Discrete Mathematics For Teachers

Ed Wheeler
Gordon College

Jim Brawner
*Armstrong Atlantic
State University*

INFORMATION AGE PUBLISHING, INC.
Charlotte, NC • www.infoagepub.com

ISBN: 978-1-61735-026-9 (paperback)
       978-1-61735-027-6 (e-book)

Printed in the United States of America

Table of Contents

Preface

MORE MATHEMATICS NEEDED

There is increasingly a national consensus that teachers who teach middle-grades and elementary mathematics need deeper and broader exposure to mathematics in both their undergraduate and in their graduate studies. *The Mathematics Education of Teachers*, published in 2001 by The Conference Board on the Mathematical Sciences, recommends 21 semester hours of mathematics for prospective teachers of middle-grades mathematics. The same publication insists that mathematics courses for teachers of all levels "should develop the habits of mind of a mathematical thinker." In Georgia, pre-service teachers preparing to teach middle-grades mathematics and pre-service teachers preparing to teach elementary school must complete 9 semester hours of mathematics content at the junior-senior level. In Texas, the Higher Education Coordinating Board is encouraging an increase in the amount of mathematics taught to elementary and middle-grades teachers. Graduate schools across the nation are developing special programs for educators who specialize in teaching mathematics to elementary school students.

However, despite the rapid development of curricula and courses to deliver more mathematics to these audiences, there is a paucity of text materials to support those efforts. For textbooks in junior-senior level courses and graduate courses, faculty members have two choices. They can teach yet another course out of one of the "Mathematics for Teachers" texts that have formed the basis of the curriculum for the last two decades. These texts tend to treat a very limited set of topics on a somewhat superficial level. Alternatively, faculty members can use mathematics textbooks written primarily for students majoring in mathematics or the sciences. Neither the topic choice nor the pedagogical style of these texts is optimal for pre-service and in-service teachers of middle-grades and elementary mathematics.

THE CHALLENGE

The challenge, then, is to develop a new generation of textbooks for pre-service teachers at the junior-senior level and graduate in-service teachers. These texts should
- extend the breadth of knowledge of mathematics in directions beyond, but related to, the topics covered in the middle-grades and elementary curricula;
- increase the depth of experience with mathematics by providing new sources of problems rich in opportunity for exploration and communication;
- present mathematics by modeling our best understanding of how mathematics should be learned and taught.
Our ability to achieve this last goal has been much enhanced by the publication of *Principles and Standards for School Mathematics* by the National Council of Teachers of Mathematics. As we complete this current project, we are very mindful of what we learn from this publication about the teaching and learning of mathematics.

WHY DISCRETE MATHEMATICS?

The topics of this new generation of textbooks must certainly go far beyond the topics covered in middle-grades and elementary mathematics. The goal is to develop teachers who not only know the mathematics they are teaching, but also understand the larger mathematical context in which the mathematics they teach has life. However, in the optimal situation, the mathematics of these textbooks will be only a few steps removed from the mathematics of the elementary and middle-grades curricula. Once this is understood, discrete mathematics is a clear choice for this new generation of textbooks. On the one hand, most of the mathematics of the K–8 curriculum is discrete, rather than continuous. Beyond this statement of the obvious is the fact that many of the classical topics of discrete mathematics arise naturally in elementary and middle-grades curricula, and can be developed robustly with relatively little prerequisite mathematical knowledge. From the *Principles and Standards for School Mathematics* we learn that the topics of discrete mathematics

> span the years from pre-kindergarten through grade 12. ... Three important areas of discrete mathematics are integrated within these Standards: combinatorics, iteration and recursion, and vertex-edge graphs. These ideas can be systematically developed from pre-kindergarten through grade 12.

Because discrete mathematics provides a rich and varied source of problems for exploration and communication, expands knowledge of mathematics in directions related to elementary and middle-grades curricula, and is easily presented using our best understanding of the ways that mathematics is learned and taught, it is an ideal topic for this new generation of textbooks.

ABOUT THIS TEXT

A quick survey of the table of contents of this text will reveal coverage of logic, sets, functions, sequences, mathematical reasoning, graph theory, trees, combinatorics, and probability, as well as some relevant applications to voting methods and social choice. It is clear that many of these topics have roots in the K–8 curriculum. Further, the topics are central ideas in the development of higher mathematics; many of them appear in textbooks tailored to the needs of future mathematicians and computer scientists.

More careful examination reveals that the material has been organized in self-contained chapters that allow the material to be selectively used in any order desired. Instructors teaching a class made up of pre-service or in-service middle-grades mathematics teachers may wish to move from Chapter 1 to Chapter 8, selecting those topics that most effectively develop the reasoning skills of their students. Instructors teaching classes containing both middle-grades and early childhood teachers may wish to use a somewhat different organization. Still other instructors may wish to select certain chapters to enrich a course taught with other texts. Whatever approach is taken, careful examination reveals that, in the spirit of *Principles and Standards for School Mathematics*, topics are presented with careful attention to the best traditions of problem solving, reasoning and proof, communication, connections with other disciplines and other areas of mathematics, and varied modes of representation.

PROBLEM SOLVING

One of the most attractive features of discrete mathematics is its rich store of problems, many of which are amenable to "hands-on" exploration and discovery. This text capitalizes on this strength in several ways. Many sections are introduced by a problem that motivates the study of the material in the section. More important, the text provides two exercise sets with each section. The first exercise set of each section is entitled *Exploratory Exercises*. Each exploratory exercise is appropriate for collaborative study; indeed the directions often instruct students to form groups to work on the problem. In many cases, the first few exploratory exercises of a section are introductory in nature, using directed inquiry to lead students to investigate a problem whose solution provides a framework for understanding the concepts of the section. Subsequent exploratory exercises may enable students to explore engaging topics that are closely related to material just covered but are not addressed specifically in the text.

Not only are the exploratory exercises designed for collaborative study, but they also facilitate the reasoning and communication that are essential to good problem solving. Many exploratory exercises could easily result in a written paper as the student discusses what the group learned in exploration. In addition, the last exercise in each exploratory exercise set is designed specifically to be a *writing exercise*, in which students are encouraged to communicate newly learned skills in written form. The second exercise set is a more traditional set of problems designed to give students an opportunity to practice working with the concepts introduced in the section and to extend those concepts in study at home.

REASONING AND PROOF

Those who teach and who aspire to teach mathematics to young people must be able to reason mathematically. This text addresses this need in a number of ways. A topic that has been given less and less attention in the traditional "Mathematics for Teachers" courses is the topic of logic. *Discrete Mathematics for Teachers* addresses this critical topic early so that careful use of mathematical language may be employed throughout the text. Beyond elementary logic, however, teachers need to understand that mathematical principles are established by proof. Chapter 3 of the text introduces the basic ideas of mathematical proof, including proof by induction and proof by contradiction. This chapter is unique in the mathematical literature in that it stresses the importance of careful preparation prior to attempting a proof. Students are carefully led to consider motivating examples, to review definitions and previously known facts, and to reflect on the resulting connections with the theorem at hand. Only then do the authors and the students engage in the task of writing a proof.

Instructors who choose to include Chapter 3 in their syllabi will find their students are able to appreciate arguments presented in the rest of the text and are able to contribute to arguments outlined in problem sets. However, despite this early attention to mathematical proof, the text does not drown in esoteric arguments. No attempt has been made to prove all results presented; proofs that are presented are chosen carefully for clarity, and arguments are often presented for those special cases in which the heart of the matter is revealed. Indeed, instructors who choose not to address the issue of proving theorems will still find the text a very pleasant pilgrimage for their students. Whatever approach is taken relative to the issue of formal proofs, instructors will find that the materials support their efforts to make mathematical reasoning more transparent to the student.

COMMUNICATION

The instructor of a class determines the extent to which students learn to appreciate the importance of communication in learning mathematics. However, the instructor who wishes to emphasize communication will have an ally in this text. The *Exploratory Exercise Sets* provide useful vehicles for promoting communication. Most of these exercises are designed for collaborative work in groups of two or more. The exercises often lead to a specific discovery or discoveries that can be the basis for a short paper written by the student. After all groups have had an opportunity to try these exercises, it is quite appropriate that solutions be shared orally with the entire class. In order to develop students' written communication skills, the last exercise in each exploratory exercise set is a *writing exercise*.

CONNECTIONS AND REPRESENTATIONS

Discrete mathematics provides many connections to other disciplines, and this text exploits those connections in multiple ways. Major topics are often introduced using these connections, and the connections are referenced in the examples and problem sets. In addition, discrete mathematics provides many connections to other topics within mathematics, such as the applications of the Pigeonhole Principle from combinatorics to fields such as geometry. Finally, the chapter *Discrete*

Applications in Political Theory provides a specific in-depth study of discrete techniques applied in the areas of political science and economics.

Discrete mathematics is rich in its many connections, but it is equally rich in its many modes of representation. For instance, one of the most powerful aspects of graph theory is the fact that the mathematical objects, graphs, can be represented as sets, as diagrams, as lists, and as matrices. The wealth of modes of representation in discrete mathematics is one of the themes of this text.

FLEXIBILITY

Many diverse audiences will find the materials in *Discrete Mathematics for Teachers* to be useful. For this reason, this text has been designed to give maximum flexibility to instructors in terms of both the particular sections covered and the order in which those sections are covered. An instructor who knows that his or her class has an adequate background in sets and functions from earlier courses and who desires a more intuitive course can skip Chapters 1, 2, and 3 and begin with Chapter 4 on graph theory, taking care to omit only those exercises labeled (*Proof Exercise*). The same instructor, desiring to emphasize careful mathematical thought, might give careful attention to Chapters 1 and 3 and then spend some time on exercises labeled (*Proof Exercise*) during the rest of the course. The chapters on graph theory and combinatorics have been written in such a way that the two topics can be taught in either order. Instructors can use the following table to tailor the course to the needs of their students and their own particular interests. Suggested course outlines are included in the *Instructor's Resource Manual*.

Chapter	Depends Upon
1: The Language of Logic	
2: Sets, Functions, and Sequences	
3: An Introduction to Mathematical Proof	Familiarity with language of logic or Chapter 1; familiarity with sets or Section 2.1
4: Graph Theory	Familiarity with sets and functions or Sections 2.1, 2.2, and 2.3
5: Trees and Directed Graphs	Sections 4.1, 4.2, 4.3, and 4.4
6: Combinatorics	Familiarity with sets or Section 2.1
7: Probability, Sections 7.1 and 7.2	Sets and functions or Sections 2.1, 2.2, and 2.3
7: Probability, Sections 7.3 and 7.4	Sections 7.1, 7.2, 6.1, 6.2, and 6.3
8: Discrete Applications in Political Theory, Sections 8.1, 8.2, 8.4, and 8.5	
8: Discrete Applications in Political Theory, Sections 8.3 and 8.4	Sections 6.1 and 6.2

OTHER RESOURCES

An *Instructor's Resource Manual* and a *Student's Solution Manual* are available as supplements for the text. The instructor's resource manual contains suggested course outlines, teaching notes, complete solutions of all exercises, and either solutions or teaching notes on exploratory exercises. The student's solution manual contains complete solutions of odd-numbered exercises.

Of course, answers to odd-numbered exercises are to be found in the back of the text for each exercise set. However, a conscious decision was made not to include answers for exploratory exercise sets in the back of the text. In some measure this decision was made because many of the exploratory exercises have no single answer. However, the answers are also omitted because there is no better way to discourage active learning, communication, and exploration than to prematurely provide an answer to a group of students who are engaged with a problem.

MANY THANKS

A textbook does not find its way to publication without important efforts by many persons beyond those listed as authors. Faculty members who taught from draft versions of the text and then provided valuable comments include Professors Cathie Aust of Clayton College and State University, Mary Ann Lee of Minnesota State University, Mankato, Dale Kilhefner of Armstrong Atlantic State University, Gwen Barber of Coastal Georgia Community College, and William Jasper of Sam Houston State University. Certainly we are very indebted to persons who reviewed the text and provided insightful comment that enabled us to polish and shape the material. This team of reviewers included:

Atif Abueida, University of Dayton, Ohio
Dr. Paul Ache, III, Kutztown University of Pennsylvania
Catherine Aust, Clayton College and State University, Georgia
Dr. David Buhl, Northen Michigan University
Mike Daven, Saint Mary College, New York
Mary C. Enderson, Middle Tennessee State University
Jerrold W. Grossman, Oakland University, Michigan
Leigh Hollyer, University of Houston, Texas
Dr. William A. Jasper, Sam Houston State University, Texas
Isa S. Jubran, PHD, SUNY College at Cortland, New York
Kathy Culver Nickell, College of DuPage, Illinois
C.L. Pinchback, University of Central Arkansas

In addition, the authors are indebted to several generations of students at Armstrong Atlantic State University who learned with us using draft versions of the manuscript. We appreciate the contributions of Lauren Mason, Mary Thornton, Diane Sellers, Fern Illidge, and many others in the Armstrong Atlantic State community who made contributions to the preparation of the text. We are very grateful to John Bakken who labored with us for three years and provided many of the figures that grace the text and to Tim Banks who created the illustrations that introduce the chapters. Finally, we very much appreciate the invaluable efforts and many kindnesses of the team at Houghton Mifflin in bringing this project to print. Specifically we must commend the fine work of Lauren Schultz, Jennifer King, and Tamela Ambush.

A FINAL NOTE

In presenting this material to pre-service and in-service teachers, it is desirable to emphasize quality more than quantity. The pace chosen should depend on the mathematical sophistication of the students. On some sections the instructor will want to linger, completing the *Exploratory Exercises*, allowing time for student presentation of some of their work, and finally sending students home to polish their skills on the standard exercise sets. It is the belief and experience of the authors that choosing topics that address the interests of both students and teacher and adjusting the pace to ensure depth as well as breadth of coverage will lead to one of the most enjoyable mathematics courses experienced by both students and teacher.

Ed Wheeler
James Brawner

Discrete Mathematics for Teachers

The Language of Logic

Chapter

1

1.1 An Introduction to Logic

Uncle Fred called ahead to announce he was bringing turkey *and* ham to the Thanksgiving celebration. When he arrived bearing only a turkey, celebrants were more than a little unhappy. Aunt Buosone called indicating that she was bringing spring rolls *or* fried rice to the meal. When she arrived with a plate of spring rolls, she was greeted with enthusiasm. We will understand the differing greetings received by Fred and Buosone when we understand the function of the words *and* and *or* in everyday discourse and in the language of logic.

At the heart of the language of mathematics lies the language associated with elementary logic.

Elementary logic provides the terminology and structures that are used in defining mathematical terms and creating mathematical arguments. Beyond its essential role in mathematics, logic provides insight in the design of computer programs and electronic circuits. In this section we take the first steps toward mastering the fundamentals of logic.

Sentences that can be identified as either true or false, but not both, are called **statements** or **propositions**.

Example 1

Determine which of the following sentences is a statement. If a sentence is a statement, determine whether it is true or false.
(1)   Atlanta is the capital of Georgia.
(2) $2 + 3 = 7$.
(3)   John Smith was the first president of the United States.
(4)   For all integers n, $n^2 + n = n(n+1)$.
(5)   Keep your eyes on the road.
(6)   What time is it?
(7)   Oh, what a beautiful sunset!
(8) $x + 2 = 6$.
(9)   She is the captain of the soccer team.
(10) This sentence is not true.

Solution

Sentences (1), (2), (3), and (4) are statements. Statements (1) and (4) are true; (2) and (3) are false. Sentence (5) (a command), sentence (6) (a question), and sentence (7) (an exclamation) are not statements because they cannot be assigned a truth value. We should pay careful attention to statements (8) and (9). They fail to be statements because we do not know the identity of "x" in sentence (8) or that of "she" in sentence (9). On the other hand, if we knew the value of "x" in (8) and the identity of "she" in (9), we could assign a truth value to both sentences. Sentence (10) is not a statement since it is self-contradictory.

Sentences like (8) and (9) that become statements when a variable has a specific value are called **open sentences** or **propositional functions**. The sentence "$x + 1 = 5$" is an open sentence. If $x = 10$, the resulting statement is not true. If $x = 4$, the statement clearly is true.

Our primary concern in this text will be sentences that are statements or open sentences. We will represent such sentences using lower-case letters such as p, q, and r. When a statement is true, we will denote its truth value by T, and when a statement is false, by F.

Example 2

Determine the truth value of statements p and q found below.
p: President Lincoln was born in Texas.
q: $2 + 3 = 5$.

Solution

Statement p is false and hence has truth value F. Statement q is true and hence has truth value T.

In ordinary speech and writing we often build more complex sentences from simpler ones. For example, we often take two sentences and build a new sentence using connectives such as *and* and *or*.

1. The birth rate in Freedom County has increased, *or* there are more young married couples in the county.

2. The number of majors in sociology at Bellevue College has increased, *and* the number of majors in physics has decreased.

Alternatively, we take a single statement or open sentence and alter it by forming its negation.

Definition: The negation of p is "It is not true that p" (denoted by $\sim p$). Suppose p is a statement. When p is true, $\sim p$ is false; when p is false, $\sim p$ is true.

The negation of "New York is the capital of the United States" is "It is not true that New York is the capital of the United States." The use of the grammatical form "It is not true that p" sometimes results in awkward sentence structures. For many statements, the negation may be more simply expressed by negating the predicate. For instance, the negation of the preceding statement could be written as "New York is not the capital of the United States."

When we create a new statement from simpler statements using *and*, *or*, or negation, the new statement is called a **compound statement**. As we noted in the definition of negation, the truth value of a compound statement is determined by the truth values of the statement or statements from which it is built. By using truth tables, we can systematically record the truth values of compound statements in terms of the possible truth values of the component statements.

TRUTH TABLES

By definition, the negation of a statement is false when the statement is true, and true when the statement is false. Table 1 summarizes the two possibilities for p and $\sim p$. This table emphasizes the fact that p and $\sim p$ cannot be simultaneously true or simultaneously false.

Table 1: Truth Values for Negation	
p	$\sim p$
T	F
F	T

When a new statement is formed from statements p and q using the connective *and*, the result is called a **conjunction**. When Uncle Fred announces that "I will bring a turkey and I will bring a ham to Thanksgiving dinner," he has created substantial expectations. The family that has gathered for dinner will be unhappy if he shows up with only a ham, with only a turkey, or with neither. The assembled diners will be happy only when both statements, "I will bring a turkey" and "I will bring a ham," are true. In similar fashion we understand the truth values of a conjunction.

Definition: "p and q" (denoted by $p \wedge q$) is called a **conjunction**. Suppose p and q are statements. The conjunction "p and q" is true if both p and q are true; otherwise "p and q" is false.

Because p is either true or false and q is either true or false when p and q are statements, there are four possibilities for the combined truth values of p and q. Table 2 illustrates these possibilities. The third column of the table identifies the truth value of $p \wedge q$ for each of the four possibilities. For example, in the last row of the table we record the fact that if statement p is false and statement q is false, then the statement $p \wedge q$ is false. In the fourth column of the table we give an example of each possibility.

Table 2: Truth Values for Conjunction			
p	q	$p \wedge q$	Examples
T	T	T	2 + 3 = 5, and 1 is less than 2.
T	F	F	2 + 3 = 5, and 4 is less than 3.
F	T	F	December has 34 days, and Christmas is on December 25.
F	F	F	December has 32 days, and January has exactly 22 days.

The words *but* and *nevertheless* are sometimes used in place of the word *and*. Thus the statement "Takira is 16, but she does not yet drive" is a conjunction.

When two statements are joined by the connective *or*, the resulting statement is called a **disjunction**. When Aunt Buosone communicates, "I will bring spring rolls or I will bring fried rice to the Thanksgiving meal," she establishes limited expectations. Certainly those assembled to eat will be very happy if she delivers a steaming plate of both delicacies, but they will also be satisfied if she delivers only the fried rice or only the spring rolls. Unhappiness will prevail, however, if she shows up with neither. Similar reasoning drives the definition of disjunction.

Definition: "p or q" (denoted by $p \vee q$) is called a **disjunction**. Suppose p and q are statements. The disjunction "p or q" is true if p is true, if q is true, or if both p and q are true.

Table 3 illustrates the truth values for the statement "p or q" for all possible combinations of truth values of p and of q.

Table 3: Truth Values for Disjunction			
p	q	$p \vee q$	Examples
T	T	T	Sugar is sweet, or December has 31 days.
T	F	T	Sugar is sweet, or the earth is flat.
F	T	T	2 + 3 = 7, or 2 is less than 7.
F	F	F	2 + 3 = 7, or 2 is greater than 5.

The *or* described by this definition is sometimes called an *inclusive or*. This means that statements using the connective *or* are false only when both parts are false.

Example 3

Suppose we have the following two statements:
p: I exercise daily.
q: I eat spinach every week.
Translate the following statements into words.

(a) $p \wedge q$ (b) $p \vee q$ (c) $\sim p$
(d) $\sim (p \vee q)$ (e) $(\sim p) \wedge q$

Solution

(a) I exercise daily, and I eat spinach every week.
(b) I exercise daily, or I eat spinach every week.
(c) I do not exercise daily.
(d) It is not the case that I exercise daily or that I eat spinach every week.
(e) I do not exercise daily, and I eat spinach every week.

Example 4

Let p be false (F), and let q be true (T). Find the truth values of the following:

(a) $p \wedge q$ (b) $p \vee q$

Solution

(a) Since p has truth value F and q has truth value T, then $p \wedge q$ has truth value F. Thus, the statement is false in this case.
(b) Since p has truth value F and q has truth value T, then $p \vee q$ has truth value T. Thus, the statement is true in this case.

Example 5

Let p be T, and let q be F. Find the truth values of the following.

(a) $\sim (p \wedge q)$ (b) $(\sim p) \wedge q$

Solution

(a) The statement $\sim (p \wedge q)$ is true.
(b) The statement $(\sim p) \wedge q$ is false. Notice that $\sim (p \wedge q)$ and $(\sim p) \wedge q$ have different truth values for the same truth values of p and q.

Example 6

Construct a truth table for $\sim (p \wedge \sim q)$.
Solution

The required truth table appears in Table 4. In building the truth table, we proceed from simple to complex. We first list the possible truth values for p and for q. Then we build a column for $\sim q$. Finally, we build successive columns for $p \wedge \sim q$ and then its negation.

Table 4: Truth Values for $\sim (p \wedge \sim q)$				
p	q	$\sim q$	$p \wedge \sim q$	$\sim (p \wedge \sim q)$
T	T	F	F	T
T	F	T	T	F
F	T	F	F	T
F	F	T	F	T

When constructing truth tables, we must be careful to list all possible cases. In statements involving p, q, and r, for example, p can be T or F, q can be T or F, and r can be T or F. The eight possible combinations of truth values are found in the first two columns of Table 5. We will investigate how to determine all possible cases that must be considered in forming a truth table in the first exploratory exercise of this section.

Example 7

Make a truth table for $p \wedge (q \vee r)$.

Solution

The required truth table appears in Table 5.

Table 5: Truth Values for $p \wedge (q \vee r)$				
p	q	r	$q \vee r$	$p \wedge (q \vee r)$
T	T	T	T	T
T	T	F	T	T
T	F	T	T	T
T	F	F	F	F
F	T	T	T	F
F	T	F	T	F
F	F	T	T	F
F	F	F	F	F

Careful consideration of all possible cases is important in constructing truth tables. In fact, consideration of all possible cases can be an important problem-solving strategy in many different kinds of problems. Use this strategy in completing the problem in Example 8.

Example 8

Arno, Terry, Raul, Mike, and Chris are starters on the Windsor Forest basketball team. Two shoot with their left hand, and three shoot with their right hand. Two are over 6 ft tall. Arno and Raul shoot with the same hand; Mike and Chris use different hands to shoot. Terry and Chris are in the same height range, but Raul and Mike are in different height ranges. The player who plays center is over 6 ft tall and is left-handed. Who is he?

Solution

As you work on this problem, you may find it helpful to make notes in a table like Table 6.

Table 6					
	Arno	Terry	Raul	Mike	Chris
Left-handed					
Right-handed					
Taller than 6 ft					
Shorter than 6 ft					

continued

   (1) What can you conclude, from these facts, about which hand Arno and Raul shoot with?
       Two shoot with their left hand, and three shoot with their right hand.
       Arno and Raul shoot with the same hand.
       Mike and Chris use different hands to shoot.
   (2) With which hand does Terry shoot?
   (3) What can you conclude about the height of Terry and Chris from these facts?
       Two are over 6 ft tall, and three are under 6 ft.
       Terry and Chris are in the same height range.
       Mike and Raul are in different height ranges.
   (4) In which height range does Arno fall?
   (5) We still need to determine which hand Mike and Chris use in shooting and which height
       range matches Mike's and Raul's height. Examine the four possible cases for Mike (LH and
       tall, LH and short, RH and tall, RH and short), eliminate the impossible cases, and explain
       why Mike must be the tall, left-handed center.

Exploratory Exercise Set 1.1

1.   One perplexing question relative to the formation of truth tables is the question of how to identify all
    cases that must be considered. We will use a geometric diagram (called a tree diagram) to help us
    with this important task.
    (a) When the truth table involves only a single statement, the matter is simple. The statement is either
       true or false. We represent this by a tree that has a single branch for each of the possibilities and
       then use the results to fill in the second and third columns of the truth table for $\sim (\sim p)$.

p	$\sim p$	$\sim (\sim p)$
T		
F		

    (b) When the truth table involves two statements p and q, and p is true, q can be either true or false.
       Similarly, when p is false, q can be true or false. In the tree diagram below we record each
       possibility for p and then the corresponding possibilities for q. Complete the tree, record each of
       the cases in the first two columns of the truth table for $p \wedge q$, and then complete the truth table.

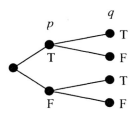

p	q	$p \wedge q$
T	T	

(c) The tree diagram below helps us identify all cases for truth tables of compound statements involving three statements p, q, and r. Complete the tree diagram and then transfer the cases to the table for $(p \wedge q) \wedge r$.

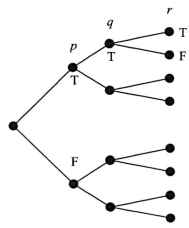

p	q	r	$(p \wedge q)$	$(p \wedge q) \wedge r$
T	T	T		

(d) Use a tree diagram to determine the cases that must be considered in constructing truth tables for compound statements involving each of the four component statements p, q, r, and s.

(e) Observe that a truth table for a compound statement involving one component statement requires two cases (the statement is either true or false). Complete this table, which records the number of cases required to construct a truth table for a compound statement with a component statement.

Number of Component Statements	Number of Cases
1	2
2	
3	
4	
5	
n	

2. Sets of attribute pieces are often found in elementary classrooms. One common set consists of 32 pieces. There are eight pieces for each color: red, blue, green, and yellow. For each color there are four different shapes: square, circle, rhombus, and triangle. For each shape and color, there are two sizes: large and small. A set of attribute pieces is given to each student, and each student is asked to place pieces that satisfy certain criteria on his or her desk. Work in pairs to determine how many attribute pieces should be on the desk for each of the following descriptions.

(a) Large and Red
(b) Red or Yellow
(c) Green or Small
(d) Square and Triangle
(e) Circle or Square
(f) Circle and Green, or Yellow

3. As we have observed, consideration of all possible cases can be an important problem-solving strategy in many different kinds of problems. Use this strategy in attempting the following problems.

(a) Five students – Leon, Sarah, Russo, Sue, and Sharon – participate in a debate tournament in which each team must have at least one affirmative debater and one negative debater. Two of the students attend Harp College, and three attend Sloan. Three are affirmative debaters, and two are negative. Leon and Sue attend the same college. Russo and Sharon attend different schools. Sarah and Russo represent the same side in the debate (both are "affirmative" or both are "negative"); Sue and Sharon represent opposite sides. A negative debater from Harp College was selected as the outstanding debater. Who was that person?

Category	Leon	Sarah	Russo	Sue	Sharon
Affirmative					
Negative					
Harp College					
Sloan College					

(b) Three soccer teams – the Cats, the Dogs, and the Ants – play a three-game tournament. We have the following information: Each team played two games. The Cats won two games; the Dogs tied one game. The Dogs scored a total of two goals; the Ants scored three goals. One goal was scored against the Cats, four against the Dogs, and seven against the Ants. Find the scores of all three games.

4. *(Writing Exercise)* You have been asked to give a short talk to a sixth-grade class about correct use of the words *and*, *or*, and *not*. Write out your lecture, being sure to use examples not given in this text.

Exercise Set 1.1

1. Determine which of the following are statements, and classify each statement as either true or false.
   (a) President George Washington was born in Alabama.
   (b) Good morning.
   (c) $5 + 4 = 9$.
   (d) $x + 5 = 8$.
   (e) Close the door.
   (f) $2 \cdot 3 = 7$.
   (g) $3x = 6$.
   (h) Help stop inflation.

2. If p is T and q is F, find the truth values of these statements. The form $\sim (\ )$ represents the negation of everything inside the parentheses.
   (a) $(\sim p) \wedge q$
   (b) $(\sim p) \vee (\sim q)$
   (c) $\sim (p \vee q)$
   (d) $\sim (p \wedge q)$
   (e) $\sim [(\sim p) \wedge q]$
   (f) $\sim (p \vee \sim q)$

3. Find the truth values of the statements in Exercise 2 if p is T and q is T.

4. Find the truth values of the statements in Exercise 2 if p is F and q is F.

5. In each of the parts that follow, construct two statements—one using *and* and the other using *or*—and find the truth values of each.
   (a) November has 30 days. Thanksgiving is always on November 25.
   (b) The smallest counting number is 2. 10 is not a multiple of 5.
   (c) $2 + 3 = 4 + 1$. $8 \cdot 6 = 4 \cdot 12$.
   (d) Triangles are not squares. 3 is smaller than 5.

6. Translate the following statements into symbolic form, using A, B, C, D, $\wedge$, $\vee$, and $\sim$, where A, B, C, and D denote the following statements:

   A. It is snowing.
   B. The roofs are white.
   C. The streets are not slick.
   D. The trees are green.

   (a) It is snowing, and the trees are green.
   (b) The trees are green, or it is snowing.
   (c) The streets are not slick, and the roofs are not white.
   (d) The trees are not green, and it is not snowing.

7. Using the statements of Exercise 6, translate the following symbolic statements into English sentences. (If statements are grouped by parentheses, set off by commas.)
   (a) A ∧ ~ B
   (b) (~ B) ∨ (~ C)
   (c) A ∧ (B ∨ C)
   (d) (A ∨ ~ C) ∧ D
   (e) ~ (A ∧ ~ D)
   (f) ~ (A ∧ ~ C)

8. If "It is snowing" and "The trees are green" are true, and if "The roofs are white" and "The streets are slick" are false, classify the following statements as true or as false.
   (a) It is snowing, and the trees are green.
   (b) The trees are green, or it is snowing.
   (c) It is not snowing, and the roofs are white.
   (d) It is not snowing, or the streets are slick.
   (e) The streets are not slick, and the roofs are not white.
   (f) The trees are not green, and it is not snowing.

9. Let *p* be the statement "My test grade was 80." Let *q* be the statement "My date was late to dinner." Let *r* be the statement "My telephone bill was more than $10." Write each of these statements in symbolic form.
   (a) My test grade was 80, and my date was late to dinner.
   (b) It is not true that my phone bill was more than $10.
   (c) My date was not late to dinner.
   (d) My date was late to dinner or my telephone bill was more than $10, but my test grade was not 80.

Understanding connectives is essential to being able to read contracts and other legal and government documents. In Exercises 10 and 11, write the statement from such a document in symbolic form and then answer the following questions.

10. The Higher Education Act specifies that a student is eligible to apply for a loan provided that the following statement is true.
    The student is enrolled (E) and in good standing (S), or not enrolled but accepted (A) for enrollment at an eligible institution.
    (a) Write the statement in symbolic form.
    (b) Is a student who is enrolled and on academic probation eligible to apply for a loan? Explain.
    (c) Is a student who is not enrolled but has been accepted for enrollment eligible to apply for a loan? Explain.

11. The contract for a deferred variable annuity that is part of a retirement program specifies that the owner of the policy can retire and receive payments provided that the owner is at least 55 years of age (A) or permanently disabled (D), and has been employed for 10 years (W).
    (a) Write the statement in symbolic form.
    (b) Lucretia is 56 years of age, is in excellent health, and has been employed for 20 years. Can she retire and receive payments? Explain.
    (c) Bill is 50 years old, can no longer work because of a chronic back injury, and has been employed for 7 years. Can he retire and receive payments? Explain.

12. In the game show "Truth or Despair," the contestants are presented with two doors. Behind one door there is a new car, behind the other door a skateboard. On each door there is a sign. Contestants are told that exactly one of the two signs is true. In each instance explain carefully how to determine which door leads to the car.
    (a) The sign on the first door reads, "There is a car or a skateboard behind this door." The sign on the second door reads, "There is a skateboard behind this door."
    (b) The sign on the first door reads, "Behind this door there is a skateboard and behind the other there is a car." The sign on the second door reads, "Behind one of these doors there is a car and behind the other there is a skateboard."

13. After Jack climbed the beanstalk and reached the Land of the Giants, he discovered that each giant belonged to one of two clans, the clan of the Liars or the clan of the Truthtellers. Every statement made by a Liar is false; every statement made by a Truthteller is true. Explain in each case how Jack can determine to which clan the giants belong.

   (a) Jack meets a giant couple and inquires of the husband, "To which clan do you belong?" The husband answers, "We are both Liars." To which clan does the husband belong and to which clan does the wife belong?

   (b) Jack meets a second giant couple. He asks the wife, "Are both you and your husband Liars?" She answers, "At least one of us is." To which clan does each belong?

   (c) On yet another occasion, Jack fell in with three giants. The first declared, "All three of us are Liars." The second said, "Only two of us are Liars." The third insisted, "Only one of us is a Liar." To which clan did each belong?

14. Classify the following compound statements as true or false.
   (a) $2 + 1 = 3$ or $3 + 5 = 9$.
   (b) $3 + 1 = 4$ and $4 + 3 = 6$.
   (c) $2 + 3 = 6$ and $4 + 5 = 9$.
   (d) $3 + 5 = 8$ or $2 + 3 = 6$.

15. Construct truth tables to show the truth values for the following:
   (a) $p \vee \sim q$ (b) $p \wedge \sim p$

16. The Law of Double Negation says that $\sim(\sim p)$ can be replaced by p in every logical expression. Complete the truth table below and use the results to explain the Law of Double Negation.

p	$\sim p$	$\sim(\sim p)$
T		
F		

17. Let p represent "The stock market is bullish." Let q represent "The Dow average is increasing." Finally, let r represent "The price of utilities is decreasing." Translate each of the following into words. When two statements are grouped by parentheses in the logical expression, group them with a comma in the English sentence.
   (a) $(\sim p) \wedge q$ (c) $(p \vee r) \wedge q$
   (b) $p \vee (\sim r)$ (d) $(\sim r \vee \sim p) \wedge q$

18. Choose three appropriate statements, p, q, and r, and then translate the following sentences into symbols:
   (a) Either I invest my money in stocks, or I put my money in a savings account.
   (b) I do not invest in stocks or deposit my money in a savings account, but I buy gold.
   (c) I do not buy gold, and I do not deposit my money in a savings account.

19. Let p represent the statement "The price of automobiles will rise," and let q represent the statement "The inflation rate will increase." Translate each of the following sentences into symbols:
   (a) The price of automobiles will rise, and the inflation rate will increase.
   (b) The price of automobiles will rise, or the inflation rate will increase.
   (c) The price of automobiles will not rise, but the inflation rate will increase.
   (d) The price of automobiles will not rise, or the inflation rate will not increase.

20. Construct truth tables to show the truth values of the following:
   (a) $(p \wedge q) \vee q$ (b) $(p \vee q) \wedge p$

21. Find the truth value of each of the following when q is true and p and r are both false.
    (a) $(p \vee q) \wedge \sim (\sim p \vee r)$
    (b) $(p \wedge \sim q) \vee (r \wedge \sim q)$
    (c) $\sim (r \wedge p) \wedge \sim (q \vee r)$
    (d) $(r \wedge p) \vee (\sim q \wedge r)$

22. Find the truth value of each expression in Exercise 21 if q and r are false and p is true.

23. Construct truth tables to show the truth values for the following:
    (a) $(p \vee \sim q) \wedge r$
    (b) $p \wedge (q \vee r)$
    (c) $(p \vee \sim p) \wedge q$
    (d) $\sim (p \vee q) \wedge r$

1.2 Conditionals and Equivalent Statements

Aaron is quite excited. After sitting through a tongue-lashing from his father concerning his falling grades, Aaron overhears his father tell his mother,

"If Aaron makes an A in senior English, then I will buy him a Jeep Cherokee."

Aaron is suddenly quite interested in the truth value of this statement.

Statements of the form "If p, then q" are called conditionals and are among the most important kinds of statements we use both in the language of mathematics and in daily discourse. Let p represent "Aaron makes an A in senior English," and let q represent "His father will buy him a Jeep Cherokee" "If Aaron makes an A in senior English, then his father will buy him a Jeep Cherokee" is denoted by "If p, then q." p is called the **hypothesis** (or antecedent) and q the **conclusion** (or consequent).

In order to decide when to classify a conditional as true and when to classify it as false, we shall think of "If p, then q" as a promise. Aaron has overheard the promise "If Aaron makes an A in senior English, then his father will buy him a Jeep Cherokee." In this situation there are four possibilities, as laid out in Table 1.

Table 1		
p: Aaron Makes A	q: Gets Cherokee	Promise Kept?
Yes	Yes	Yes
Yes	No	No
No	Yes	Yes
No	No	Yes

The only case in which the promise is broken occurs when Aaron makes an A, but his father does not buy the car. In the last two cases, the promise is not broken because Aaron does not make an A; hence, whether the father does or does not buy the car does not affect the promise. This example suggests the following definition of a conditional.

Definition: A **conditional** has the form "If p, then q" where p and q are statements or open sentences. It is denoted by $p \rightarrow q$. Suppose p and q are statements. The conditional $p \rightarrow q$ is false when p is true and q is false, and is true otherwise.

The truth values of conditional statements are illustrated in Table 2.

p	q	$p \rightarrow q$	Examples
Table 2: Truth Values of Conditional Statements			
T	T	T	If air quality in Atlanta is below federal standards, then low-polluting gasoline must be used.
T	F	F	If George Bush is president, then a Democrat lives in the White House.
F	T	T	If 5 < 3, then bicycles have two wheels.
F	F	T	If 5 < 3, then bicycles have three wheels.

In ordinary language we use conditionals to describe circumstances in which there is a relationship between the hypothesis and the conclusion. In the conditional "If it rains, then I will not water the lawn," there is a clear connection between the hypothesis, "It rains," and the conclusion, "I will not water the lawn." In logic, an "if-then" statement does not necessarily imply relationship between the hypothesis and conclusion. Because in logic we are concerned with the form of statements rather than with the truth of the subject matter, we permit conditionals like "If 3 + 4 = 10, then Tony Blair is president of the United States." Moreover, this conditional is true since the hypothesis is false. (Note the difference between a conditional that is true and a conditional with a true conclusion.) In logic any two statements can be connected with the conditional connective, and the result can then be classified as true or false.

Example 1
    (a) Determine whether statements *p* and *q* are true or false.
        p: Baseball is played with a racket.
        q: Football is played with a bat.
    (b) Determine the truth value of the conditional "If baseball is played with a racket, then football is played with a bat."
Solution
    (a) Neither statement is true.
    (b) The conditional is true because a conditional with a false hypothesis and a false conclusion is true.

Conditionals are used so often both in ordinary language and in mathematics that they occur in many different forms. The conditional "If *p*, then *q*" can also be written as
    q if *p*.
    p implies *q*.
    When *p* then *q*.
    p only if *q*.

Example 2
    Rewrite the following conditional in four additional forms.
        If it snows, then the streets are slick.
Solution
    (a) The streets are slick if it snows.
    (b) Snowy weather implies the streets are slick.
    (c) When it snows, then the streets are slick.
    (d) It snows only if the streets are slick.

RELATED CONDITIONALS

There are three related conditionals that can be formed from the conditional $p \to q$: its converse, its inverse, and its contrapositive.

> **Definition:** The **converse of the conditional** "If p, then q" is the conditional "If q, then p." That is, the converse of $p \to q$ is $q \to p$.

"If it is cold, then it is snowing" is the converse of "If it is snowing, then it is cold." This example shows that the truth of a conditional in no way ensures the truth of its converse. "If it is snowing, then it is cold" is a true conditional, but the fact that it is cold does not imply that it is snowing. Similarly, the truth of a conditional certainly does not require that its converse be false. "If $2 + 2 + 2 = 6$, then $3 \cdot 2 = 6$" and its converse, "If $3 \cdot 2 = 6$, then $2 + 2 + 2 = 6$," are both true.

> **Definition:** The **inverse of a conditional** is the conditional that results when p and q are replaced by their negations; that is, the inverse of $p \to q$ is $(\sim p) \to (\sim q)$.

The inverse of "If a polygon is a rectangle, then it is a parallelogram" is "If a polygon is not a rectangle, then it is not a parallelogram."

Although the truth of a conditional does not ensure the truth of the converse or inverse of the conditional, much advertising assumes that you believe that it does. Consider the conditional "If you are a great basketball player, then you wear expensive Swoosh basketball shoes." Certainly a quick check of shoes on the feet of the major basketball stars will convince you that this is a true statement. Advertisers hope that you will assume that the converse is true: "If you wear expensive Swoosh basketball shoes, you will be a great basketball player."

> **Definition:** The **contrapositive of the conditional** "If p, then q" is "If $\sim q$, then $\sim p$"; that is, the contrapositive of $p \to q$ is $\sim q \to \sim p$.

The contrapositive of "If it is snowing, then it is cold" is "If it is not cold, then it is not snowing." Common sense suggests that these two statements say the same thing. The fact that the truth of a conditional ensures the truth of the associated contrapositive, and vice versa, will be verified later in this section.

To summarize the definitions of converse, inverse, and contrapositive, consider the following example:

Example 3
Find the converse, inverse, and contrapositive of the following conditional:
If it rains, then I buy a new umbrella.
Solution
Converse: If I buy a new umbrella, then it is raining.
Inverse: If it does not rain, then I do not buy a new umbrella.
Contrapositive: If I do not buy a new umbrella, then it is not raining.

LOGICALLY EQUIVALENT STATEMENTS

When two statements r and s have the same truth values in every possible situation, we say that r is logically equivalent to s.

> **Definition:** In logic, two statements r and s are **equivalent statements** if they have identical truth tables.

The observation that two statements are logically equivalent is very important. It means that both in mathematical discourse and in ordinary conversation the two statements can be used interchangeably without changing the meaning of the communication. If statements that are not logically equivalent are substituted for one another, however, there is a risk that the meaning will be changed.

Example 4

Show that a conditional and its contrapositive are equivalent.

Solution

In Table 3 we record all possible truth values for both $p \rightarrow q$ and $\sim q \rightarrow \sim p$. Observe that the truth values for $p \rightarrow q$ and $\sim q \rightarrow \sim p$ are the same.

Table 3					
p	q	$\sim p$	$\sim q$	$p \rightarrow q$	$\sim q \rightarrow \sim p$
T	T	F	F	T	T
T	F	F	T	F	F
F	T	T	F	T	T
F	F	T	T	T	T

Because a conditional and its contrapositive are equivalent, the statement "If you wash your hair with Shed shampoo, then your dandruff disappears" can be used interchangeably with the statement "If your dandruff does not disappear, then you do not wash your hair with Shed shampoo."

Example 5

Using a truth table, show that the converse of a conditional is not equivalent to the conditional.

Solution

Table 4 shows that the truth values for $p \rightarrow q$ and $q \rightarrow p$ are not the same for some truth values of p and of q. Therefore, a conditional and its converse are not equivalent.

Table 4			
p	q	$p \rightarrow q$	$q \rightarrow p$
T	T	T	T
T	F	F	T
F	T	T	F
F	F	T	T

OTHER EQUIVALENT STATEMENTS

In Exercise 17 you will use truth tables to verify De Morgan's Laws.

De Morgan's Laws: For any statements p and q,
   (a) $\sim (p \vee q)$ is equivalent to $\sim p \wedge \sim q$.
   (b) $\sim (p \wedge q)$ is equivalent to $\sim p \vee \sim q$.

Example 6

Use one of De Morgan's Laws to write the negation of "I will eat a hamburger, or I will not eat a bowl of soup."

Solution

Let p be "I will eat a hamburger," and let q be "I will not eat a bowl of soup." Then $\sim p$ is "I will not eat a hamburger," and $\sim q$ is "I will eat a bowl of soup." We want the negation of $p \vee q$, which is $\sim (p \vee q)$. By the first of De Morgan's Laws, $\sim (p \vee q)$ is equivalent to $(\sim p) \wedge (\sim q)$. The negation is "I will not eat a hamburger, and I will eat a bowl of soup."

Example 7

Use one of De Morgan's Laws to complete the parts of this example.
(a) Consider the statement "It is March, and azaleas are in bloom." Identify statements p and q so that this statement can be understood as "p and q."
(b) Write the negation of the statement in (a) using one of De Morgan's Laws.
(c) Use one of De Morgan's Laws to write the negation of this statement:
     Both Robert and Patrick scored 20 points tonight.

Solution

(a) p: It is March. q: Azaleas are in bloom.
(b) Either it is not March or the azaleas are not in bloom.
(c) Either Robert did not score 20 points tonight or Patrick did not score 20 points tonight.

One of the most important equivalencies relates to the formation of the negation of a conditional.

Definition: The Negation of a Conditional: The negation of the conditional $p \rightarrow q$ is equivalent to $p \wedge \sim q$; that is, $\sim (p \rightarrow q)$ is equivalent to $p \wedge \sim q$.

In Exercise 13 we will verify that these conditionals are equivalent using truth tables. However, we can intuitively understand this definition by observing that if Manuel promises, "If it rains, then I will bring an umbrella," the one circumstance in which his date Maria would protest would be in the event that "It rained and Manuel did not bring an umbrella." The equivalence of $\sim (p \rightarrow q)$ and $p \wedge \sim q$ is the basis of a method of mathematical proof called Proof by Contradiction that will be discussed in Chapter 3. In Exercise 14 we will also observe that $p \vee q$ and $\sim p \rightarrow q$ are logically equivalent.

Example 8

Write the negation of this conditional in two ways: If n^2 is even, then n is even.

Solution

The standard way to form the negation of a statement is to precede it by the words "It is not true that...." Hence, an awkward form of the negation of this conditional is "It is not true that if n^2 is even then n is even." Since $\sim (p \rightarrow q)$ is equivalent to $p \wedge \sim q$, another form of the conditional is

n^2 is even and n is not even.
n^2 is even and n is odd.

To this point we have used negations, conditionals, *and*, and *or* to create complex statements from simpler components. Yet another way to connect two statements or open sentences involves the use of the phrase *if and only if*. The statement "A graph is a tree if and only if the graph is connected and has no cycles" is an example of a biconditional.

Definition: The **biconditional** formed from p and q is "p if and only if q" (denoted by $p \leftrightarrow q$). Suppose p and q are statements. The biconditional is true when p and q have the same truth values. It is false otherwise.

In the third column of Table 5 are found the truth values for the biconditional $p \leftrightarrow q$. Compare Column 3 and Column 6 in this table. Observe that the truth values in these two columns are the same, which proves that the biconditional $p \leftrightarrow q$ is equivalent to $(p \to q) \wedge (q \to p)$. This fact has important consequences in mathematics. To prove a theorem of the form $p \leftrightarrow q$, we prove two separate theorems, $p \to q$ and $q \to p$.

Table 5					
p	q	$p \leftrightarrow q$	$p \to q$	$q \to p$	$(p \to q) \wedge (q \to p)$
T	T	T	T	T	T
T	F	F	F	T	F
F	T	F	T	F	F
F	F	T	T	T	T

In mathematics when a statement of the form $p \to q$ is true, we describe this situation in either of the following ways:

For p to be true, it is *necessary* for q to be true.

For q to be true, it is *sufficient* for p to be true.

When the biconditional $p \leftrightarrow q$ is true, we can describe this by saying the truth of p is necessary and sufficient for the truth of q.

Exploratory Exercise Set 1.2

1. Work in pairs. Complete each part of this exercise individually, and then compare and discuss your work with your partner.
    (a) Write each of these conditional statements in the form "If p, then q."
        (1) Hai can vote if he is 18.
        (2) A muddy path implies that I will wear my boots.
        (3) When I finish studying, then I will go to the party.
        (4) Number x is a whole number only if x is an integer.
    (b) We know that the contrapositive "If $\sim q$, then $\sim p$" is equivalent to "If p, then q." Use this fact and the work we did in (a) to write a statement equivalent to each statement in (a). [For example, the statement "John pays taxes if he works" can be rewritten as "If John works, then he pays taxes." Thus the contrapositive "If John does not pay taxes, then he does not work" is equivalent to the original statement: "John pays taxes if he works."]

2. Work in pairs. Complete this exercise individually and then compare and discuss your work with your partner. One of De Morgan's Laws states that for any statements p and q, $\sim(p \vee q)$ is equivalent to $(\sim p) \wedge (\sim q)$.
    (a) Identify statements p and q so that the following statement can be understood as "p or q."
        Amelia made an A in mathematics, or Amelia made a B in English.
    (b) One way to write the negation of the statement in (a) is
        It is not true that Amelia made an A in mathematics or Amelia made a B in mathematics.
        Write the negation another way using one of De Morgan's Laws.
    (c) Use one of De Morgan's Laws to write the negation of this statement:
        Ardie can pay me now, or Ardie will work for me later.

3. Work in pairs. Complete each part of this exercise individually, and then compare and discuss your work with your partner. In the section we learned that the negation of $p \to q$ is equivalent to $p \wedge \sim q$.
    (a) Identify statements p and q so that the following statement can be understood as "$p \to q$."

If the month is March, then the azaleas are in bloom.
(b) One way to write the negation of the statement in (a) is
It is not true that if the month is March, then the azaleas are in bloom.
Write the negation another way.
(c) Use this same equivalence to write the negation of this statement:
If figure *ABCD* is a rectangle, then it is a parallelogram.

4. *(Writing Exercise)* Work with a partner to solve the following puzzle problem involving conditionals. You may find the suggested three-step outline helpful in your solution. After arriving at a solution, each of you should write a careful exposition of the solution, including any charts or tables that you used in the process.

Three star basketball players truthfully reported their grades to their coach.

Anders: If I passed math, then so did Barker.
I passed English if and only if Carlos did.

Barker: If I passed math, then so did Anders.
Anders did not pass Spanish.

Carlos: Either Anders passed Spanish or I did not pass it.
If Barker did not pass English, then neither did Anders.

Each of the three players passed at least one subject. Each subject was passed by at least one of the three. Carlos did not pass the same number of subjects as either of the other players. Determine which subjects each of the three players passed.

• Both Anders and Barker made statements about mathematics. In view of their statements, we see that either they both passed mathematics or they both failed mathematics. We have summarized these two cases in charts.

•

	A	B	C			A	B	C
M	P	P			M	F	F	P
E					E			
S					S			
	Case I					**Case II**		

In the second case, how do we know that Carlos must have passed mathematics?

• Use Anders's statement about English to place more information in the charts. Observe that this will require you to create two possible charts for Case I and two possible charts for Case II.

• Use the other information in the problem to eliminate three of the cases and determine the academic success of the three basketball players.

Exercise Set 1.2

1. Let A represent "It is snowing"; B, "The roofs are white"; C, "The streets are slick"; and D, "The trees are covered with ice." Write the following in symbolic notation:
(a) If it is snowing, then the trees are covered with ice.
(b) If it is not snowing, then the roofs are not white.
(c) If the streets are not slick, then it is not snowing.
(d) If the streets are slick, then the trees are not covered with ice.

2. In Exercise 1, assume that A is true, that B and C are both false, and that D is true. Classify the conditional statements as either true or false.

3. State the converse, the inverse, and the contrapositive of each of the following conditionals.
(a) If a triangle is a right triangle, then one angle has a measure of 90°.
(b) If a number is a prime, then it is odd.
(c) If two lines are parallel, then alternate interior angles are equal.
(d) If Joyce is smiling, then she is happy.

4. Find the truth value of each of the following statements.
   (a) If the moon orbits the earth, then $2 + 2 = 5$.
   (b) If Lincoln was the first United States president, then $2 + 2 = 5$.
   (c) If Mexico is the southern neighbor of the United States, then $3 + 3 = 6$.
   (d) If Canada is in Asia, then $3 + 3 = 6$.

5. Write the converse, inverse, and contrapositive of each statement in Exercise 4.

6. Using the notation of Exercise 1, translate the following into sentences:
   (a) $\sim A \to B$
   (b) $\sim C \to \sim B$
   (c) $(\sim B \wedge \sim C) \to A$
   (d) $(A \vee B) \to \sim C$
   (e) $\sim (A \to \sim B)$
   (f) $(\sim C \vee \sim A) \to \sim B$

7. In the game show "Truth or Despair," the contestants are presented with two doors. Behind one door there is a new car; behind the other door there is a skateboard. On each door there is a sign. Contestants are told that exactly one of the two signs is true. Explain carefully how to determine which door leads to the new car if these are the signs seen by the contestant:
   The sign on the first door reads, "If there is a car behind this door, then there is a skateboard behind the other door."
   The sign on the second door reads, "There is a car behind this door."

8. After Jack climbed the beanstalk and reached the Land of the Giants, he discovered that each giant belonged to one of two clans, the clan of the Liars or the clan of the Truthtellers. Every statement made by a Liar is false; every statement made by a Truthteller is true. Explain in each case what Jack can learn about which clan the giants to whom he is talking belong.
   (a) Jack meets a giant couple and asks the husband to tell him something about the couple. The giant husband answers, "If I am a Truthteller, then so is my wife."
   (b) Jack meets another giant couple and again asks the husband to tell him something about the couple. The giant husband answers, "I am a Truthteller if and only if my wife is a Truthteller."

9. Commercials and advertisements are often based on the assumption that naive audiences will accept that the converse, inverse, contrapositive, and original statement are all true. Write the converse, inverse, and contrapositive of the following statements. As you write, think about whether they have the same truth values.
   (a) If you brush your teeth with White-as-Snow, then you have fewer cavities.
   (b) If you like this book, then you love mathematics.
   (c) If you are strong, then you eat Barlies for breakfast.
   (d) If you use Wave, then your clothes are bright and colorful.

Understanding conditionals is essential to being able to read contracts and other legal and government documents. In Exercises 10 and 11, answer the following questions based on a conditional statement in such a document.

10. From an insurance policy: If the recomputed premium exceeds the premium stated on the declaration page, then you must pay the excess to Guarantee Auto.
    (a) Suppose the recomputed premium is $800 while the premium on the declaration page is $600. What must the insured person do?
    (b) Suppose the recomputed premium is $550 while the premium on the declaration page is $600. What does this policy statement indicate that the insured person must do?

11. From IRS instructions: If your return is more than 60 days late, the minimum penalty will be $100 or the amount of any tax you owe, whichever is smaller.
    (a) Suppose the return is 65 days late and you owe $75 in taxes. What is the minimum penalty?
    (b) Suppose the return is 59 days late. What does this policy say about the penalty?

12. Write the following in "if-then" form:
    (a) Carelessness leads to accidents.
    (b) Whenever I see June, my heart throbs.
    (c) A triangle is isosceles if two sides are congruent.
    (d) I will be happy if I pass.

13. (a) Use truth tables to show that $p \wedge \sim q$ is equivalent to $\sim (p \rightarrow q)$.
    (b) Use the result from (a) to rewrite "Nakisha is a music major, but she does not sing well."
    (c) Use the result from (a) to write the negation of the conditional "If my dog has fleas, then he scratches often."

14. (a) Use truth tables to show that $p \vee q$ and $\sim p \rightarrow q$ are logically equivalent.
    (b) Use the result from (a) to rewrite the statement "Either I do not fail English, or I go to summer school."
    (c) Use the result from (a) to rewrite "If I stay up all night, then I will skip the luncheon."

15. Use the result from Exercise 13(a) to rewrite these statements:
    (a) Ed drives a truck, and he does not wear socks.
    (b) Ana lives in Memphis, and she rides a bus to work.
    (c) It is not true that if the black dog sits, then he wants to eat.
    (d) It is not true that if the phone rings, then trouble follows.

16. Use the result from Exercise 14(a) to rewrite these statements:
    (a) Either I do not eat supper, or I go to bed early.
    (b) Ed sleeps late, or he is very sluggish.
    (c) If the frog does not leave the pond, then he will be lonesome.
    (d) If the cat catches a mouse, then she is quite content.

17. Use truth tables to verify De Morgan's Laws.
    (a) $\sim (p \vee q)$ is equivalent to $\sim p \wedge \sim q$.
    (b) $\sim (p \wedge q)$ is equivalent to $\sim p \vee \sim q$.
    (c) Use (a) to write the negation of "I will take history, or I will take physics."
    (d) Use (b) to write the negation of "I am on the baseball team, and I am on the basketball team."

18. In this section we saw that the biconditional "p if and only if q" is equivalent to the statement "If p then q and if q then p." Use this fact to rewrite the following statements.
    (a) A triangle is equilateral if and only if the triangle is equiangular.
    (b) If a dancer is graceful then he is athletic, and if a dancer is athletic then he is graceful.
    (c) A mathematician is charming if and only if she is asleep.
    (d) If a connected graph with n vertices is a tree, then it has $n - 1$ edges, and if a connected graph with n vertices has $n - 1$ edges, then it is a tree.

19. Write each of the following statements in "if-then" form.
    (a) Triangles are not squares.
    (b) Birds of a feather flock together.
    (c) Honest politicians do not accept bribes.

20. Write the converse and contrapositive of each part of Exercise 19.

21. Using truth tables, determine whether the following pairs of statements are equivalent.
    (a) $p \vee q; \sim p \rightarrow q$
    (b) $\sim (p \wedge q); \sim p \wedge \sim q$
    (c) $\sim (p \vee q); \sim p \vee \sim q$
    (d) $p \rightarrow q; \sim p \rightarrow \sim q$

22. Dale, Sigmund, and Bob each own a car. One of them owns a Saturn, one owns a Honda, and one owns a Plymouth. One car is white, one car is blue, and one car is gray. Given the following information, determine the color and owner of each car.

   Dale does not own the Saturn, and his car is not blue.

   If the Honda does not belong to Dale, then it is white.

   If the blue car either is the Saturn or belongs to Sigmund, then the Plymouth is white.

   If the Honda is either gray or white, then Sigmund does not own the Saturn.

23. Use one of De Morgan's Laws in writing the contrapositive of the following statements.
   (a) If two lines do not intersect, then the lines are either parallel or skew.
   (b) If p is true and q is true, then the conjunction $p \wedge q$ is true.

24. A *contradiction* is a compound statement that is false for all possible truth values of the simple statements that compose it. A contradiction is said to be logically false. Use truth tables to show that the following statements are contradictions.
   (a) $p \wedge \sim p$.   (b) $(p \rightarrow q) \wedge (p \wedge \sim q)$.

25. A *tautology* is a compound statement that is true for all possible values of the simple statements that compose it. A tautology is said to be logically true. Use truth tables to determine which of the following statements are tautologies.
   (a) $p \rightarrow p$ (c) $(p \rightarrow q) \rightarrow p$
   (b) $[p \wedge (p \rightarrow q)] \rightarrow \sim q$ (d) $[(\sim p \vee q) \wedge p] \rightarrow q$

26. Using a truth table, determine whether $r \vee (s \wedge t)$ and $(r \vee s) \wedge (r \vee t)$ are logically equivalent.

1.3 Quantifiers, Venn Diagrams, and Valid Arguments

All serious environmentalists support Senate Bill 137. Senator Snodgrass supports Senate Bill 137. Therefore, it is clear that Senator Snodgrass is a serious environmentalist.

If something about this reasoning disturbs you, then you are wise. We need to understand more about quantified statements (statements using words like *all* and *some*) and about the characteristics of a valid argument. Then we will be able to better evaluate Senator Snodgrass' commitment to environmental concerns.

In Section 1.1 we saw that open sentences like "He is a politician" and "$x + 2 = 5$" are not statements. However, we can modify these sentences to make statements by providing the identity of "He" in the first sentence and of "x" in the second sentence. "George Bush is a politician" is clearly a true statement, whereas "For $x = 2$, $x + 2 = 5$" is a false statement.

Another way to make sentences such as "He is a politician" and "$x + 2 = 5$" into statements is to use words like *all* and *some*. "All men are politicians" is false, whereas "For some number x, $x + 2 = 5$" is clearly true. Words such as *all*, *some*, and *no* are called **quantifiers** and must be given special attention when used in building statements. Quantifiers give information about "how many" in the statements in which they occur. Statements involving quantifiers are called **quantified statements.**

Examples of quantified statements include the following:

Some women have red hair.	No professors are bald-headed.
All bananas are yellow.	Some students do not work hard.

UNIVERSAL QUANTIFIERS

The words *all*, *every*, and *each* are called **universal quantifiers** because when these words are added to an open sentence to make it a statement, the sentence must be true in all possible instances in order for the statement to be true.

All prime numbers greater than 2 are odd.
Every automobile pollutes the atmosphere.
For each number x, $x + 3 = 3 + x$.
All men have black hair.

Statements written using universal quantifiers have the property that with a little thought, they can also be written as conditional statements.   "All illegal acts are immoral" can be written as "If an act is illegal, then it is immoral."

Example 1

　Rewrite (a) as a conditional statement and rewrite (b) as a universally quantified statement.
　(a) All teachers own a computer.
　(b) If an animal is a dog, then the animal loves to bury bones.

Solution

　(a) If a person is a teacher, then he or she owns a computer.
　(b) All dogs love to bury bones.

EXISTENTIAL QUANTIFIERS

Other quantified statements are intended to indicate that there exists at least one case in which the statement is true. Such statements generally involve one of the **existential quantifiers**: *some, there exist*, or *there exists at least one*.

> Some men have black hair.
> There exist students who do not work hard.
> There exists at least one student who does not work hard.

Example 2

Make a statement out of the sentence "The man is tall" using
(a) an existential quantifier
(b) a universal quantifier

Solution

(a) Some men are tall. (At least one man is tall.) [This is a true statement.]
(b) All men are tall. [This is a false statement.]

NEGATION OF QUANTIFIERS

Sometimes confusion arises when we try to write the negation of a quantified statement. In order to understand the source of this confusion, suppose we have a box in which we have placed a collection of letters from the alphabet. Consider the statement "All letters in the box are vowels." How would you write the negation of this statement? Before we determine the correct negation, let's look at some possible contents of the box.

	Contents of Box	**Description**
Case 1	*a e i o*	All of the letters are vowels.
Case 2	*a e t s*	Some letters are vowels; some are not.
Case 3	*r s t b*	None of the letters are vowels.

Possible negations of the statement "All letters in the box are vowels" would include the following, two of which are erroneously used as negations.
(a) No letters in the box are vowels.
(b) All letters in the box are not vowels.
(c) Some letters in the box are not vowels.

To be a negation, the newly formulated statement must have truth values that are the opposite of the truth values of the original statement in every possible situation. Let's examine the three candidates for this negation in view of the three cases identified above. Table 1 presents a truth table for these cases.

Table 1: Truth Table			
Statement and Possible Negations	**Case 1** {*a, e, i, o*}	**Case 2** {*a, e, t, s*}	**Case 3** {*r, s, t, b*}
Statement: All letters are vowels.	T	F	F
(a) No letters are vowels.	F	F	T
(b) All letters are not vowels.	F	F	T
(c) Some letters are not vowels.	F	T	T

Only the third possible negation (c) has truth values that are the opposite of the truth values of the original statement in all cases. We can use "Some letters in the box are not vowels" as a negation of "All letters in a box are vowels."

The other two possibilities are not negations of the original statement, because they yield the same truth value as the original statement in Case 2.

As the preceding discussion suggests, it is sometimes difficult to state correctly the negations of statements involving quantifiers. The following are patterns that can be used for forming the negation of quantified statements:

Statement	Negation
All *p* are *q*.	Some *p* are not *q*.
Some *p* are *q*.	No *p* is *q*. All *p* are not *q*.
Some *p* are not *q*.	All *p* are *q*.
No *p* is *q*.	Some *p* are *q*.

Example 3

Write the negation of each statement.
(a) Some women have red hair. (At least one woman has red hair.)
(b) All bananas are yellow.
(c) No professor is bald-headed.
(d) Some students do not work hard. (At least one student does not work hard.)

Solution

(a) "No woman has red hair" or "All women have hair that is not red."
(b) Some bananas are not yellow.
(c) Some professors are bald-headed. (At least one professor is bald-headed.)
(d) All students work hard.

To show that a universally quantified statement is false, you need find only one case for which the statement is false; that is, you need identify only one **counterexample**. However, in showing that an existentially quantified statement is false, you must show that it is false for all possibilities. Similarly, an existentially quantified statement is true if you can find one case for which it is true, but a universally quantified statement is true only if it is true for all cases.

VENN DIAGRAMS

A diagram in which the interiors of simple closed curves such as circles are used to represent collections of objects (or sets) is called a **Venn diagram**. Venn diagrams provide geometrical representations of the relationships indicated by statements involving quantifiers.

Example 4

Draw Venn diagrams for the following statements.
(a) All dogs (*D*) are animals (*A*).
(b) Some kindergarten students (*K*) are able to read (*R*).
(c) No cat (*C*) is a dog (*D*).

continued

Solution

(a) We will represent the collection or set of all dogs by the circular region labeled *D*. Similarly we represent the collection of animals by the circular region labeled *A*. Since all dogs are animals, the circular region *D* lies entirely within the circular region *A*. (See Figure 1(a).) If part of the circular region *D* were outside the circular region *A*, this would imply that some dogs are not animals. Similarly, if the circular region *A* were entirely within the circular region *D*, this would imply that all animals are dogs. Note that in Figure 1(a), the dot labeled *x* represents an animal that is not a dog.

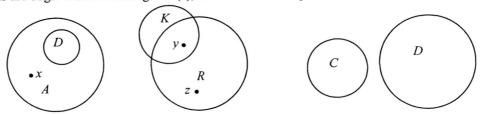

Figure 1(a)   **Figure 1(b)**   **Figure 1(c)**

(b) We use the circular region *R* to represent the collection of all readers and the circular region *K* to represent all kindergarten students. The only thing we know for certain is that at least one point in *K* is in *R*. Ordinarily we illustrate this with a diagram like the one shown in Figure 1(b). Note that in Figure 1(b), the dot labeled *y* represents a child who is both in kindergarten and can read, and *z* represents a child who can read but is not in kindergarten.

(c) The circular region *C* representing cats shares no points with the circular region *D* representing dogs, because not one cat is a dog. (See Figure 1(c).)

VALID ARGUMENTS

Consider the following three statements:

    All leghorns are chickens.
    All chickens are fowls.

    Therefore, all leghorns are fowls.

These three statements form what is called an **argument** in logic. The two statements above the line are called the **premises**, and the statement below the line is called the **conclusion**. We say the argument is **valid** if in all cases where the premises are true, the conclusion is true. When quantified statements are used in the argument, we quite often can use Venn diagrams to determine whether the reasoning is valid.

Example 5

    Consider the following two arguments. Use Venn diagrams to determine whether either is valid.

(a) All leghorns are chickens.   (b) All leghorns are chickens.
    All chickens are fowls.   All chickens are fowls.
    -------------------------------------   -------------------------------------
    Therefore, all leghorns are fowls.   Therefore, all fowls are leghorns.

Solution

Figure 2

continued

(a) In Figure 2, let the interior of the small circular region represent leghorns. According to   the first premise, all leghorns (*L*) are chickens (*C*), so the circular region of leghorns lies entirely *within* the circular region of chickens. Likewise, the circular region of chickens (*C*) is within the rectangle of fowls (*F*). Thus anything within the circular region of leghorns (*L*) is automatically within the square of fowls (*F*). Therefore, all leghorns are fowls. Thus the argument is valid.

(b) Figure 2 includes a fowl (at *X*) that is not a leghorn. Thus the second argument is not valid.

In Examples 6 and 7 we will illustrate again how to use Venn diagrams to show that an argument is valid or not valid.

Example 6

(a) Represent the following statement with a Venn diagram.
All football players are excellent students.

(b) Suppose Manuel is a football player. Locate him in the Venn diagram.

(c) Explain why the following is a valid argument.
All football players are excellent students.
Manuel is a football player.

Therefore, Manuel is an excellent student.

Solution

(a) See the Venn diagram in Figure 3, where *F* represents football players and *S* represents good students.

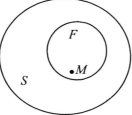

Figure 3

(b) Manuel can be located anywhere within the circular region *F*. See the placement of *M* in Figure 3.

(c) Since all of circular region *F* is inside of circular region S, Manuel is inside of *S* no matter where he is placed. Therefore the argument is valid.

Example 7

(a) Represent the following statement with a Venn diagram.
Some football players are excellent students.

(b) Suppose Thomas is a football player. Find two different places that he could be located in the Venn diagram.

(c) Explain why the following is not a valid argument.
Some football players are excellent students.
Thomas is a football player.

Therefore, Thomas is an excellent student.

continued

Solution

(a) See the Venn diagram in Figure 4, where F represents football players and S represents good students.

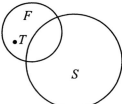

Figure 4

(b) Thomas can be located anywhere within the circular region F. In particular, we have indicated one possible location for Thomas with the letter T in Figure 4. Notice that in this case, Thomas is not in the circular region S.

(c) It is possible to place Thomas in the Venn diagram in such a way that he is not in circular region S and hence is not a good student. Therefore, the argument is not valid.

Let's turn now to the problem posed in the first paragraph of this section.

Example 8

Use a Venn diagram to determine whether the following argument is valid or invalid.
All serious environmentalists support Senate Bill 137. Senator Snodgrass supports Senate Bill 137.
Therefore, it is clear that Senator Snodgrass is a serious environmentalist.

Solution

Let E represent people who are serious environmentalists. All of these people support Senate Bill 137 (C). But as Figure 5 indicates, Senator Snodgrass (S) may support Senate Bill 137 (C) without necessarily being inside E. Therefore, the argument is not valid.

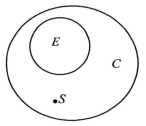

Figure 5

Exploratory Exercise Set 1.3

1. Work in pairs. Each partner should write a statement using a universal quantifier and a second statement using an existential quantifier.

(a) Exchange the statements written with a universal quantifier.

(1) Write a negation for your partner's statement.

(2) Write your partner's statement in the form "If p, then q."

(3) Compare your work with that of your partner and discuss the correctness of your answers.

(b) Exchange the statements written with an existential quantifier.
  (1) Write a negation for your partner's statement.
  (2) Compare your work with that of your partner and discuss the correctness of your answers.

2. Consider the contents of these three baskets of flowers.
   Basket 1: Red roses, red peonies, red carnations
   Basket 2: Red roses, yellow daisies, pink carnations
   Basket 3: Yellow daisies, lavender orchids, pink carnations
Complete the following table in which you record the truth values of the four statements relative to the three baskets.

Statements	Basket 1	Basket 2	Basket 3
1. Some flowers are red.			
2. Some flowers are not red.			
3. No flowers are red.			
4. All flowers are not red.			

From the evidence of this table, which of Statements 2, 3, and 4 can be used as the negation of Statement 1?

3. Lewis Carroll, the name that appears as the author of the children's classic *Alice in Wonderland*, was the pen name of Charles Dodgson, an Oxford mathematician whose specialty was logic. In addition to writing *Alice in Wonderland*, he wrote many logical puzzles for children and other nonmathematicians. Work in pairs to determine whether you can find a conclusion that uses each of the premises in the following puzzles.
  (a) Clever use of Venn diagrams and thoughtful use of quantifiers can be helpful in finding the conclusion that suits the puzzle. Consider these three premises and the Venn diagram beside it. In the Venn diagram, *I* represents "Illogical" and *D* represents "Despised."

Babies are illogical.
Nobody is despised who can manage a crocodile.
Illogical persons are despised.

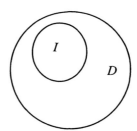

  (1) Write the phrase "Babies are illogical" using the universal quantifier *all*.
  (2) Write the phrase "Illogical persons are despised" using the universal quantifier *all*.
  (3) Place a circular region for "Persons who can manage a crocodile" and a circular region for "Babies" in the Venn diagram, and explain why "Babies cannot manage crocodiles" is a conclusion that satisfies the premises in the puzzle.
  (b) For each of the following puzzles by Lewis Carroll, find a conclusion that uses all the premises. Thoughtful use of Venn diagrams may be helpful.
  (1) No ducks waltz.
     No officers ever decline to waltz.
     All my poultry are ducks.
     ----
  (2) No experienced person is incompetent.
     Jenkins is always blundering.
     No competent person is always blundering.
     ----

(3) All puddings are nice.
This dish is a pudding.
No nice things are wholesome.

----

(4) Nobody who really appreciates Beethoven fails to keep silence while the Moonlight   Sonata is being played.
Guinea pigs are hopelessly ignorant of music.
No one who is hopelessly ignorant of music ever keeps silence
while the Moonlight Sonata is being played.

----

4. *(Writing Exercise)* Write a mini-lecture for your class on the appropriate way to negate a universal quantifier. You may wish to refer to your work in Exploratory Exercise 1 and the discussion prior to Example 3.

Exercise Set 1.3

1. Make a Venn diagram showing each of the following:
    (a) All ims are ems.
    (b) Some men are over six feet tall.
    (c) No cams are dills.
    (d) All men are people, and some men are lazy.

2. Make a Venn diagram showing
    (a) the relationship among cats, dogs, and animals
    (b) the relationship among boys, girls, and blond people
    (c) some x's are y's; some x's are z's; no y's are z's
    (d) some y's are x's; no z's are x's; some z's are y's
    (e) all x's are y's; some y's are z's; no x's are z's

3. Assign a value for x that makes each open sentence a true statement.
    (a) $x + 4 = 7$ (c) $x^2 = 16$
    (b) $x - 7 = 4$ (d) $x + 3 = 3 + x$

4. If possible, assign a value for x that makes each open sentence in Exercise 3 a false statement.

5. Use a quantifier to make each open sentence in Exercise 3 into a true statement.

6. Use a quantifier to make each open sentence in Exercise 3 into a false statement.

7. Write the negation of each statement without using the expression "It is not true that."
    (a) All athletes over 7 feet tall play basketball.
    (b) Some students work hard at their studies.
    (c) All men use Bob-Bob hair oil.
    (d) Some professors are not intelligent.
    (e) No man weighs more than 500 pounds.
    (f) All Martians are green and have three eyes, or my name is not Captain Midnight.
    (g) Every mystery novel involves either a murder or a bungled robbery.

8. Write the negation of each statement. Do not use "It is not true that."
    (a) There exists a counting number greater than 50.
    (b) Not all counting numbers are greater than 5.
    (c) There exists an x such that $x + 2 = 7$.
    (d) For all x, $(x + 3) + 2 = 3 + (2 + x)$.
    (e) For all x, $x(x + 2) = x^2 + 2x$.
    (f) For some a, $a(a + 1) = a^2 + a$.

9. State the premises and the conclusion for each of these arguments.
    (a) All donors are generous.   (b) All diligent students succeed.
         Some young persons are donors.   All education majors are diligent.
         ------------------------------------   ---
         Therefore, some young persons are generous.   Therefore, all education majors succeed.

10. Write each of these statements as a conditional.
    (a) All frogs are green.
    (b) All cats are intelligent.
    (c) Each college professor is brilliant.
    (d) No people with long legs are athletes.

11. Show that the following statements are false by finding a counterexample.
    (a) All counting numbers are even.
    (b) All governors of states have been men.

12. Write each of these conditionals as a universally quantified statement.
    (a) If a figure is a square, then it is a rectangle.
    (b) If a number is divisible by 4, then it is divisible by 2.

13. (a-b) Write a negation for each part of Exercise 12 by using the fact that $\sim(p \rightarrow q)$ is equivalent to $p \wedge \sim q$.
    (c-d) Write a negation for each part of Exercise 12 by using the rules for forming negations of universally
          quantified statements.

14. Write a quantified statement to describe the relationship represented by each of the Venn diagrams.

(a)    (b)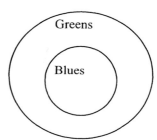

15. Describe Dr. X as completely as possible.

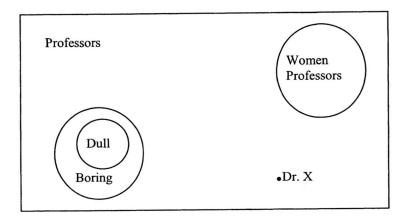

In Exercises 16 through 21, use Venn diagrams to determine whether the reasoning is valid. That is, determine whether the third statement (the conclusion) must be true if it is assumed that the first two statements (the premises) are true.

16. All fast runners are athletes. No people with long legs are athletes. Therefore, no people with long legs are fast runners.

17. All spheres are round. All circular regions are round. Therefore, all circular regions are spheres.

18. Carelessness always leads to accidents. Mrs. Yeager had an accident. Therefore, Mrs. Yeager was careless.

19. All cats are intelligent animals. Fido is an intelligent animal. Therefore, Fido is a cat.

20. Some ems are red. All ims are ems. Therefore, some ims are red.

21. All athletes are fast runners. Some people with short legs are not fast runners. Therefore, no athletes have short legs.

22. Write a valid conclusion for the following premise: All sales must come from production, and consumption must come from sales.

In Exercises 23 through 25, use Venn diagrams to select valid conclusions from the given possibilities.

23. All x's are y's.
    Some x's are z's.
    Some x's are w's.
    Therefore,
    (a) all w's are y's.
    (b) all z's are y's.
    (c) some y's are w's.
    (d) some x's are not y's.

24. All whole numbers are integers.
    All integers are rational numbers.
    No irrational numbers are rational numbers.
    All rational numbers are real numbers.
    All irrational numbers are real numbers.
    x is an integer.
    Therefore,

    (a) x is a whole number.   (c) x is an irrational number.
    (b) x is a real number.   (d) x is not a rational number.

25. All authors are intelligent.
    Some people from New York are authors.
    Some people from Philadelphia are not intelligent.
    Therefore,

    (a) no person from Philadelphia is an author.
    (b) some people from New York are intelligent.
    (c) no authors are from New York.

26. Use Venn diagrams to test whether the conclusion is a necessary result of the given facts.
    (a) All students who study hard make A's.
        All math students study hard.
        Fred is a math student.

        Therefore, Fred makes an A.
    (b) Some business majors find jobs.
        All marketing majors are business majors.

        Therefore, some marketing majors find jobs.

27. In mathematics we often use the symbol "∃" to represent an existential quantifier and the symbol "∀" to represent a universal quantifier. Hence,
    "$\exists x, x + 2 = 7$" is read "For some $x, x + 2 = 7$" and
    "$\forall x, (x + 3) + 2 = 3 + (2 + x)$" is read "For all $x, (x + 3) + 2 = 3 + (2 + x)$."
    Rewrite the following statements using the symbols ∃ and ∀.
    (a) For some $x, x^2 = 4$.
    (b) For all $x, x^2 = 4$.
    (c) For some triangles, the sum of the measures of the interior angles is 180°.
    (d) For all triangles, the sum of the measures of the interior angles is 180°.

1.4 Thinking About Valid Arguments

Consider the following two sets of statements.

Argument A:
If Daniel studies, then he will make an A.
If Daniel makes an A, then he will make the
Dean's List.
Daniel studies.

Argument B:
If there is a path connecting each pair of vertices
in a graph, then the graph is connected.
In graph G there is a path connecting each pair of
vertices.

Can you imagine a conclusion that follows from the statements in Argument A? Did you choose the statement "Daniel makes the Dean's List"? Were you able to deduce a conclusion from Argument B? Did you conclude, "Graph G is connected"? If so, then you are reasoning clearly. Further, you are skilled at two of the classic patterns of reasoning that form the heart of deductive reasoning.

As we observed in Section 3, the collection of statements

> If Daniel studies, then he will make an A.
> If Daniel makes an A, then he will make the Dean's List.
> Daniel studies.
> ---
> Therefore, Daniel makes the Dean's List.

is an example of a **deductive argument**. Both in mathematics and in everyday affairs, arguments arise in which we need to deduce a correct conclusion from a given set of statements. In Section 3, we noted that an argument consists of two parts: a set of two or more statements called **premises** and a single statement called the **conclusion**. An argument is **valid** if the conclusion is true in every circumstance that the conjunction of the premises is true. If, in some case, the conjunction of the premises is true and the conclusion is false, then the argument is **invalid**. Invalid arguments are sometimes called **fallacies**. Note that *a valid argument may have a conclusion that is false if any one of the premises fails to be true.*

Arguments are often written symbolically by naming the statements that form the argument:

> If inflation occurs, then the price of automobiles increases. $p \rightarrow q$
> Inflation occurs. p
> ---   --------
> Therefore, the price of automobiles increases. $\therefore q$

The three dots $\therefore$ are read as *therefore*

When verbal arguments are converted into symbolic form, we find that the same patterns often occur. For example, the pattern in the previous paragraph,

$$(p \rightarrow q) \wedge p; \therefore q$$

occurs frequently. In Example 1 we will see that that arguments that follow this pattern are always valid. This pattern of reasoning is called the **rule of detachment**, affirming a true hypothesis, or, classically, *modus ponens*.

Example 1

Use a truth table to verify the rule of detachment.

Solution

Remember that in order for an argument to be valid, the conclusion must be true in every circumstance in which the conjunction of the premises is true. Table 1 presents the required truth table.

Table 1				
p	q	$p \to q$	$(p \to q) \land p$	q
T	T	T	T	T
T	F	F	F	F
F	T	T	F	T
F	F	T	F	F

Notice that for the one case in which the conjunction of the premises $(p \to q) \land p$ is true, the conclusion q is also true. Thus the argument is valid.

Rule of Detachment: Because in every case where $(p \to q) \land p$ is true, q is true, the argument

$$p \to q$$
$$p$$
$$\text{-------}$$
$$\therefore q$$

is valid. As we have noted, this pattern of argument is called the rule of detachment.

Example 2

Determine the validity of the following argument.

If John is a thief, then he is a lawbreaker.	$p \to q$
John is a thief.	p
-------------------------------------	-------
Therefore, John is a lawbreaker.	$\therefore q$

Solution

This argument is valid because of the rule of detachment. Observe that because we recognize this previously established pattern, we do not have to do a truth table to show that the argument is valid.

Another often-used pattern of argument, which can be stated as a theorem and proved by means of a truth table, is the **chain rule** or, classically, *hypothetical syllogism*.

CHAIN RULE

The following argument is valid and is called the *chain rule*.

$$p \rightarrow q$$
$$q \rightarrow r$$

$$\therefore p \rightarrow r$$

Table 2 verifies that this argument is valid. Observe that in each of the four cases in which the conjunction of the premises is true, the conclusion is true.

Table 2						
p	q	r	$p \rightarrow q$	$q \rightarrow r$	$(p \rightarrow q) \wedge (q \rightarrow r)$	$p \rightarrow r$
T	T	T	T	T	T	T
T	T	F	T	F	F	F
T	F	T	F	T	F	T
T	F	F	F	T	F	F
F	T	T	T	T	T	T
F	T	F	T	F	F	T
F	F	T	T	T	T	T
F	F	F	T	T	T	T

Example 3

Friends of Joyce know that the two conditionals

If Joyce is smiling, then she is happy.
If Joyce is happy, then she is polite.

are both true statements. Thus they know, by the chain rule, that "If Joyce is smiling, then she is polite," is a true statement.

An extension of the chain rule gives
$$[(p \rightarrow q) \wedge (q \rightarrow r) \wedge (r \rightarrow s) \wedge \ldots \wedge (w \rightarrow z)] \rightarrow (p \rightarrow z)$$

Example 4

Show that the following argument is valid.

If a woman is a good speaker, then she is a good teacher.	$p \rightarrow q$
If a woman is a good teacher, then she is friendly.	$q \rightarrow r$
If a woman is friendly, then she is polite.	$r \rightarrow s$
If a woman is polite, then she is well liked.	$s \rightarrow t$
--	---------
Therefore, if a woman is a good speaker, she is well liked.	$\therefore p \rightarrow t$

Solution

This argument is valid by the extension of the chain rule.

In the next example, we use our knowledge of equivalent statements to test the validity of the argument.

Example 5

Show that the following argument is valid.

If John is telling the truth, then Earl is guilty.
Earl is not guilty.

Therefore, John is not telling the truth.

Solution

Because a conditional and its contrapositive are equivalent, $p \rightarrow q$ may be replaced by $\sim q \rightarrow \sim p$. Thus, the argument becomes $(\sim q \rightarrow \sim p) \wedge \sim q$; therefore, $\sim p$. This argument is valid by the rule of detachment.

In Example 5, we used the fact that a conditional can always be replaced by its contrapositive to arrive at a valid argument. This same reasoning can be used to derive another common pattern of argument from the rule of detachment. This pattern is called the **rule of contraposition**, denying the conclusion, or, classically, *modus tollens*.

Rule of Contraposition: The argument

$$p \rightarrow q$$
$$\sim q$$

$$\therefore \sim p$$

is valid and is called the *rule of contraposition*.

Of course, we can also justify the rule of contraposition with truth tables and will do so in Exercise 9.

Example 6

Show that the following argument is valid.

If I go bowling, I will not study.	$p \rightarrow \sim q$
If I do not study, I will take a nap.	$\sim q \rightarrow r$
I will not take a nap.	$\sim r$
--------------------------------------	-----------
Therefore, I will not go bowling.	$\therefore \sim p$

Solution

Now $p \rightarrow \sim q$ and $\sim q \rightarrow r$ lead to the conclusion $p \rightarrow r$ by the chain rule. By the rule of contraposition, $p \rightarrow r$ and $\sim r$ yield the valid conclusion, $\sim p$.

Example 7

Show that the following argument is not valid.

If someone is a college graduate, then he or she is intelligent.	$p \rightarrow q$
Juan is intelligent.	q
--	-------
Therefore, Juan is a college graduate.	$\therefore p$

Solution

We will examine the validity of this argument by constructing the truth table in Table 3.

Table 3				
p	q	$p \to q$	$(p \to q) \wedge q$	p
T	T	T	T	T
T	F	F	F	T
F	T	T	T	F
F	F	T	F	F

Observe that when p is F and q is T, the conjunction of the premises, $(p \to q) \wedge q$, is T whereas p is F. Thus the argument is not valid.

We need to remember that there is a difference between a valid argument and a true conclusion. **A conclusion drawn from an argument is true if the argument is valid and if all the premises are true.**

Example 8

Determine the validity of the following argument and the truth of its conclusion.

If a person is a college professor, then he or she is brilliant.
Dr. Wheeler is a college professor.

Therefore, Dr. Wheeler is brilliant.

Solution

The argument is valid, but many generations of students have wondered about the truth of the first premise. Therefore, the truth of the conclusion may be contested.

Exploratory Exercise Set 1.4

1. Choose a partner. In this exercise we will explore the rules of detachment and contraposition and the chain rule by writing arguments that use them. Working individually, answer all questions in (a), and then discuss the results with your partner. Again working individually, answer all questions in (b) and then discuss your work with your partner.
   (a) From each of the following sets of premises, write a conclusion that will create a valid argument, and indicate which of the rules of logic you used.
      (1) If you build it, then they will come.
          If they come, then you will meet Shoeless Joe.
          ----
      (2) If Amelia buys a dress, then she will buy shoes.
          Amelia buys a dress.
          ----
      (3) If Sergio sells his quota, then he gets a bonus.
          Sergio does not get a bonus.

   (b) Consider the statement "If my car is dirty, then I must wash the car."
      (1) Using this statement as the first premise, create a valid argument that follows the pattern of the rule of detachment.
      (2) Using this statement as the first premise, create a valid argument that follows the pattern of the chain rule.
      (3) Using this statement as the first premise, create a valid argument that follows the pattern of the rule of contraposition.

2. In this exercise we will explore the process of using a truth table to show that an argument is valid.
    (a) Let p be the statement "I will buy a skateboard." Let q be the statement "I will buy a scooter."
        Write the following argument in symbolic form.

        I will buy a skateboard, or I will buy a scooter
        I will not buy a skateboard.

        Therefore, I will buy a scooter.

    (b) Consider the symbolic pattern for the argument from (a). (An argument that follows this pattern is
        known classically as a *disjunctive syllogism*.)

        $$p \vee q$$
        $$\sim p$$

        $$\therefore q$$

        (1) Complete this truth table, which shows that the argument is valid.

p	q	$\sim p$	$p \vee q$	$(p \vee q) \wedge \sim p$	q
T	T	F	T	F	
T	F	F			
F	T	T			
F	F	T			

        (2) Explain why this truth table shows that the argument is valid.
    (c) Consider the following argument.

        $$p \rightarrow q$$
        $$\sim p$$

        $$\therefore \sim q$$

        (1) Complete this truth table to show that the argument is not valid.

p	q	$\sim p$	$p \rightarrow q$	$(p \rightarrow q) \wedge \sim p$	$\sim q$
T	T	F	T	F	
T	F	F			
F	T	T			
F	F	T			

        (2) Explain why the truth table shows that the argument is not valid.
    (d) Use your work in (c) to help you evaluate the validity of the following argument. Explain your
        thinking carefully.

        If it snows, then I get cold.
        It does not snow.

        Therefore, I do not get cold.

3. In Exploratory Exercise 3 of Section 3 we solved some of the logical puzzles of Lewis Carroll using
    Venn diagrams and statements with quantifiers. Work in pairs to use some of the reasoning patterns of
    this section to help decipher some more of Carroll's puzzles.
    (a) In this Carroll puzzle, rewrite each premise as a conditional.
        A. No kitten that loves fish is unteachable.
        B. No kitten without a tail will play with a gorilla.
        C. Kittens with whiskers always love fish.
        D. No teachable kitten has green eyes.
        E. No kittens have tails unless they have whiskers.
        ----
        (1) Rewrite A as the conditional "If a kitten loves fish, then...."

(2) Rewrite B as the conditional "If a kitten has no tail ...."

(3) Rewrite each of C, D, and E as a conditional.

(4) Familiarize yourself with the contrapositives of the conditionals from (1) – (3).

(5) Use the chain rule to combine the five conditionals (or their contrapositives) from (1) to (4) to arrive at a conclusion.

(b) For each of the following puzzles by Lewis Carroll, find a conclusion that uses all the premises. As an intermediate step, rewrite the premises as conditionals and examine the contrapositives of the conditionals.

(1) Babies are illogical.
Nobody is despised who can manage a crocodile.
Illogical persons are despised.
----

(2) All unripe fruit is unwholesome.
All these apples are wholesome.
No fruit, grown in the shade, is ripe.
----

(3) The only books in this library that I do not recommend for reading are unhealthy in tone.
The bound books are all well written.
All the romances are healthy in tone.
I do not recommend that you read any of the unbound books.
----

4. *(Writing Exercise)* We have seen that universally quantified statements can be written as conditionals, and vice versa. We have used Venn diagrams to check the validity of arguments using universally quantified statements, and we have determined the validity of arguments using conditional statements by identifying rules of logic or using truth tables. Please write a letter to a puzzled nephew, explaining the connection between the two ways of understanding valid arguments. In the letter please include

- An example of an argument using conditional statements that you rewrite using universally quantified statements. Once you have written the argument using universally quantified statements, check its validity using Venn diagrams.

- An example of an argument using universally quantified statements that you rewrite using conditional statements. Once you have written the argument using conditional statements, determine the validity of the arguments using either rules of logic or truth tables.

Exercise Set 1. 4

In Exercises 1 through 3, rewrite the argument in symbols, using the symbols that are suggested.

1. If you eat your squash (*s*), then you may go out and play (*p*).
You eat your squash.

Therefore, you may go out and play.

2. If you studied Latin *(L)*, then Spanish is easy *(S)*.
Spanish is not easy.

Therefore, you did not study Latin.

3. If prices increase *(p)*, then consumers will complain *(c)*.
If consumers complain, then managers will fret *(m)*.

Therefore, if prices increase, then managers will fret.

4. (a – c) For each of the arguments in Exercises 1 – 3, identify which of the rules of logic ensures that the argument is valid.

5. Identify the rule or rules of logic used to find the valid conclusion in each of the following:

   (a) $p \to \sim q$ (b) $a \to b$
        q $b \to c$
        --------   --------
        $\therefore \sim p$ $\therefore a \to c$

   (c) If two sides of a triangle are congruent, then the angles opposite these sides are congruent. Side *BC* is congruent to side *AB* in triangle *ABC*.

        Therefore, the angles opposite *BC* and *AB* are congruent.
   (d) If the movie is not over, then they will buy popcorn. They do not buy popcorn.

        Therefore, the movie is over.
   (e) If Kahleel wins, he is happy. If Kahleel is happy, he treats his sister well.

        Therefore, if Kahleel wins, he treats his sister well.

6. Find a valid conclusion to each of the following sets of premises, and indicate the rule or rules of logic that you use.
   (a) If the sun shines, the flowers grow. If the flowers grow, the garden flourishes.
   (b) If Fred hits a home run, the game is won. Fred hits a home run.
   (c) If Devron is not sick, he will be at the wedding. Devron is not at the wedding.
   (d) If Jermonte studies logic, then he will make an A in geometry. If he makes an A in geometry, then he will qualify for physics. Jermonte studies logic.

7. Draw a valid conclusion from each of the following sets of statements. Explain why your conclusion is valid.
   (a) If Susan is a freshman, then Susan takes mathematics. Susan is a freshman.
   (b) You will fail this test if you do not study. You do not fail the test.
   (c) You cry if you are sad. You are sad.
   (d) You will receive your allowance if you cut the grass. You do not receive your allowance.

8. Arrange the following statements in order to create a valid argument with the conclusion "If Joan is tall, then Joan teaches seventh grade."
   (a) If Joan has red hair, then she teaches seventh grade.
   (b) If Joan is tall, then she wears contacts.
   (c) Therefore, if Joan is tall, then she teaches seventh grade.
   (d) If Joan wears contacts, then she has red hair.

9. (a) Complete this truth table, which will show that the rule of contraposition,
$((p \rightarrow q) \wedge \sim q); \sim p$, is a valid argument.

p	q	$\sim q$	$p \rightarrow q$	$(p \rightarrow q) \wedge \sim q$	$\sim p$
T	T	F			
T	F	T			
F	T	F			
F	F	T			

(b) Explain why the truth table you completed shows that this argument is valid.

In each of Exercises 10 through 15, determine whether the argument is valid by using truth tables. Specifically, check whether the conclusion is true for each case in which the conjunction of the premises is true.

10.
$$p \rightarrow q$$
$$p \rightarrow r$$

$$\therefore q \rightarrow r$$

11.
$$p \vee q$$
$$p$$

$$\therefore \sim q$$

12.
$$p \rightarrow \sim q$$
$$q \vee r$$
$$p$$

$$\therefore r$$

13.
$$p \vee \sim q$$
$$\sim p \vee r$$

$$\therefore q \rightarrow r$$

14.
$$q \rightarrow \sim p$$
$$q \wedge r$$

$$\therefore \sim p \rightarrow \sim r$$

15.
$$q \rightarrow r$$
$$\sim p \vee q$$
$$p$$

$$\therefore r$$

In Exercises 16 through 24, write the arguments in symbolic form. Then determine whether any of the three rules of logic from this section can be used to determine that the arguments are valid. If not, check validity with truth tables.

16. If a man is a good speaker, then he is a good teacher.
Mr. Faulkner is a good teacher.

Therefore, he is a good speaker.

17. If I can't go to town, then I will go to the shopping center.
I can go to town.

Therefore, I will not go to the shopping center.

18. When the movie is over, we must go home.
The movie is over.

Therefore, we must go home.

19. You will fail this course if you do not study enough.
    You are not studying enough.

    Therefore, you will fail this course.

20. If R, then not S.
    If not S, then T.

    Therefore, if R, then T.

21. If a quadrilateral is a parallelogram, then its opposite sides are parallel.
    The opposite sides of quadrilateral $ABCD$ are parallel.

    Therefore, $ABCD$ is a parallelogram.

22. If q, then not p.
    q and r are true.

    Therefore, if not p, then not r.

23. If $a's$ are $b's$, then $c's$ are $d's$.
    $a's$ are not $b's$.

    Therefore, $c's$ are not $d's$.

In everyday discourse, we sometimes make errors by violating our understanding of elementary logic (for example, using two statements as equivalent when, in fact, they are not). Other common errors have their source in misinterpretation and misrepresentation of human experience. In Exercises 26 through 29 we describe four such errors. Explain what is wrong with the fallacies in these exercises.

24. **Fallacy of False Experts.** Tiger Woods addressed a Senate panel on the issue of aid to the country of Zimberia.

25. **Fallacy of the Loaded Question.** (a) Have you stopped beating your wife? (b) You look good in this suit; will it be cash or charge?

26. **Fallacy of Composition.** (a) Your essay is too long; therefore, each sentence in your essay is too long. (b) The choral presentation was too loud; therefore, the tenors were too loud.

27. **Fallacy of False Cause.** I saw a black cat; I lost my pocketbook. Seeing a black cat brings bad luck.

28. Write two additional examples of each of the following:
    (a) Fallacy of False Experts
    (b) Fallacy of the Loaded Question
    (c) Fallacy of Composition
    (d) Fallacy of False Cause

Sets, Functions, and Sequences

Chapter
2

2.1 Sets

Over a hundred years ago, the German mathematician Georg Cantor borrowed the everyday word *set* to describe a specific mathematical concept. The language of sets that he introduced was particularly useful in finding a precise way to talk about infinite quantities. His work provided much of the grammar and vocabulary of modern mathematics.

The "new math" that was introduced in the 1960s emphasized the language of sets to such a degree that many critics feared children and adults beginning their study of mathematics would be lost in a sea of abstract symbols. Over time, however, it has become clear that the language of sets can be a very useful mode of representation that clarifies ideas across a broad spectrum of introductory mathematics. Perhaps much of the backlash in response to the "new math" could have been avoided if more emphasis had been placed on showing how set language and other abstractions can be useful tools for clarifying or drawing connections among more "traditional" mathematical ideas. The basic vocabulary of sets explored in this section, probably familiar already to many students, should not be considered as a topic in isolation. Rather, it will be used throughout the rest of the book to provide a clearer picture of the topics that follow.

DEFINING SETS

What is a set? We define a **set** to be a collection of objects; the objects themselves are called the **elements** (or **members**) of the set. We typically use a capital letter such as A, B, or C to represent a set and use a lower-case letter such as a, b, or c to represent an element of the set. We use the notation $a \in A$ to mean that the object a **belongs to** (or **is an element of**) the set A. If a is *not* an element of the set A, we write $a \notin A$.

There are a few sets that we will encounter frequently in our discussion and use as a basis for defining other sets. We can describe a set by listing its elements, separated by commas, and enclosing the list in braces. When it is impossible or inconvenient to list all of the elements of a set, we can use an ellipsis (...) to indicate "and so forth" when the pattern is clear.

> The set of **natural numbers**: $\{1, 2, 3, \ldots\}$, also called the set of **counting numbers**
> The set of **integers**: $\{\ldots, -3, -2, -1, 0, 1, 2, 3, \ldots\}$

Example 1
> Let A be the set of all natural numbers less than 5.
> Let B be the set of all even integers strictly between 0 and 10.
> Let C be the set of all U.S. presidents elected before 2003.

In these examples, $3 \in A$, but $3 \notin B$. Notice that the elements of a set need not be numbers, as illustrated in set C.

The only stipulation in our definition of a set is that there must be a clear way to determine whether a particular object is an element of the set; in that case, we say that the set is **well defined**. If we try to define a set D as the set of all nice people in the world, there is no definitive way to decide whether a particular person belongs to that set. In this case, we would say that D is not a well-defined set.

DESCRIBING SETS

There are several different ways to describe a set. The sets A, B, and C in Example 1 above are all described **verbally**. It is sometimes possible to describe a set by simply listing all of its elements. By convention, we enclose the list in those curlicue symbols called braces. For example, set A in the example above can be written as $A = \{1, 2, 3, 4\}$. This method of describing a set is called the **listing** method (or *tabulation method*, or *roster method*).

It is possible, but not very convenient, to describe the set C using the listing method, since there are 42 elements in this set. It would be impossible, of course, to list all of the elements of an infinite set. We can get around both of these problems by using an ellipsis (...), provided that there is an obvious pattern in the list. We could write

$$C = \{\text{G. Washington, J. Adams, T. Jefferson, ..., G. W. Bush}\}.$$

In addition, if we write $D = \{1, 4, 9, 16, ...\}$, it seems apparent that D is the set of squares of natural numbers. Care must be used, however, to avoid confusion over which missing numbers are implied by the ellipsis. For example, $\{1, 3, ..., 15\}$ could represent the odd counting numbers less than or equal to 15, but it could also represent the set $\{1, 3, 6, 10, 15\}$. This is the set of the first five **triangular** numbers, which we will explore later, in Chapter 6.

A third method of describing a set, called **set-builder notation**, describes the criteria for membership in the set, as in the verbal method, but uses set notation to do so. Set A from Example 1 could be written as

$$A = \{n \mid n \text{ is a natural number } < 5\}$$

This sentence would be read aloud as "A is the set of all elements n such that n is a natural number less than 5." The vertical bar corresponds to the words *such that*. Notice that after the vertical bar, n represents one typical element of the set, so the singular rather than the plural verb form is used.

EQUALITY AND SUBSETS

With three different methods of describing sets, how can we tell when two sets are the same? We say that two sets are **equal** if they contain exactly the same elements. Note that the order in which the elements are listed is not important. $\{1, 3, 5\}$ and $\{1, 5, 3\}$ are equal as sets because they have exactly the same elements. Similarly, $\{I, M, P, S\}$ and $\{M, I, S, S, I, S, S, I, P, P, I\}$ are equal as sets, even though the second set has some elements listed more than once.

Definition: If all of the elements of a set A are also elements of another set B, we say that A **is a subset of** B, and we write $A \subseteq B$. In this case, we also say that A **is contained in** B.

Example 2
>If $F = \{2, 4, 8, 16\}$ and $G = \{2, 4, 6, 8, 10, 12, 14, 16\}$, then $F \subseteq G$, since every element of F is also an element of G. Notice that G is not a subset of F (we write $G \not\subseteq F$), since not every element of G is also an element of F. For example, $6 \in G$, but $6 \notin F$.

Once we have defined the notion of subset, we have a particularly nice way of defining what it means for two sets A and B to be equal. This will be especially useful in Chapter 3, when we attempt to prove that two sets are equal.

Definition: Two sets are **equal** if each is contained in the other. In symbols, if $A \subseteq B$ and $B \subseteq A$, then $A = B$.

THE EMPTY SET AND PROPER SUBSETS

There is one set that contains no elements at all. It is called the **empty set**, or the **null set**, and is written as $\varnothing$ or { }. The empty set has the following interesting property:

Subset Property of $\varnothing$: If B is any set, then $\varnothing$ must be a subset of B.

If this seems strange at first glance, consider the alternative. If $\varnothing$ is *not* a subset of B, then there must be some element of $\varnothing$ that is not in B. But this is impossible, since $\varnothing$ has no elements to begin with. At the other extreme, every set B is a subset of itself; that is, $B \subseteq B$, because every element of B is contained in B.

It is often useful to give attention to those subsets of B other than the entire set B itself. Such a set is called a **proper subset** of B. If A is a proper subset of B, we write $A \subset B$. For example, the set of all Republican U.S. presidents elected before 2003 is a proper subset of set C of Example 1, since not all U.S. presidents elected before 2003 were Republicans. On the other hand, the set of all male U.S. presidents elected before 2003 is not a proper subset of C. Notice the similarity between the subset symbols, $\subseteq$ and $\subset$, and the symbols we use for describing inequalities of numbers, $\leq$ and $<$. In each case, one of the symbols allows equality, and the other does not.

Example 3
List all possible subsets of $\{a, b, c\}$.
Solution
{ }, $\{a\}$, $\{b\}$, $\{c\}$, $\{a, b\}$, $\{a, c\}$, $\{b, c\}$, $\{a, b, c\}$.
It is important to remember the two extremes, the empty set and the whole set itself.

We introduce here the helpful notation $n(A)$ for describing the **number of elements** of a set A. We use this notation, however, only for a set with a finite number of elements. Thus $n(\{a, b, c\}) = 3$, $n(\{M, I, S, S, I, S, S, I, P, P, I\}) = 4$, and $n(\{x \mid x \text{ is an integer and } x^2 < 10\}) = 7$, but $n(\{1, 4, 9, 16, \ldots\})$ is undefined because the set described has infinitely many elements.

BASIC SET OPERATIONS

When we do arithmetic with numbers, we begin with the operations of addition, subtraction, multiplication, and division. With sets, our basic operations are union, intersection, and complement.

If A and B are two sets, then the **union** of A and B, written $A \cup B$, is the set of all elements that are in A or in B or in both. The **intersection** of A and B, written $A \cap B$, is the set of all elements that are in both A and B. If A and B have no elements in common, we say that A and B are **disjoint**. Notice that in this case, $A \cap B = \varnothing$.

Example 4
Let $A = \{1, 3, 5, 7, 9\}$ and let $B = \{6, 7, 8, 9\}$. Then the union of A and B is $A \cup B = \{1, 3, 5, 6, 7, 8, 9\}$. Notice that we do not need to list elements 7 and 9 more than once in the description of $A \cup B$ using the listing method. The intersection of A and B is $A \cap B = \{7, 9\}$.

Venn diagrams, named after the English logician John Venn, often give a helpful visual picture that describes the relationship among two or more sets. Note the flexibility of Venn diagrams to illustrate different concepts; we used them in Section 1.3 to represent logical statements involving quantifiers. Each set is represented by a simple closed curve, usually a circle or rectangle. The elements of the set A, for example, are represented by points inside the circle labeled as A. The following two diagrams illustrate $A \cup B$ and $A \cap B$, respectively.

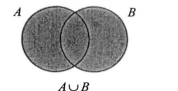

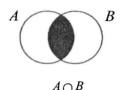

$A \cup B$ $A \cap B$

The third basic set operation is the complement. Notice the spelling; this is not to be confused with the "My, your hair looks nice today" kind of compliment. The idea is that the complement of set A contains exactly those elements that are *not* in A. In order to make sense of this idea, however, we have to have some fixed set of elements to choose from. We call this set the **universal set**, or simply the **universe**, often denoted U. All sets under discussion will be subsets of this universal set. The **complement** of a set A, written $\overline{A}$, is then defined to be the set of all elements in the universal set that are not in A. In set-builder notation, $\overline{A} = \{x \in U \mid x \notin A\}$.

Example 5

If the universal set is $\{1, 2, 3, 4, 5, 6, 7, 8, 9, 10\}$ and $A = \{1, 3, 5, 7, 9\}$, then the complement of A is $\overline{A} = \{2, 4, 6, 8, 10\}$. If we had not specified the universal set, we would not have known whether to include 0 or –1 or even π or George Washington in the complement of A.

In a Venn diagram, the universal set is generally represented by the region inside a rectangle, which encloses everything else in the diagram. In the diagram below, the shaded area represents the complement of A.

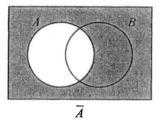

$\overline{A}$

It is sometimes useful to describe the complement of a set A relative to a set B that is smaller than the universal set. The **complement of A relative to B** is the set of all elements in B that are not in A. This is sometimes called the **set difference** of B and A and is written $B - A$. In set notation, we write $B - A = \{x \in B \mid x \notin A\}$. Verbally, $B - A$ is the set of all the elements of B that are not in A.

Example 6

If $A = \{1, 3, 5, 7, 9\}$ and $B = \{6, 7, 8, 9\}$, then the complement of A relative to B is $B - A = \{6, 8\}$, and the complement of B relative to A is $A - B = \{1, 3, 5\}$.

Notice that $A - A = \emptyset$ for every set A. If A and B are disjoint, then $A - B = A$ and $B - A = B$.

PROPERTIES OF SET OPERATIONS

Let's make a brief digression to recall a few familiar properties of addition and multiplication of numbers.

(1) (Identity) $a + 0 = a$ and $a \cdot 1 = a$ for all real numbers a.

(2) (Commutative) $a + b = b + a$ and $a \cdot b = b \cdot a$ for all real numbers a and b.

(3) (Associative) $(a + b) + c = a + (b + c)$ and
                              $(ab)c = a(bc)$ for all real numbers a, b, and c.

(4) (Distributive) $a(b + c) = ab + ac$ for all real numbers a, b, and c.

In Property (1) we say that 0 is an identity element for addition and 1 is an identity element for multiplication. The commutative and associative laws of addition and multiplication are given in (2) and (3), and the property given in (4) is called the distributive law of multiplication over addition. We have similar properties of sets for the operations of union and intersection.

Properties of Set Operations: For all sets A, B, and C,

(1) (Identity Properties) $A \cup \varnothing = A$ and $A \cap U = A$.

(2) (Commutative Properties) $A \cup B = B \cup A$ and $A \cap B = B \cap A$.

(3) (Associative Properties) $(A \cup B) \cup C = A \cup (B \cup C)$ and
                                      $(A \cap B) \cap C = A \cap (B \cap C)$.

(4) (Distributive Properties) $A \cap (B \cup C) = (A \cap B) \cup (A \cap C)$ and
                                      $A \cup (B \cap C) = (A \cup B) \cap (A \cup C)$.

(5) (DeMorgan's Laws) $\overline{(A \cup B)} = \overline{A} \cap \overline{B}$ and $\overline{(A \cap B)} = \overline{A} \cup \overline{B}$.

Notice that the empty set serves as the identity for unions, and the universal set U is the identity for intersections. The properties in (5) are known as **De Morgan's laws,** named after the English mathematician Augustus De Morgan. Notice the similarity to De Morgan's laws of logic from Section 1.1.

In Chapter 1, we saw how Venn diagrams can be used to illustrate logical statements involving quantifiers. We can also use Venn diagrams to illustrate the properties of set operations by shading the sets described on each side of the equation and showing that they are the same. It often requires several steps to shade each set. An illustration of the first of De Morgan's laws is given below.

Example 7

Use a series of Venn diagrams to illustrate $\overline{(A \cup B)} = \overline{A} \cap \overline{B}$.

Solution

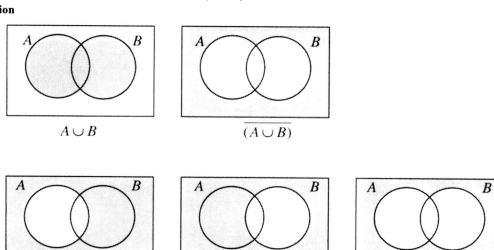

$A \cup B$ $\overline{(A \cup B)}$

$\overline{A}$ $\overline{B}$ $\overline{A} \cap \overline{B}$

We construct the left-hand side with two diagrams and the right-hand side with three. Notice that the final diagrams in each row look identical.

To illustrate properties involving three sets, the arrangement of three circles shown below is generally used.

Example 8

Shade the diagrams below to demonstrate the first of the distributive properties:
$A \cap (B \cup C) = (A \cap B) \cup (A \cap C)$.

Solution

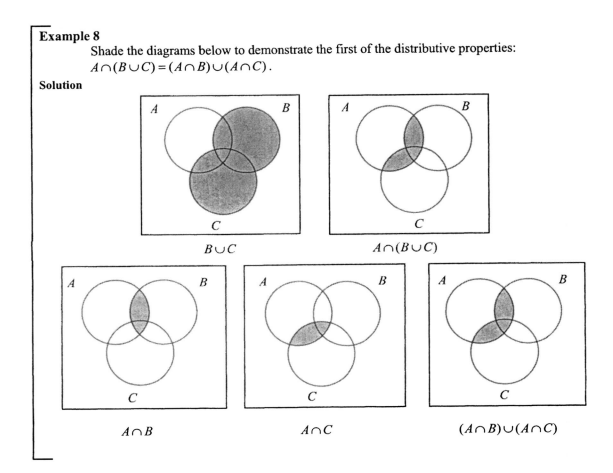

Just for Fun: Frequently described as jovial and witty, the English mathematician Augustus De Morgan, when asked his age, would reply, "I was x years old in the year x^2." Given that he lived entirely in the nineteenth century, can you figure out when DeMorgan was born?

Exploratory Exercise Set 2.1

1. Split up into pairs. Each person in the pair then writes down a description of a set using words. Exchange descriptions and describe your partner's set using the listing method. Then describe your partner's set using set-builder notation. Swap back and discuss whether your partner accurately described your set.

2. Recall that in Example 3, we found eight subsets of the set $\{a, b, c\}$.
   (a) List all of the subsets of $\{a, b\}$.
   (b) List all of the subsets of $\{a, b, c, d\}$.
   (c) Make a conjecture (a guess) about the number of subsets of a set with five elements.
   (d) Check your guess in part (c) by listing all of the subsets of $\{a, b, c, d, e\}$.
   (e) Make a conjecture about the number of subsets of a set with n elements. (We will prove this fact in Chapter 3, using induction.)

3. A town board consists of six people: Al, Bo, Cy, Di, Ed, and Flo. To be approved, a resolution must receive the votes of a majority (more than half) of the board members. List all possible successful voting coalitions as subsets of the set of town board members.

4. Split up into pairs.

   (a) Consider the second of DeMorgan's laws, $\overline{A \cap B} = \overline{A} \cup \overline{B}$. Have one partner describe the left-hand side of the equation with a Venn diagram or a series of Venn diagrams, labeling each diagram as in Example 7. Have the other partner do the same for the right-hand side of the equation. When you are both finished, compare your final diagrams. Do they look the same?

   (b) Consider the second distributive property, $A \cup (B \cap C) = (A \cup B) \cap (A \cup C)$, and repeat the procedure of part (a). How do your two diagrams compare?

5. (*Writing Exercise*) There is a curious paradox that arises when considering sets of sets. Can a set be a member of itself? Bertrand Russell posed the following question, which is known as Russell's paradox: "Is the set of all sets that are not members of themselves a member of itself?"

   (a) Discuss how answering either "yes" or "no" to Russell's paradox leads to a contradiction.

   (b) Explain how the following riddle is related to Russell's paradox:
   The barber in a certain town shaves all the men in the town who do not shave themselves, and no others. Who shaves the barber?

   (c) There is a trick answer to the riddle in part (b) that avoids the contradictions of Russell's paradox. What do you think it might be?

Exercise Set 2.1

In these exercises, we refer to the following sets:
   The counting numbers: {1, 2, 3, ...},
   The integers: {..., –3, –2, –1, 0, 1, 2, 3, ...}
   The square numbers: {0, 1, 4, 9, 16, ...}

1. Let $A = \{1, 2, 3, 4, 5\}$. Classify each of the following statements as true or false.

   (a) $2 \in A$.   (b) $\{2\} \in A$.

   (c) $0 \in A$.   (d) $\varnothing \in A$.

   (e) $\varnothing \subset A$.   (f) $\{1, 2, 3\} \in A$.

   (g) $\{1, 2, 3\} \subseteq A$.   (h) $\{1, 2, 3\} \subset A$.

   (i) $4 \subset A$.   (j) $\{1, 2, 3, 4, 5\} \subset A$.

2. Let $A = \{2, 4, 8\}$. In each of the following statements, fill in the blank with one of the symbols $\in, \notin, \subset,$ or $\not\subset$ in order to make the statement true.

   (a) 0 ___ A.   (b) $\{0\}$ ___ A.

   (c) 2 ___ A.   (d) 6 ___ A.

   (e) $\{8\}$ ___ A.   (f) $\{2, 4\}$ ___ A.

   (g) $\{2, 4, 6\}$ ___ A.   (h) $\{2, 4, 8\}$ ___ A.

   (i) $\varnothing$ ___ A.   (j) $\{\varnothing\}$ ___ A.

3. Describe the set B in Example 1 in this section using the listing method.

4. Write the sets B and C of Example 1 in this section using set-builder notation.

5. Let $E = \{n \mid n \text{ is an integer and } n^2 < 10\}$. Describe E using the listing method.

6. Describe each of the following sets using the listing method.
   (a) The set of all single-digit odd counting numbers
   (b) The set of all living former U.S. presidents
   (c) The set of all U.S. state capitals that were named for U.S. presidents
   (d) The set of all square numbers less than 50
   (e) The set of all even counting numbers

7. Describe each of the sets in the previous exercise using set-builder notation.

8. Give a verbal description of each of the following sets:
    (a) $\{a, e, i, o, u, y\}$
    (b) $\{1, 2, 3, 4, 5, 6, 7, 8, 9\}$
    (c) $\{$Sunday, Monday, Tuesday, ..., Saturday$\}$
    (d) $\{..., -5, -3, -1, 1, 3, 5, ...\}$
    (e) $\{$Mercury, Venus, Earth, Mars, Jupiter, Saturn, Uranus, Neptune, Pluto$\}$

9. Describe each of the following sets using the listing method:
    (a) $\{x \mid x$ is an odd counting number with one digit$\}$
    (b) $\{x \mid x^2 = 9\}$
    (c) $\{x \mid x$ is a two-digit counting number that looks the same upside down as right-side up$\}$
    (d) $\{x \mid x$ is a state in the United States that begins with the letter A$\}$

10. Which of the following sets are well defined?
    (a) The set of all cute children
    (b) The set of all counting numbers less than one million
    (c) The set of all nice teachers
    (d) The set of all large integers
    (e) The set of all negative integers

11. Which of the following represent equal sets?
    $A = \{1, 2, 3, 4, 5\}$
    $B = \{a, b, c, d, e\}$
    $C = \{1, 4, 2, 5, 3\}$
    $D = \{1, 1, 2, 3, 4, 4, 5, 5, 5\}$
    $E = \{x \mid x$ is a counting number less than 5$\}$

12. Which of the following represent equal sets?
    $A = \{ \}$
    $B = \{\varnothing\}$
    $C = \varnothing$
    $D = 0$
    $E = \{0\}$

13. Let $A = \{a, b, c, e\}$, $B = \{b, d, e, f\}$, $C = \{a, c, e\}$, and $D = \{d, f, g\}$. Describe each of the following sets using the listing method:
    (a) $A \cup B$ (b) $A \cap B$
    (c) $A \cap D$ (d) $B \cap C$
    (e) $A \cap C$ (f) $A \cup C$
    (g) $C \cap D$ (h) $A - B$
    (i) $B - A$ (j) $A - D$

14. Let $U = \{1, 2, 3, 4, 5, 6, 7, 8, 9, 10\}$, $A = \{2, 4, 6, 8, 10\}$, $B = \{1, 2, 3, 4, 5\}$, and $C = \{1, 3, 5, 7, 9\}$.
    Describe each of the following sets using the listing method:
    (a) $A \cup B$ (b) $A \cap B$
    (c) $B \cap C$ (d) $B \cup C$
    (e) $A \cap C$ (f) $\overline{A}$
    (g) $\overline{B}$ (h) $\overline{A \cup B}$
    (i) $\overline{A} \cap \overline{B}$ (j) $\overline{A} \cup \overline{B}$

15. Let U be the set of months of the year; let A be the set of months that have exactly 30 days; let B be the set of months that begin with the letter "J"; let C be the set of months that contain the letter "r." Describe each of the following sets using the listing method.
    (a) $A \cup B$
    (b) $B \cap C$
    (c) $\overline{C}$
    (d) $B \cup \overline{C}$
    (e) $\overline{B} \cap C$

16. Shade the portion of the Venn diagram below that illustrates each of the following sets:
    (a) $A \cup B$
    (b) $B \cap C$
    (c) $(A \cup C) \cap B$
    (d) $B \cup \overline{C}$
    (e) $\overline{B} \cap C$
    (f) $A \cap \overline{B} \cap C$

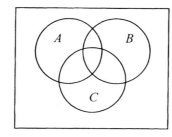

17. In each of the following Venn diagrams, describe the shaded region both verbally and in terms of unions, intersections, and complements of the sets A, B, and C.
    (a)
    (b)

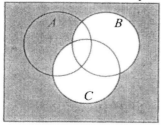 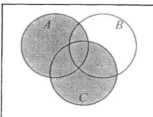

18. Let $A = \{1, 2, 3, 4\}$.
    (a) List all of the subsets of A that have four elements. How many are there?
    (b) List all of the subsets of A that have three elements. How many are there?
    (c) List all of the subsets of A that have two elements. How many are there?
    (d) List all of the subsets of A that have one element. How many are there?
    (e) List all of the subsets of A that have no elements. How many are there?
    (f) How many subsets does A have?

19. In the Venn diagram below, sets A, B, and C are subsets of a universal set U. Label each of the following eight disjoint sets in the diagram below. Mark the first set with a 1, the second with a 2, and so on.
    (1) $A \cap B \cap C$
    (5) $\overline{A} \cap \overline{B} \cap C$
    (2) $\overline{A} \cap B \cap C$
    (6) $\overline{A} \cap B \cap \overline{C}$
    (3) $A \cap \overline{B} \cap C$
    (7) $A \cap \overline{B} \cap \overline{C}$
    (4) $A \cap B \cap \overline{C}$
    (8) $\overline{A} \cap \overline{B} \cap \overline{C}$

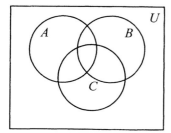

20. Suppose you want to draw a Venn diagram depicting four sets A, B, C, and D. If the sets A, B, C, and D do not have to be represented by circles, draw a Venn diagram in which the universal set U is the union of the following 16 regions, each of which is connected, and no two of which overlap.

(1) $A \cap B \cap C \cap D$

(2) $\bar{A} \cap B \cap C \cap D$

(3) $A \cap \bar{B} \cap C \cap D$

(4) $A \cap B \cap \bar{C} \cap D$

(5) $A \cap B \cap C \cap \bar{D}$

(6) $\bar{A} \cap \bar{B} \cap C \cap D$

(7) $\bar{A} \cap B \cap \bar{C} \cap D$

(8) $\bar{A} \cap B \cap C \cap \bar{D}$

(9) $A \cap \bar{B} \cap \bar{C} \cap D$

(10) $A \cap \bar{B} \cap C \cap \bar{D}$

(11) $A \cap B \cap \bar{C} \cap \bar{D}$

(12) $\bar{A} \cap \bar{B} \cap \bar{C} \cap D$

(13) $\bar{A} \cap \bar{B} \cap C \cap \bar{D}$

(14) $\bar{A} \cap B \cap \bar{C} \cap \bar{D}$

(15) $A \cap \bar{B} \cap \bar{C} \cap \bar{D}$

(16) $\bar{A} \cap \bar{B} \cap \bar{C} \cap \bar{D}$

2.2 Relations

Who are your relations? Are you related to Mike? Most of us are familiar with the everyday context of the words *relation* and *related* to express a family connection. We can be more specific and say that "Betty is Suzie's mother" or "Maggie and Keisha are cousins." In each case there is a relation between two people. We can extend this idea beyond family connections to express relations between other people or even between objects. In the statements "Smudge is the pet of Frank" and "Bill's car is the same color as Rhonda's car," two elements are related by a connecting expression. The expressions "is the mother of" and "is the cousin of" define relations on the set of all people, and the expression "is the same color as" defines a relation on the set of all objects with color. The expression "is the pet of" defines a relationship between the set of all animals and the set of all people.

ORDERED PAIRS

To define this concept of relation more precisely, we need to introduce some mathematical terms that will apply to many other contexts in addition to the examples given above. An **ordered pair** consists of two components, with one specified as coming first, and the other second. The components could be people, animals, numbers, objects, or just about anything. We write an ordered pair by listing the two components in the specified order, separated by a comma, and we enclose the pair in parentheses. For example, (Betty, Suzie) and (Maggie, Keisha) are ordered pairs of people, whereas (Smudge, Frank) is an ordered pair where the first component is an animal and the second is a person. The important thing to remember about ordered pairs is that the order is important. (Suzie, Betty) and (Betty, Suzie) are different ordered pairs because the components are listed in a different order. In general, we say that two ordered pairs are **equal** if their first components are identical and their second components are identical.

You may have encountered ordered pairs of numbers or letters, such as (2, 5) or (x, y), in an algebra class. We use such ordered pairs to describe the location of points in the xy-plane, which is also called the Cartesian plane after the French mathematician and philosopher René Descartes. Monsieur Descartes also lends his name to an operation on sets that is helpful in describing sets of ordered pairs.

CARTESIAN PRODUCTS

Example 1

A restaurant serves three flavors of ice cream: vanilla, chocolate, and strawberry. Toppings of chocolate sauce and caramel sauce are available. The different ways of having ice cream with a topping are given by the following set of ordered pairs:
{(vanilla, chocolate), (vanilla, caramel),
(chocolate, chocolate), (chocolate, caramel),
(strawberry, chocolate), (strawberry, caramel)}

If A = {vanilla, chocolate, strawberry} is the set of possible ice cream flavors, and B = {chocolate, caramel} is the set of possible topping flavors, then the set listed above is called the **Cartesian product** of A and B and is denoted $A \times B$. Notice that in each ordered pair, the ice cream flavor is listed first and the topping second. If we wanted to reverse this order and list the topping first, we could use the Cartesian product $B \times A$.

Definition: The **Cartesian product** of sets A and B, denoted $A \times B$, is the set of all ordered pairs (a, b), where the first component a is an element of the set A, and the second component b is an element of the set B.

Using set-builder notation, we would write $A \times B = \{(a, b) \mid a \in A, b \in B\}$. You might have grasped the reason for the name Cartesian because of the connection with ordered pairs in the Cartesian plane, but why is this set of possible ordered pairs called a *product*? After all, we're not multiplying flavors together in Example 1. The answer becomes evident if we look at the number of elements in the Cartesian product. In Example 1, there are 3 choices for the flavor of the ice cream, 2 choices for the flavor of the topping, and $3 \cdot 2 = 6$ elements in the Cartesian product. In general, if A and B are finite sets, then the number of elements in the Cartesian product $A \times B$ is the product of the number of elements in A times the number of elements in B. We can state this more succinctly using the notation $n(A)$ introduced in the previous section:

Theorem 1: If A and B are finite sets, then $n(A \times B) = n(A) \cdot n(B)$.

Theorem 1 is a simplified version of the multiplicative rule for counting, which we will explore further in Chapter 6 on combinatorics. We will give a careful proof of this theorem after we have discussed mathematical proofs in Chapter 3. Let us see how we can use this theorem in a few examples.

Example 2

Let $A = \{1, 2, 3\}$ and $B = \{2, 4, 6, 8\}$.
(a) How many elements are in $A \times B$? List them.
(b) How many elements are in $B \times A$? List them.
(c) How many elements are in $A \times A$? List them.

Solution

(a) Theorem 1 tells us there are $3 \cdot 4 = 12$ elements in $A \times B$. They are
(1, 2), (1, 4), (1, 6), (1, 8), (2, 2), (2, 4), (2, 6), (2, 8), (3, 2), (3, 4), (3, 6), (3, 8).
(b) Theorem 1 tells us there are $4 \cdot 3 = 12$ elements in $B \times A$. They are
(2, 1), (2, 2), (2, 3), (4, 1), (4, 2), (4, 3), (6, 1), (6, 2), (6, 3), (8, 1), (8, 2), (8, 3).
(c) From Theorem 1, there are $n(A) \cdot n(A) = 3 \cdot 3 = 9$ elements in $A \times A$. They are
(1, 1), (1, 2), (1, 3), (2, 1), (2, 2), (2, 3), (3, 1), (3, 2), and (3, 3).

Notice that in the previous example, the Cartesian products $A \times B$ and $B \times A$ are not equal, since (1, 2), for example, is an element of $A \times B$ but not an element of $B \times A$. However, since $3 \cdot 4 = 4 \cdot 3$, $A \times B$ and $B \times A$ do contain the same number of elements. In fact, Theorem 1 guarantees that for any two sets A and B, $n(A \times B) = n(B \times A)$, since $n(A) \cdot n(B) = n(B) \cdot n(A)$.

RELATIONS

Let us now revisit the idea of relations that we discussed at the beginning of this section. Recall that the expressions "is the mother of" and "is a cousin of" define relations between people. The ordered pairs (Queen Elizabeth II, Prince Charles), (Eve, Cain), and (Barbara Bush, George W. Bush) are included in the relation "is the mother of." Similarly, the ordered pair (Maggie, Keisha) satisfies the relation "is a cousin of" if Maggie is a cousin of Keisha. Clearly some, but not all, ordered pairs of people satisfy each of these relations. If P represents the set of all people, then we can think of the relation "is the mother of" as a subset of the Cartesian product $P \times P$. If A represents the set of all animals, then the relation "is a pet of" can be thought of as a subset of $A \times P$. We can now formulate the general definition.

Definition: A **relation** R from a set A to a set B is a subset of $A \times B$. In the special case where R is a subset of $A \times A$, we say that R is a **relation on** A.

Example 3

Let P represent the set of all people and A the set of all animals, as above, and let R be the relation "is a pet of" from set A to set P. To say that (Smudge, Frank) $\in R$ means that Smudge is the pet of Frank. If Q is the relation "is the mother of" on set P, then (Betty, Suzie) $\in Q$ means Betty is the mother of Suzie.

Example 4

Let A = {dollar, euro, peso} and let B = {Canada, France, U.S.A.}. If $a \in A$ and $b \in B$, then the relation "a is a monetary unit in b" is defined by {(dollar, Canada), (dollar, U.S.A.), (euro, France)}, which is a subset of the Cartesian product $A \times B$ = {(dollar, Canada), (dollar, France), (dollar, U.S.A.}, (euro, Canada), (euro, France), (euro, U.S.A.}, (peso, Canada), (peso, France), (peso, U.S.A.)}.

Example 5

Let R be the relation "is greater than" on the set A = {1, 2, 3, 4}. We can describe R by listing the ordered pairs in $A \times A$ where the first component is greater than the second. Thus R = {(2, 1), (3, 1), (3, 2), (4, 1), (4, 2), (4, 3)}. Notice that, among others, (1, 2) and (3, 3) are elements of $A \times A$, but not elements of R, since in each case the first component is not greater than the second.

We can give a visual representation of relations by drawing a dot for each component of the ordered pair and an arrow from the first component to the second. In the diagram on the left, suppose Betty is the mother of Suzie, Jo is the mother of Ed, and Gert is the mother of Ham. In the diagram on the right, we illustrate the relation "is greater than" from Example 5.

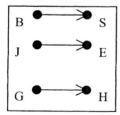

Figure 1: "is the mother of"

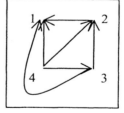

Figure 2: "$>$"

SPECIAL TYPES OF RELATIONS

Some, but not all, relations on a set A (i.e., subsets of $A \times A$) satisfy one or more special properties that we are familiar with from our knowledge of the mathematical symbols "=" and ">". It is often helpful to classify a relation in terms of whether it is *reflexive*, *symmetric*, and/or *transitive*.

Definition: Let R be a relation on a set A.

(1) To say R is **reflexive** means that $(a, a) \in R$ for all $a \in A$.

(2) To say R is **symmetric** means that $(a, b) \in R$ implies $(b, a) \in R$.

(3) To say R is **transitive** means that $[(a, b) \in R$ and $(b, c) \in R]$ implies $(a, c) \in R$.

Example 6

Determine whether the following relations on the set of all people are reflexive, symmetric, or transitive:
(a) "is the mother of"
(b) "is a cousin of"
(c) "has the same color eyes as"

Solution

(a) This relation is not reflexive (nobody can be her own mother), not symmetric (if Betty is Suzie's mother, then Suzie cannot be Betty's mother), and not transitive (if Betty is Suzie's mother and Suzie is Margaret's mother, then Betty is Margaret's grandmother, not her mother).

(b) This relation is symmetric, but not reflexive (you are not generally considered to be your own cousin). Neither is it transitive, which we can see from the following example. If Keisha and Maggie are cousins, and Carol is Keisha's sister, then (Keisha, Maggie) $\in R$ and (Maggie, Carol) $\in R$, but (Keisha, Carol) $\notin R$, since Keisha and Carol are not cousins but sisters.

(c) Check that this relation is reflexive, symmetric, and transitive.

Notice that we do not try to determine whether "is a pet of" has any of these three properties, since it is a relation from one set, A, to a different set, P.

Example 7

Determine whether the following relations on the set of real numbers are reflexive, symmetric, or transitive:
(a) "greater than"
(b) "greater than or equal to"
(c) "equals"

Solution

These relations should be familiar from algebra. We summarize:
(a) "greater than" is not reflexive (3 is not greater than 3) or symmetric (2 > 0 but 0 is not greater than 2), but it is transitive (if $a > b$ and $b > c$, then $a > c$).
(b) "greater than or equal to" is reflexive and transitive but not symmetric.
(c) "equals" is reflexive, symmetric, and transitive.

Definition: A relation on a set A is called an **equivalence relation** if it is reflexive, symmetric, and transitive.

Example 8

Determine whether $R = \{(1, 1), (1, 2), (1, 3), (2, 1), (2, 2), (3, 1), (3, 3)\}$ is an equivalence relation on the set $A = \{1, 2, 3\}$.

Solution

R is reflexive, since $(1, 1)$, $(2, 2)$, and $(3, 3)$ are all in R. Notice that if you reverse the order of any ordered pair in R, the new ordered pair is again in R. This means that R is symmetric. To check the transitive property, notice that $(2, 1)$ and $(1, 3)$ are in R but $(2, 3)$ is not in R. Therefore, R is not transitive and hence is not an equivalence relation. This relation is illustrated in the diagram below.

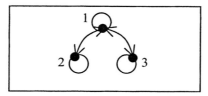

Figure 3: The relation in Example 8

Among the relations we have considered in the previous examples, "has the same eye color as" is an equivalence relation on the set of all people, and "=" is an equivalence relation on the set of all real numbers. A nice thing happens when we have an equivalence relation on a set A.

Example 9

Suppose Al, Flo, and Hannah have brown eyes, Bo and Gary have blue eyes, Cindy and Dawn have green eyes, and Ephraim has hazel eyes. The diagram below illustrates the equivalence relation "has the same eye color as" on the set $P = \{$Al, Bo, Cindy, Dawn, Ephraim, Flo, Gary, Hannah$\}$.

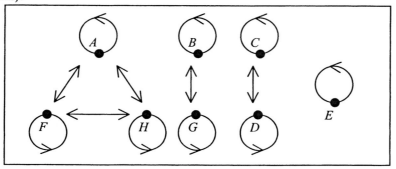

Figure 4: The eye color equivalence relation

Notice that the equivalence relation divides the set P into four disjoint subsets, $\{$Al, Flo, Hannah$\}$, $\{$Bo, Gary$\}$, $\{$Cindy, Dawn$\}$, and $\{$Ephraim$\}$, and that the union of these four subsets is the entire set P. Within each subset, all members share the same eye color. We call these subsets **equivalence classes**. Every equivalence relation on a set A divides the set A into disjoint subsets called equivalence classes. Within each equivalence class, all members are related to each other.

We close this section with an example of a relation that is central to the branch of mathematics called number theory. In reading this example, remember that an even integer is a multiple of 2 and that the sum of two even integers is even. In Chapter 3 we explore how to write proofs of such statements.

Example 10

Let A be the set of all integers, and let R be the relation defined by

$R = \{(a,b) \in A \times A \mid a - b \text{ is an even number}\}$.

In other words, an integer a is related to another integer b if the difference $a - b$ is an even integer. Is R is an equivalence relation on A? If so, describe the equivalence classes.

Solution

To see whether R is an equivalence relation, we check the three properties.

(1) If a is any integer, then $(a, a) \in R$, since $a - a = 0$, which is even.

(2) If $(a, b) \in R$, then $a - b = c$ is even, so $b - a = -c$ is also even, and $(b, a) \in R$.

(3) If $(a, b) \in R$ and $(b, c) \in R$, then $a - b$ and $b - c$ are even, so $a - c = (a - b) + (b - c)$ is also even, since the sum of two even numbers is even. (We will prove this carefully as Theorem 1 in Chapter 3.) Thus $(a, c) \in R$.

Since R satisfies the reflexive, symmetric, and transitive properties, it is an equivalence relation. This equivalence relation partitions the set of all integers into two equivalence classes: the even integers and the odd integers.

Exploratory Exercise Set 2.2

1. Break up into groups of 4 to 6 people.
   (a) Within each group, draw a diagram illustrating each of the following relations on the set of people in your group:
      (1) is taller than
      (2) is the same sex as
      (3) was born in the same month as
   (b) Describe each relation in part (a) by listing all ordered pairs of people in your group that are elements of the relation.

2. Test each of the three relations in the previous exercise to see whether they are reflexive, symmetric and/or transitive. Which are equivalence relations? Describe the equivalence classes for each equivalence relation.

3. Working in pairs, consider the set $A = \{1, 2, 3\}$.
   (a) How many elements are in $A \times A$? List them.
   (b) List all possible subsets of $A \times A$ that are equivalence relations on A.
   (c) Draw a diagram and give the equivalence classes for each of the equivalence relations you found in part (b).

4. Recall the commutative and associative properties of set operations from the previous section. We will explore whether the Cartesian product satisfies these properties.
   (a) Let $A = \{1, 2\}$ and $B = \{3, 4\}$.
      (1) Compute $A \times B$.
      (2) Compute $B \times A$.
      (3) Explain why the Cartesian product is not commutative.
      (4) What must be true about the sets A and B in order for $A \times B = B \times A$?
   (b) Let $A = \{1, 2\}$, $B = \{3\}$, and $C = \{4, 5\}$.
      (1) Compute $(A \times B) \times C$.
      (2) Compute $A \times (B \times C)$.
      (3) Explain why the Cartesian product is not associative.

5. Let $A = \{2, 4\}$ and $B = \{1, 2, 3\}$ be subsets of the universal set $U = \{1, 2, 3, 4, 5\}$. Define $\overline{A \times B}$ to be the subset of all elements of $U \times U$ that are not in $A \times B$. With this definition, determine whether $\overline{A \times B} = \overline{A} \times \overline{B}$.

6. Working in groups of three, we will explore the connection between arrow diagrams and relations on a set A. Each member of the group should complete (1) and (2) for parts (a), (b), and (c). Then, after discussing individual solutions, the group should complete (3) together.
   (a) Let $A = \{1, 2, 3\}$.
       (1) Draw an arrow diagram that corresponds to a reflexive relation on A.
       (2) Draw an arrow diagram that corresponds to a nonreflexive relation on A.
       (3) Describe a feature that all diagrams of reflexive relations share.
   (b) Let $A = \{1, 2, 3\}$.
       (1) Draw an arrow diagram that corresponds to a symmetric relation on A.
       (2) Draw an arrow diagram that corresponds to a nonsymmetric relation on A.
       (3) Describe a feature that all diagrams of symmetric relations share.
   (c) Let $A = \{1, 2, 3, 4\}$.
       (1) Draw an arrow diagram that corresponds to a transitive relation on A.
       (2) Draw an arrow diagram that corresponds to a nontransitive relation on A.
       (3) Describe a feature that all diagrams of transitive relations share.

7. Divide into pairs. Have one member do part (a) and the other member part (b).
   (a) The symbols $\subseteq$ ("is a subset of") and $\subset$ ("is a proper subset of") are relations on the set of all subsets of a given set U.
       (1) Determine whether $\subseteq$ is reflexive, symmetric, or transitive.
       (2) Determine whether $\subset$ is reflexive, symmetric, or transitive.
   (b) The symbols $\leq$ ("is less than or equal to") and $<$ ("is less than") are relations on the set of real numbers.
       (1) Determine whether $\leq$ is reflexive, symmetric, or transitive.
       (2) Determine whether $<$ is reflexive, symmetric, or transitive.
   (c) Compare your results from parts (a) and (b).

8. (*Writing Exercise*) Your younger brother is having difficulty understanding the reflexive, symmetric, and transitive properties of relations. Write a paragraph explaining to him, using words rather than symbols, what each of these three properties means. As an example, include an explanation of which properties the relation "is a brother of" satisfies on the set of all people.

Exercise Set 2.2

1. Let $A = \{1, 3, 5\}$ and $B = \{2, 3\}$. List the elements of each of the following Cartesian products:
   (a) $A \times B$ (b) $A \times A$ (c) $B \times B$

2. Let $A = \{x, y, z\}$ and $B = \{1, 2, 3\}$. List the elements of each of the following Cartesian products:
   (a) $A \times B$ (b) $B \times A$

3. Determine the number of elements in the Cartesian product $A \times B$ if
   (a) $A = \{2, 4, 6, 7\}$ and $B = \{1, 3, 4, 5, 6\}$
   (b) $A = \{1, 2, 3, 4, 5\}$ and $B = \{a, b, c, ..., z\}$

4. If $A = \{1, 2, 3, 4\}$, $B = \{5, 6, 7\}$, and $C = \{8, 9\}$, determine the number of elements in each of the following Cartesian products:
   (a) $A \times B$ (b) $A \times C$
   (c) $B \times C$ (d) $(A \times B) \times C$

5. Describe in words a relation between the elements of the ordered pairs in each of the following sets:
    (a) {(Yankees, New York), (Padres, San Diego), (Mariners, Seattle)}
    (b) {(South Dakota, Pierre), (Kentucky, Frankfort), (Nevada, Carson City)}
    (c) {(Mark Twain, Samuel Clemens), (George Eliot, Mary Ann Evans), (Lewis Carroll, Charles Dodgson)}

6. Describe each of the following relations on the set $A = \{1, 2, 3\}$ as a subset of $A \times A$.
    (a) is less than
    (b) is less than or equal to
    (c) equals
    (d) is not equal to

7. Draw an arrow diagram to illustrate each of the relations in the Exercise 6.

8. Describe each of the following relations on the set $A = \{\varnothing, \{1\}, \{2\}, \{1,2\}\}$ as a subset of $A \times A$.
    (a) is a subset of
    (b) is a proper subset of
    (c) is not a proper subset of
    (d) equals

9. Draw an arrow diagram to illustrate each of the relations in Exercise 8.

10. Determine whether each of the following relations is reflexive, symmetric, or transitive on the set of all people:
    (a) is younger than
    (b) was born in the same city as
    (c) likes
    (d) is a relative of

11. Determine whether each of the following relations is reflexive, symmetric, or transitive on the given set A:
    (a) {(1, 1), (1, 2), (2, 1), (2, 2), (3, 3)} on $A = \{1, 2, 3\}$
    (b) {(a, a), (a, b), (b, b), (b, c), (c, b), (c, c)} on $A = \{a, b, c\}$
    (c) {(1, 1), (1, 2), (1, 3), (2, 1), (2, 2), (3, 1), (3, 3), (4, 4)} on $A = \{1, 2, 3, 4\}$

12. (a) Draw an arrow diagram to represent each of the relations in Exercise 11.
    (b) Determine which of the relations in Exercise 11 are equivalence relations.

13. The arrow diagram below represents a relation on the set $\{1, 2, 3, 4, 5\}$.

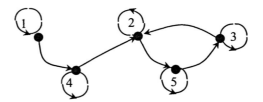

Figure 5: Arrow diagram for Exercise 13

    (a) List the ordered pairs in the relation.
    (b) Determine whether the relation is reflexive, symmetric, or transitive.

14. Give an example of a relation on the set $A = \{a, b, c\}$ that is
    (a) reflexive and transitive, but not symmetric
    (b) transitive, but not reflexive or symmetric
    (c) symmetric and transitive, but not an equivalence relation

15. Consider the set $A = \{\text{one, two, three, four, five, six, seven, eight, nine, ten}\}$.
Identify the equivalence classes in each of the following relations on A:
(a) begins with the same letter of the alphabet as
(b) has the same number of letters as
(c) has the same remainder when divided by three as

16. Draw an arrow diagram illustrating each of the equivalence relations in Exercise 15.

17. Let $A = \{-1, 0, 1\}$, and let R be the relation on A that consists of all ordered pairs
(x, y) in $A \times A$ such that $x^2 + y^2 = 1$.
(a) List the ordered pairs in the relation R.
(b) Determine whether R is reflexive, symmetric, or transitive.

18. Let A be the set of all integers, and let R be the relation on A defined by
$$R = \{(a,b) \in A \times A \mid a + b \text{ is a multiple of } 3\}$$
Determine whether the relation R is reflexive, symmetric, and/or transitive.

19. Let A be the set of all integers, and let R be the relation defined by
$$R = \{(a,b) \in A \times A \mid a - b \text{ is a multiple of } 4\}$$
(a) (*Proof*) Show that the relation R is an equivalence relation.
(b) Identify the equivalence classes for this equivalence relation.

20. Suppose A is a set with n elements and R is an equivalence relation on A.
(a) What is the least number of elements that R can have?
(b) What is the greatest number of elements that R can have?

2.3 Functions

In the previous section we investigated relations from a set A to a set B. Recall that in a relation R, it is possible for an element of the set A to be paired with several different elements of the set B. For example, we saw that (Maggie, Keisha) and (Maggie, Carol) are both elements of the relation "is a cousin of" if Keisha and Carol are both cousins of Maggie. In this section we want to concentrate on a special type of relation called a **function**, in which each element of the set A is paired with exactly one element of the set B.

WHAT'S MY RULE?

One fun way to introduce students to the concept of functions is by playing the game "What's My Rule?" One round of the game might go like this:

When Angie said 5, I said 11. When Bill said 1, I said 3. When Chanda said 3, I said 7. What's my rule?

We can use this game to define the basic notation that we use to describe functions. The numbers that Angie, Bill, and Chanda give are called **inputs**. The responses that the Rule Maker gives are called **outputs**. The set of all possible inputs is called the **domain**, and the set of all possible outputs is called the **range**. We can describe the inputs and outputs of this game as a set of ordered pairs, with the input given first and the output second: {(5, 11), (1, 3), (3, 7)}. It is sometimes helpful to organize these ordered pairs in the form of a table, as shown below.

Input	Output
5	11
1	3
3	7

Can you think of a rule that would give these outputs? Once we know the rule, then we can determine the output for any possible input. One possible rule can be described verbally as "Double the input and add 1." If this is the rule, then we can easily determine the output for any given input. If the input is 4, we double 4 and add 1 to get an output of 9. To describe this rule algebraically, we simply consider what happens when the input is an arbitrary number x. If the input is x, then we double x and add 1 to get an output of $2x + 1$.

We have just described a relation between a set of inputs and a set of outputs. Notice that if Dawn and Eddie both call out the same input number, the output number will be the same each time. This type of relation, where each input is paired with exactly one output, is called a **function**. We are now ready to make the following definition.

Definition: A **function** from a set A to a set B is a relation from A to B in which each element of A (called an **input**) is paired with exactly one element of B (called an **output**). The set of all inputs is called the **domain** of the function; it is simply the set A. The set of all outputs is called the **range** of the function; the range is a subset of B.

Example 1

Which of the following relations from $A = \{1, 2, 3\}$ to $B = \{1, 2, 3, 4, 5\}$ describe a function from A to B?

(a) $\{(1, 2), (2, 3), (3, 5)\}$
(b) $\{(1, 1), (2, 1), (3, 1)\}$
(c) $\{(1, 3), (1, 5), (3, 5)\}$
(d) $\{(1, 2), (2, 4)\}$

Solution

In relations (a) and (b), each input is paired with exactly one output, so these relations do describe a function. Relation (c) has two different outputs paired with the input 1. This cannot describe a function. In relation (d), the inputs 1 and 2 are both paired with one element of B, but the input element 3 of the set A is not paired with any output, so this is not a function from A to B.

Example 2

Draw an arrow diagram for each relation in Example 1; then find the domain and range for each of the functions in Example 1.

Solution

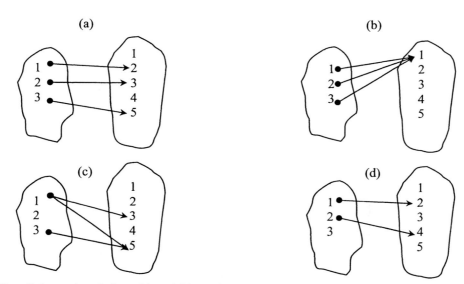

Recall that only relations (a) and (b) are functions from A to B. In (a) and (b), the domain is $A = \{1, 2, 3\}$. In (a), the range is $\{2, 3, 5\}$ and in (b) the range is $\{1\}$. Notice that in each case, the range is a subset of the set $B = \{1, 2, 3, 4, 5\}$. Although relation (d) is not a function from A to B, it could be considered a function from $\{1, 2\}$ to B, in which case the domain would be $\{1, 2\}$ and the range would be $\{2, 4\}$. Relation (c) cannot be a function at all, since two different outputs, 3 and 5, are paired with the same input 1.

FUNCTION NOTATION

It is often helpful to visualize functions with a machine like the one pictured below. Inputs go into the funnel at the top, and outputs come out of the tube at the bottom. We typically label functions with a name like f or g.

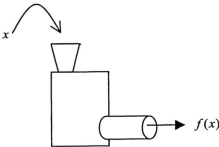

Let us give the name f to the function we described in the "What's My Rule?" game. Recall that when the input was 5, the output was 11. In function notation, we would write $f(5) = 11$. In general, if x is an input to the function f, then we write $f(x)$ to denote the output. If the rule for the function f is "Double the input and add 1," then we can describe the rule in function notation by writing $f(x) = 2x + 1$.

Example 3

Suppose that the rule for a function g is given by $g(x) = x^2 - 1$. Find $g(2)$, $g(0)$, $g(-1)$, $g(a)$, and $g(a + 1)$.

Solution

$$g(2) = 2^2 - 1 = 4 - 1 = 3$$
$$g(0) = 0^2 - 1 = -1$$
$$g(-1) = (-1)^2 - 1 = 1 - 1 = 0$$
$$g(a) = a^2 - 1$$
$$g(a+1) = (a+1)^2 - 1 = (a^2 + 2a + 1) - 1 = a^2 + 2a$$

Notice that the domain for the function g in the previous example is not given explicitly. If we specify that the domain is $A = \{-2, -1, 0, 1, 2\}$, then we can determine the range and draw an arrow diagram to represent the function g.

$$g(x) = x^2 - 1$$

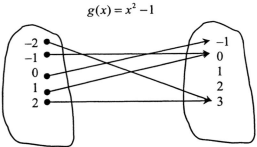

The range is the set of all outputs, which is $\{-1, 0, 3\}$.

GRAPHS OF FUNCTIONS

We have seen that a function can be represented in a number of different ways. As with any relation, it can be represented by a set of ordered pairs or, more visually, by an input-output machine, a chart, or an arrow diagram. Alternatively, a function may be given by its rule, such as $f(x) = 2x + 1$ or $g(x) = x^2 - 1$.

Although charts and arrow diagrams give us a visual description of a function, these representations can be unwieldy if the domain of the function is large. Even when the domain is relatively small, we can often get a clearer picture of a function by looking at its graph.

A graph is a geometric picture of a set of ordered pairs. First, two number lines are drawn, one horizontal (usually called the *x*-axis) and one vertical (usually called the *y*-axis). Each ordered pair can be represented by a point in the plane containing these two lines. By convention, the first element of the ordered pair gives the point's location along the horizontal line, and the second element of the ordered pair gives the point's location along the vertical line. Then we can draw the graph of any relation by plotting each ordered pair as a point.

This plane is called the Cartesian plane (in honor of René Descartes) or simply the *xy*-plane. Notice that points in the Cartesian plane are described by ordered pairs. We saw in Section 2.1 that elements of the Cartesian product of two sets are also described by ordered pairs.

Example 4

Draw the graph of each of the following relations:
(a) {(1, 1), (2, 3), (3, 3), (3, 4), (4, 5)}
(b) {(1, 3), (2, 1), (3, 3), (4, 2)}

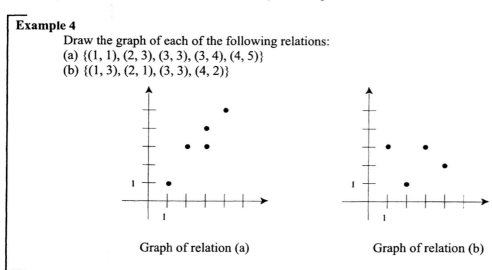

Graph of relation (a)   Graph of relation (b)

Notice that relation (a) in Example 4 cannot be a function since two distinct outputs are paired with the same input value of 3. This is shown in the graph of relation (a) by the two points on the same vertical line at $x = 3$. This observation is true in general, and it gives us a graphical way to determine whether a relation is a function.

The Vertical Line Test: If no vertical line intersects the graph of a relation in more than one point, then the relation is a function. If any vertical line passes through more than one point of the graph of a relation, then the relation is not a function.

Example 5

Determine whether each of the following graphs represents a function:

(a)   (b)

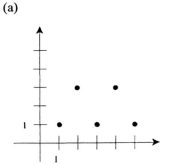   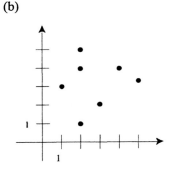

Solution

The graph in part (a) represents a function because no two points lie on the same vertical line. The graph in part (b) does not represent a function because the vertical line at $x = 2$ passes through more than one point: (2, 1), (2, 4), and (2, 5).

Example 6

An Internet service provider (ISP) charges customers $5 for the first hour online and $1 for each additional hour online. Describe this ISP's pricing function in each of the following ways:

(a) as a set of ordered pairs
(b) as a chart
(c) as an arrow diagram
(d) as a graph

Solution

(a) (1, 5), (2, 6), (3, 7), (4, 8), (5, 9), …

(b)

Hours online	1	2	3	4	5	⋯
Charge in dollars	5	6	7	8	9	⋯

(c)

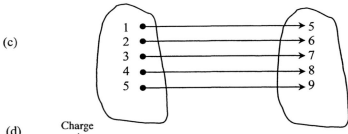

(d)

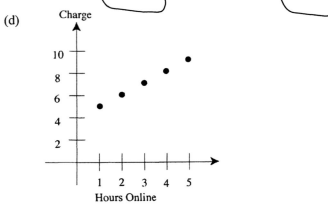

ONE-TO-ONE AND ONTO FUNCTIONS

If f is a function from A to B, we have seen that f must pair each element of A with exactly one element of B. In an arrow diagram, this means that there is exactly one arrow leaving each element of the domain, A. Must every element of B have exactly one arrow pointing toward it? Look back at the arrow diagram in Example 3 to convince yourself that the answer is no. In that example, some elements of $B = \{-1, 0, 1, 2, 3\}$ have two arrows leading to them, and some have none. Nonetheless, g is a perfectly good function from $A = \{-2, -1, 0, 1, 2\}$ to $B = \{-1, 0, 1, 2, 3\}$. In the chapters that follow, we will have occasion to refer to special types of functions that are particularly well behaved.

> **Definition:** A function from A to B is called **one-to-one** if it does not send two distinct inputs to the same output. A function from A to B is called **onto** B if every element of B is an output for some input in A.

It is helpful to look at the arrow diagram of a function to examine these two properties. A function from A to B is one-to-one if no element of B has more than one arrow pointing toward it. A function is onto B if every element of B has at least one arrow pointing toward it.

Example 7

Determine whether each of the functions from parts (a) and (b) of Example 2 is one-to-one or onto $B = \{1, 2, 3, 4, 5\}$.

(a)

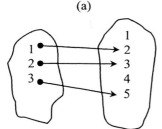

(b)

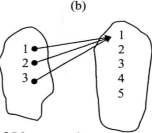

Solution

The function in part (a) is one-to-one, since no element of B has more than one arrow pointing toward it, but is not onto B, since some elements of B (namely 1 and 4) have no arrows leading to them. The function in part (b) is neither one-to-one nor onto.

Exploratory Exercise Set 2.3

1. Divide into pairs and play the "What's My Rule?" game. One student starts out as the Rule Maker, and the other is the Guesser. The Rule Maker thinks up a rule and writes it down on a piece of paper but does not show it to the Guesser. The Guesser says a number (the input), and the Rule Maker responds with the output number. The Guesser makes a chart of inputs and outputs and tries to guess the rule after each response. After the Guesser successfully guesses the rule, switch roles.

2. Divide into pairs. Have each student write down a verbal rule for a function (such as "Triple the input and then subtract 2"). Swap papers and write down a rule for the other student's function using function notation (such as "$f(x) = 3x - 2$"). Then have each student write down a (different) rule using function notation and have the other student translate it into a verbal rule.

3. In this exercise we will reformulate the definition of a one-to-one function using the function notation $f(x)$.
    (a) Let $f : X \to Y$ be a one-to-one function, and let a and b be two elements of X.

If $a \neq b$, what must be true about $f(a)$ and $f(b)$? Complete the following reformulation of the definition of what it means for a function f to be one-to-one:

A function $f : X \rightarrow Y$ is one-to-one provided that if $a \neq b$, then _____ .

(b) Recall from Chapter 1 that the contrapositive of the conditional statement "If P then Q" is "If $\sim Q$ then $\sim P$," where $\sim P$ is the negation of P. Use the contrapositive to reformulate the definition of a one-to-one function one more time:

A function $f : X \rightarrow Y$ is one-to-one provided that if _____ , then $a = b$.

4. Let $A = \{a, b, c, d\}, B = \{j, k, l, m, n\}, C = \{p, q, r\}$, and $D = \{w, x, y, z\}$.
    Draw an arrow diagram describing each of the following:
    (a) A relation from A to B that is not a function
    (b) A function from A to B that is neither one-to-one nor onto B
    (c) A function from A to B that is one-to-one but not onto B
    (d) A function from A to C that is onto C but not one-to-one
    (e) A function from A to D that is both one-to-one and onto D

5. Let A, B, C, and D be the same sets as in the Exercise 4.
    (a) Is there a function from A to B that is onto B?
    (b) Is there a function from A to C that is one-to-one?
    (c) If X and Y are both finite sets, and f is a function from X to Y that is both one-to-one and onto Y, then what must be true about X and Y?
    (d) Let $X = \{1, 2, 3, \ldots\}$ be the set of counting numbers, and let $Y = \{2, 4, 6, \ldots\}$ be the set of even counting numbers. Is there a function from X to Y that is both one-to-one and onto Y? If so, describe its rule in function notation: $f(x) = $ _____ .

6. (*Writing Exercise*) Many students have difficulty remembering the definition of a one-to-one function. This might be because the name suggests that such a function need only send one input to one output.
    (a) For which functions is it true that every one input must be sent to one output?
    (b) For which functions is it true that every two distinct inputs must be sent to two distinct outputs?
    (c) Some mathematicians have suggested that one-to-one functions should actually be called *two-to-two functions*. Write a letter to your local senator or math teacher defending the reasoning behind this change in terminology.

Exercise Set 2.3

1. Consider the following game of "What's My Rule?"
    When Al said 3, I said 8. When Bo said 5, I said 24. When Carol said 2, I said 3. When Diane said 1, I said 0.
    (a) Describe this "rule" as a set of ordered pairs.
    (b) If Eddie says 4, what do you think I will say?
    (c) What's my rule? In other words, write a rule in words for this game.
    (d) Use function notation to describe this rule.

2. Consider the following game of "What's My Rule?"
    When Al said 2, I said 4. When Bo said 1, I said 7. When Carol said 2, I said 3. When Diane said 4, I said 7.
    (a) Describe the "rule" as a set of ordered pairs.
    (b) Explain why this "rule" does not describe a function.

3. Which of the following sets of ordered pairs represent functions? Explain.
    (a) $\{(1, 4), (2, 3), (3, 2), (5, 3)\}$
    (b) $\{(2, 5), (4, 1), (3, 3), (2, 2)\}$
    (c) $\{(1, 3), (2, 3), (3, 3), (4, 3)\}$
    (d) $\{(4, 2), (1, 1), (0, 0), (4, -2)\}$

4. Draw an arrow diagram to represent each of the relations in the previous exercise.

5. Which of the following arrow diagrams represent a function?
    (a)   (b)

    (c)   (d)

6. Represent each of the relations in Exercise 5 as a set of ordered pairs.

7. (a) Give the domain and range for each of the functions in Exercise 5.
    (b) Of those functions, which are one-to-one?
    (c) Of those functions, which are onto the set {1, 2, 3, 4, 5}?

8. Draw an arrow diagram to represent each of the following functions on the set {1, 2, 3, 4, 5}:
    (a) $f(x) = 6 - x$
    (b) $f(x) = (x - 3)^2$
    (c) $f(x) = 2x - 3$

9. Which of the functions in Exercise 8 are one-to-one?

10. Represent each of the functions in Exercise 8 as a set of ordered pairs.

11. Which of the following tables define y as a function of x?

    (a)
x	1	2	3	2
y	4	3	2	1

    (b)
x	1	−1	3	0
y	1	3	2	1

    (c)
x	−2	−1	1	2
y	4	1	1	4

    (d)
x	1	1	1	1
y	1	2	3	4

12. Given each of the following rules for $f(x)$, compute $f(-1)$, $f(2)$, and $f(3)$.
    (a) $f(x) = 3x - 5$
    (b) $f(x) = 10 - 2x$
    (c) $f(x) = x^2 + x$
    (d) $f(x) = 3$

13. Draw a graph of each of the functions in Exercise 12, assuming the domain is {−1, 0, 1, 2, 3} in each case.

14. Determine whether each of the functions in Exercise 12, with domain {−1, 0, 1, 2, 3}, is one-to-one.

15. Which of the following graphs represent a function?

(a) 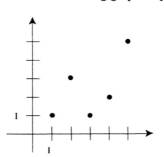   (b)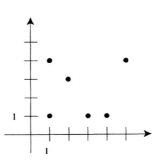

(c) 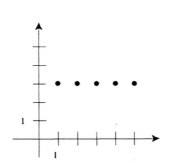   (d)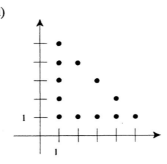

16. Which of the following graphs represent a function that is one-to-one?

(a) 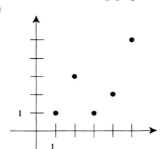   (b)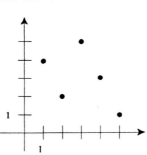

17. Which of the graphs in Exercise 16 represent a function that is onto {1, 2, 3, 4, 5}?

18. An online brokerage firm charges an annual fee of $50, plus an additional fee of $5 for each transaction.
    (a) Write the total charges in one year as a function of the number of transactions, x.
    (b) What will the total charges over one year be if there are 5 transactions? 10 transactions? 25 transactions?
    (c) If you pay this brokerage firm a total of $115 over one year, how many transactions were you charged for?

19. A long-distance phone plan charges 5 cents a minute for the first 10 minutes and 15 cents for each additional minute.
    (a) Write the long-distance charges as a function of the number of minutes, x, if x is greater than 10.
    (b) What will the long-distance charges be for a 20-minute phone call? for a 60-minute phone call?
    (c) If you are charged $3.50 for a long-distance call, how long were you on the phone?

20. Guess the rule for each of the following functions.

(a)
x	1	2	3	4
y	4	6	8	10

(b)
x	1	2	3	4
y	3	6	12	24

(c)
x	1	2	3	4	5
y	1	4	9	16	25

(d)
x	1	2	3	4
y	4	3	2	1

2.4 Sequences

Much of mathematical thinking involves investigating patterns. One way that many students learn to detect patterns and develop conjectures is by studying sequences. In general, a **sequence** is a list that comes equipped with a precise order. It should be clear which item of the list is first, or second, or fifth. We refer to the items of the list as the **terms** of the sequence. We will focus our discussion on sequences of numbers.

Example 1

In each of the following sequences, describe the pattern you see in words, and give the likely next term in the sequence.
(a) 1, 4, 7, 10, 13, ...
(b) 3, 6, 12, 24, 48, ...
(c) 1, 0, 1, 0, 1, ...
(d) $\frac{2}{1}, \frac{3}{2}, \frac{4}{3}, \frac{5}{4}, \frac{6}{5}, \ldots$

Solution

(a) The terms are increasing by 3. The likely next term is $13 + 3 = 16$.
(b) The terms are doubling. The likely next term is $2 \cdot 48 = 96$.
(c) The terms are alternating between 1 and 0. The likely next term is 0.
(d) Numerators and denominators are both increasing by 1. The likely next term is $\frac{7}{6}$.

It is often useful to refer to the individual terms of a sequence. We use the symbol a_1 to denote the first term of a sequence. Similarly, a_2, a_3, and a_4 denote the second, third, and fourth terms, respectively, and a_n denotes the **nth term**, or the **general term** of the sequence.

Example 2

Find a_1, a_4, and likely candidates for a_{10} and the general term a_n in each of the sequences in the previous exercise.

Solution

(a) The first term is $a_1 = 1$ and the fourth term is $a_4 = 10$. If the pattern we identified in the previous example continues, namely that the terms increase by 3, then the tenth term will be $a_{10} = 1 + 3 \cdot 9 = 28$, and the general term will be $a_n = 1 + 3(n-1) = 3n - 2$. The following table illustrates our computations.

Number of Term (n)	1	2	3	4	...	10	...	n
Term (a_n)	1	4	7	10	...	28	...	$1 + (n-1) \cdot 3$
Computation Used	1	1+3	1+2·3	1+3·3	...	1+9·3	...	$1 + (n-1) \cdot 3$

(b) $a_1 = 3, a_4 = 24$; if the pattern continues, we will double 5 more times after 48 to get $a_{10} = 48 \cdot 2^5 = 1536$. For the general term, we will start with 3 and double it $(n-1)$ times to get $a_n = 3 \cdot 2^{n-1}$

(c) $a_1 = 1, a_4 = 0$, and since 10 is even, a_{10} will be 0 if the pattern continues. The general term a_n will be 1 if n is odd and 0 if n is even. One way to describe this general term is to say that a_n is the remainder when n is divided by 2.

continued

(d) $a_1 = \frac{2}{1} = 2, a_4 = \frac{5}{4}$; if the pattern continues, the tenth term will be $a_{10} = \frac{11}{10}$ and the general term will be $a_n = \frac{n+1}{n}$.

You may have noticed that finding a formula for the nth term of a sequence is similar to the work we did in the previous section finding a rule for a function $f(x)$. In fact, a sequence is sometimes defined as a function whose domain is the set of counting numbers $\{1, 2, 3, \ldots\}$.

Notice that if we are given only the first few terms of a sequence, we cannot know for certain that we have identified the pattern that actually describes the sequence. (See Exploratory Exercise Set 2.4 for a vivid illustration of this uncertainty.) If we are given more terms, we may get a better idea of the pattern. How many terms must we be given before we can be sure of the pattern? Sadly, we can never be given enough terms to be sure of the pattern. In fact, the pattern may not be apparent at all.

On the other hand, if we are given a formula for the nth term, then we can say precisely what any term of the sequence will be. If, for example, a sequence begins 3, 3, 5, 4, 4, ... , we can only guess what the pattern is. Here are two plausible guesses for the continuation of the sequence:

(a) 3, 3, 5, 4, 4, 6, 5, 5, 7, 6, 6, 8, 7, 7, 9, ...
(b) 3, 3, 5, 4, 4, 5, 5, 5, 5, 6, 6, 5, 7, 7, 5, ...

Even if we are given ten or one hundred terms of the sequence, we still cannot say for certain what the pattern is. If, however, we are given a formula such as

a_n = the number of letters in the English name for n,

then we know that the first twelve terms of the sequence are 3, 3, 5, 4, 4, 3, 5, 5, 4, 3, 6, 6, ...

Example 3
Determine the following terms of the sequence with the formula
a_n = the number of letters in the English name for n.
(a) a_{20} (b) a_{47} (c) a_{119}

Solution
Twenty has 6 letters, so $a_{20} = 6$. Similarly, $a_{47} = 10$ and $a_{119} = 18$.

ARITHMETIC SEQUENCES

The first sequence we examined in this section has a particularly nice property. The apparent pattern in the sequence, which begins 1, 4, 7, 10, 13, ..., is that the terms increase by 3. If this is indeed the pattern, then notice that if we subtract any term from the next term, we always get a difference of 3: $4 - 1 = 3$; $7 - 4 = 3$; $10 - 7 = 3$; .... We say that consecutive terms of this sequence have a **common difference** of 3. Any sequence in which consecutive terms have a common difference is called an **arithmetic sequence**. We often denote the common difference of an arithmetic sequence with the letter d.

Example 4
In each of the following examples, determine whether the sequence is arithmetic. If so, find the common difference d.
(a) 3, 8, 13, 18, 23, ...

continued

(b) 2, 3, 5, 8, 12, ...
(c) 10, 9, 8, 7, 6, ...
(d) 7, 1, –5, –11, –17, ...
(e) 1, –1 , 1, –1, 1, ...

Solution

The sequences in (a), (c) and (d) are arithmetic; the sequences in (b) and (e) are not. In (a) the common difference is $d = 5$. In (c) the common difference is $d = -1$. In (d) the common difference is $d = -6$.

A formula for the general term, a_n, of an arithmetic sequence is easy to determine once you know the common difference, d, as well as the first term, a_1, of the sequence. To find the second term a_2, we simply add d to the first term, a_1. To find the third term, we could add d to the second term, or we could add **two times** d to the first term. Similarly, recall that in the arithmetic sequence of Example 2(a), in order to find the tenth term, a_{10}, we added $9d$ to the first term. More generally, the nth term of an arithmetic sequence is the sum of the first term and $(n-1)$ times the common difference. In more succinct mathematical notation,

$$a_n = a_1 + (n-1)d .$$

We summarize with the following table:

Number of Term (n)	1	2	3	4	$\cdots$	10	$\cdots$	n
Term (a_n)	a_1	$a_1 + d$	$a_1 + 2d$	$a_1 + 3d$	$\cdots$	$a_1 + 9d$	$\cdots$	$a_1 + (n-1)d$

Example 5

Determine a formula for the general term of the following sequences:
(a) 3, 7, 11, 15, 19, ...
(b) 100, 99, 98, 97, 96, ...
(c) The arithmetic sequence with first term 2 and common difference 7
(d) The arithmetic sequence with first term 80 and second term 65
(e) The arithmetic sequence with first term 4 and tenth term 58

Solution

(a) Since $d = 4$, $a_n = 3 + 4(n-1) = 3 + 4n - 4 = 4n - 1$.

(b) Since $d = -1$, $a_n = 100 + (-1)(n-1) = 100 - n + 1 = 101 - n$.

(c) $a_n = 2 + 7(n-1) = 7n - 5$.

(d) Since $d = -15$, $a_n = 80 + (-15)(n-1) = 95 - 15n$.

(e) We first need to determine d. Since $a_1 = 4$ and $a_{10} = 4 + 9d = 58$, $9d = 54$, and $d = 6$. Thus $a_n = 4 + 6(n-1) = 6n - 2$.

Example 6

At the new Giga-Plex Cinema there are 10 seats in the first row, 13 seats in the second row, 16 seats in the third row, and, in general, 3 more seats in each row than in the previous row.
(a) Find a formula for the number of seats in the nth row.
(b) How many seats are in the 20th row?
(c) If the last row has 100 seats, how many rows are there?

continued

Solution

(a) If a_n represents the number of seats in the nth row, then $\{a_n\}$ is an arithmetic sequence with first term 10 and common difference $d = 3$. A formula for the general term is thus $a_n = 10 + 3(n-1) = 3n + 7$.

(b) The number of seats in the twentieth row is $a_{20} = 3 \cdot 20 + 7 = 67$.

(c) If $a_n = 3n + 7 = 100$, then $3n = 93$, and $n = 31$. There are 31 rows.

GEOMETRIC SEQUENCES

The sequence 3, 6, 12, 24, 48, … of Example 1(b) is not an arithmetic sequence, since consecutive terms do not share a common difference. The apparent pattern, however, is almost as nice: We simply double a term to get to the next term. Instead of a common difference, consecutive terms of this sequence share a **common ratio** of 2:

$$6 \div 3 = 2; 12 \div 6 = 2; 24 \div 12 = 2; 48 \div 24 = 2; \ldots$$

Such a sequence, in which consecutive terms have a common ratio, is called a **geometric sequence**, and the common ratio is usually denoted r.

Example 7

Determine whether each of the following sequences is geometric; if it is, find the common ratio r.

(a) $3, 2, \frac{4}{3}, \frac{8}{9}, \frac{16}{27}, \ldots$

(b) $2, -2, 2, -2, 2, \ldots$

(c) $1, \frac{1}{2}, \frac{1}{3}, \frac{1}{4}, \frac{1}{5}, \ldots$

(d) $1, \frac{1}{2}, \frac{1}{4}, \frac{1}{8}, \frac{1}{16}, \ldots$

(e) $1, 4, 9, 16, 25, \ldots$

Solution

The sequences in parts (a), (b), and (d) are geometric, with common ratios $\frac{2}{3}$, -1 , and $\frac{1}{2}$, respectively.

Similar to the case with arithmetic sequences, a formula for the general term of a geometric sequence can be determined if the common ratio and the first term of the sequence are known. Rather than repeatedly adding the common difference to the first term, we get the nth term of a geometric sequence by repeatedly multiplying the first term by the common ratio. More precisely, $a_n = a_1 \cdot r^{n-1}$.

Example 8

Find a formula for the general term of the geometric sequences in Example 7.

Solution

Recall that sequences (a), (b), and (d) are geometric.

(a) $a_n = 3\left(\frac{2}{3}\right)^{n-1} = \dfrac{2^{n-1}}{3^{n-2}}$

(b) $a_n = 2(-1)^{n-1}$

(d) $a_n = 1\left(\frac{1}{2}\right)^{n-1} = \dfrac{1}{2^{n-1}}$

Quantities that change by a fixed percentage over each interval of time provide a wealth of examples of geometric sequences. Recall that doubling, or multiplying by 2, corresponds to a 100% increase. Similarly, increasing a quantity by 30% is the same as multiplying it by 130%, or 1.3. Reducing a quantity by 25% is the same as multiplying it by 75%, or 0.75.

Example 9

Lynette's annual salary at her new job is $40,000. If she gets a 5% raise each year, what will her annual salary be for her fifth year on the job?

Solution

If Lynette's salary increases by 5%, then her new salary will be 105% of her old salary, or 1.05 times her old salary. If a_n represents her salary for her nth year on the job, then $\{a_n\}$ is a geometric series with first term $a_1 = 40,000$ and common ratio $r = 1.05$. Her salary during her fifth year will be $a_5 = \$40,000(1.05)^4 = \$48,620.25$.

Example 10

The amount of a certain medication in the bloodstream decreases by 20% each hour. If there are 15 milliliters of the medication in the bloodstream at 1:00 p.m., how much will be in the bloodstream at 4:00 p.m.?

Solution

If a_n represents the milliliters of the medication in the bloodstream at n o'clock, then $\{a_n\}$ is a geometric sequence with common ratio $r = 1 - 0.2 = 0.8$ and first term $a_1 = 15$. The amount of medication in the bloodstream at 4:00 p.m. is $a_4 = 15(0.8)^3 = 7.68$ milliliters.

Exploratory Exercise Set 2.4

1. Divide up into pairs. Have each person write down a sequence of ten numbers with a distinct pattern. On a separate sheet of paper, write down the first five numbers of your sequence and give it to your partner.
   (a) Try to describe the pattern of your partner's sequence in words.
   (b) Write down what you think are the next five terms of your partner's sequence, and then compare your answers.

2. Write down the first five terms of the sequence whose general term is given by:
   (a) $a_n = 5 - 2n$
   (b) $a_n = n^2 + n$
   (c) $a_n = \dfrac{n}{n+1}$
   (d) $a_n = 4 \cdot 3^n$

3. Find a formula for the general term a_n of the sequence whose first five terms are:
   (a) 1, 5, 9, 13, 17, …
   (b) 72, 36, 18, 9, $\frac{9}{2}$, …
   (c) 0, 3, 8, 15, 24, …
   (d) –1, 2, –3, 4, –5, …

4. Which exercise did you find more difficult, Exercise 2 or Exercise 3? Explain.

5. Neil Sloane's On-Line Encyclopedia of Integer Sequences is an excellent resource for exploring sequences of integers. It can be found on the Web at
   http://www.research.att.com/~njas/sequences/index.html

(a) Go to this web site and type in the first five terms of the sequence 1, 5, 9, 13, 17, ... and click on the Search button. How many sequences did you find?

(b) Which sequence corresponds to the one you described in Exercise 3(a)?

(c) List the first ten terms of the sequence of odd integers that can be written in the form $x^2 + y^2$ for integers x and y.

(d) Explain why it is not a good idea to define a sequence by listing its first few terms.

6. Consider a sequence whose first five terms are 4, 20, 100, 500, 2500, ...
    (a) Could this be an arithmetic sequence? If so, what is the common difference d?
    (b) Could this be a geometric sequence? If so, what is the common ratio r?
    (c) Fill in the following table, which illustrates how to compute the terms of the sequence.

Number of Term (n)	1	2	3	4	$\cdots$	10	$\cdots$	n
Term (a_n)	4	20	100	___	$\cdots$	___	$\cdots$	___
Computation Used	4	$4 \cdot 5$	$4 \cdot 5^2$	___	$\cdots$	___	$\cdots$	___

7. Complete a similar table for the general geometric series with first term a_1 and common ration r.

Number of Term (n)	1	2	3	4	$\cdots$	10	$\cdots$	n
Term (a_n)	a_1	$a_1 \cdot r$	___	___	$\cdots$	___	$\cdots$	___

8. Look for a pattern in the first three squares, and try to guess the next square.
    (a) $4^2 = 16$; $34^2 = 1156$; $334^2 = 111,556$; $3334^2 = $ _____
    (b) $7^2 = 49$; $67^2 = 4489$; $667^2 = 444,889$; $6667^2 = $ _____

9. Describe a pattern for the following sequence, and then give the next three terms:
    1, 11, 21, 1211, 111221, 312211, ... (*Hint*: Say the digits out loud.)

10. On the TI-73 ™ calculator, the CONSTANT function can be used to illustrate arithmetic and geometric sequences. Consider the arithmetic sequence 4, 7, 10, 13, 16, ... with first term $a_1 = 4$ and common difference $d = 3$.
    (a) Describe in words how to obtain a particular term from the previous term.
    (b) To set the constant function on the TI-73 ™, carry out the following steps:
        (1) Enter [2nd] [SET].
        (2) Select **Single** mode, rather than **Multiple**.
        (3) To the right of $C_1 =$, type [+] 3.
        (4) Enter [2nd] [QUIT].
    (c) Now enter 1 [CONST]. On the left side of the screen you should see $1 + 3$. Record what you see on the right side of the screen.
    (d) Now press [CONST] several more times and describe what you see on the screen.
    (e) What sequence will you obtain if you enter 4 [CONST] in part (c)? List the first five terms.

11. Repeat the previous exercise to illustrate the geometric sequence 6, 12, 24, 48, 96, ... on the TI-73™.
    (a) Describe in words how to obtain a particular term from the previous term.
    (b) To set the constant function on the TI-73 ™, carry out the following steps:
        (1) Enter [2nd] [SET].
        (2) Select **Single** mode, rather than **Multiple**.
        (3) To the right of $C_1 =$, type [×] 2.
        (4) Enter [2nd] [QUIT].
    (c) Now enter 3 [CONST]. On the left side of the screen you should see 3*2. Record what you see on the right side of the screen.
    (d) Now press [CONST] several more times and describe what you see on the screen.
    (e) What sequence will you obtain if you enter 6 [CONST] in part (c)? List the first five terms.

12. (*Writing Exercise*) Suppose you are given the first five terms of a sequence. Write a paragraph describing to your classmates how you would determine whether the sequence is arithmetic, geometric, or neither. In which cases can you be certain of your answer?

Exercise Set 2.4

1. Write down the first five terms of the sequence whose general term is given by
   (a) $a_n = n + 4$
   (b) $a_n = 5n - 3$
   (c) $a_n = \frac{1}{n}$
   (d) $a_n = \frac{1}{2^n}$
   (e) $a_n = n^2 + 1$
   (f) $a_n = 2 + (-1)^n$
   (g) $a_n = 2 \cdot (-1)^n$
   (h) $a_n = (-1) \cdot 2^n$
   (i) $a_n = 5 \cdot 2^n$
   (j) $a_n = 11 - n$

2. (a) Which of the sequences in Exercise 1 are arithmetic? Give the common difference d for each one.
   (b) Which of the sequences in Exercise 1 are geometric? Give the common ratio r for each one.

3. Describe a pattern in words, and then find the next three terms for each of the following sequences:
   (a) 3, 7, 11, 15, 19, ...
   (b) 1, –2, 3, –4, 5, ...
   (c) 8, 2, –4, –10, –16, ...
   (d) 2, 10, 50, 250, 1250, ...
   (e) $\frac{1}{3}, \frac{1}{9}, \frac{1}{27}, \frac{1}{81}, \frac{1}{243}, \ldots$
   (f) $\frac{1}{4}, \frac{2}{5}, \frac{3}{6}, \frac{4}{7}, \frac{5}{8}, \ldots$
   (g) 1, 6, 5, 10, 9, ...
   (h) 11, –22, 44, –88, 176, ...
   (i) 1, 3, 6, 10, 15, ...
   (j) 1, 11, 121, 1331, 14641, ...

4. (a) Which of the sequences in Exercise 3 could be arithmetic? Give the common difference d for each one.
   (b) Which of the sequences in Exercise 3 could be geometric? Give the common ratio r for each one.

5. For each of the following sequences, determine whether the sequence could be arithmetic, geometric, or neither. If it is arithmetic, give the common difference d. If it is geometric, give the common ratio, r.
   (a) 1, 0, 1, 0, 1, ...
   (b) 9, 7, 5, 3, 1, ...
   (c) 1, –1, 1, –1, 1, ...
   (d) 4, 20, 100, 500, 2500, ...
   (e) 2, –4, 6, –8, 10, ...

6. For each of the following sequences, find a formula for the apparent general term a_n:
   (a) –1, 2, –3, 4, –5, ...
   (b) 1, 4, 9, 16, 25, ...
   (c) 1, 3, 6, 10, 15, ...
   (d) 3, 10, 17, 24, 31, ...
   (e) 3, 6, 12, 24, 48, ...

7. (a) Which of the sequences in Exercise 6 could be arithmetic? Give the common difference d for each one.
   (b) Which of the sequences in Exercise 6 could be geometric? Give the common ratio r for each one.

8. Write down the first five terms of the arithmetic sequence with
    (a) first term 7 and common difference 10
    (b) first two terms −4 and 5
    (c) first term 2 and fourth term 20
    (d) third term 14 and fifth term 8

9. Determine a formula for the general term a_n of each of the sequences in Exercise 8.

10. Find a formula for the general term a_n of the following arithmetic sequences:
    (a) 3, 7, 11, 15, 19, ...
    (b) 22, 17, 12, 7, 2, ...
    (c) 4, 13, 22, 31, 40, ...
    (d) 50, 49, 48, 47, 46, ...
    (e) 13, 24, 35, 46, 57, ...

11. Using a TI-83™ calculator, enter 1, 2, 3, 4, 5 into the list L_1 and the first five numbers of each sequence from Exercise 10 into the list L_2. Then use the LinReg command to find a formula for the nth term of the sequence. Relate this formula to the one you found by hand in Exercise 10.

12. A school auditorium has 12 seats in the front row, 13 seats in the second row, 14 seats in the third row, and so on. If there are 30 rows in the auditorium, how many seats are in the last row?

13. Miguel's starting annual salary at his new job is $35,000. If he receives a $2000 raise each year, what will his annual salary be after 6 years?

14. Write down the first five terms of the geometric sequence with
    (a) first term 1 and common ratio $r = 3$
    (b) fourth term 4 and common ratio $r = \frac{1}{2}$
    (c) first term 2 and second term − 40
    (d) third term 28 and fourth term 56

15. Determine a formula for the general term a_n of each of the sequences in Exercise 14.

16. Find a formula for the general term a_n of the following geometric sequences:
    (a) 8, 24, 72, 216, 648, ...
    (b) −5, 10, −20, 40, −80, ...
    (c) 400, 200, 100, 50, 25, ...
    (d) 2500, −500, 100, −20, 4, ...
    (e) $\frac{4}{9}, \frac{4}{3}$, 4, 12, 36, ...

17. Can a sequence be both arithmetic and geometric? If so, find one. If not, explain why not.

18. Suppose a sequence has first term 2 and third term 18.
    (a) If the sequence is arithmetic, find the common difference d, and list the first five terms of the sequence.
    (b) If the sequence is geometric, find two possible values for the common ratio r, and list the first five terms of the geometric sequence in each case.

19. There are 1600 students currently enrolled in the Idaville School District. If the enrollment increases by 5% each year, approximately how many students will be enrolled 5 years from now?

20. Alice starts a new job with an annual salary of $40,000 and a 5% raise each year. At the same time, Bob starts a new job with an annual salary of $50,000 and a $2000 raise each year. Will Alice's annual salary ever be greater than Bob's? If so, when?

21. A chessboard has 64 squares. Suppose a cent is placed on the first square, two cents on the second square, four cents on the third square, eight cents on the fourth square, and twice as many cents on each successive square.
    (a) Which will be the first square to have more than a million dollars on it?
    (b) How much money will be on the last square?

2.5 Recursive Sequences

With many of the sequences we studied in the previous section, we can describe the apparent pattern by saying how to get from one term to the next. For example, we could tell a student how to write down the arithmetic sequence 1, 4, 7, 10, 13, … with only two instructions:

> 1. Start with 1.
> 2. Add 3 to any term to get the next term.

Notice that both instructions are necessary if we want the student to write down the particular sequence 1, 4, 7, 10, 13, … . Without the first instruction, the student might write down 3, 6, 9, 12, 15, … or 8, 11, 14, 17, 20, … or indeed any arithmetic sequence with a common difference of 3. And without the second instruction, of course, the student could write down any sequence that begins with 1. We can rephrase these instructions in the mathematical language of sequences by saying

> 1. $a_1 = 1$
> 2. $a_n = a_{n-1} + 3$, for $n \geq 2$

Recall that a_1 denotes the first term of the sequence, a_n denotes the nth term of the sequence, and a_{n-1} represents the term that comes immediately before the nth term. You may have noticed that the second instruction does not make sense if n happens to be 1, since we have not defined $a_{1-1} = a_0$. Luckily, this does not cause a problem because the first instruction tells us how to find a_1.

When we define a sequence with two such instructions, we say that the sequence is **defined recursively**. The first instruction, which tells us how to get started, is called an **initial condition**. The second instruction is called a **recurrence relation**. It tells us how to get to the next term if we know the previous term or terms.

Example 1

Define the following sequences recursively, giving two instructions in mathematical language.
 (a) 3, 6, 12, 24, 48, …
 (b) 10, 5, 0, –5, –10, …
 (c) 2, 4, 7, 11, 16, 22, …
 (d) 2, 3, 5, 9, 17, …

Solution

 (a) Observe that we double each term to get to the next term. The instructions would be $a_1 = 3$ and $a_n = 2a_{n-1}$ for $n \geq 2$.

 (b) We subtract 5 to get the next term, so $a_1 = 10$ and $a_n = a_{n-1} - 5$ for $n \geq 2$.

 (c) Although this sequence is neither arithmetic nor geometric, notice that the differences between consecutive terms form a nice pattern: 2, 3, 4, 5, 6, … . In other words, you add 2 to get to the second term, add 3 to get to the third term, and so on. In mathematical language, $a_1 = 2$ and $a_n = a_{n-1} + n$ for $n \geq 2$.

 (d) Notice that each term after the first is 1 less than twice the previous term. That means $a_1 = 2$ and $a_n = 2a_{n-1} - 1$ for $n \geq 2$. Alternatively, you might notice that the differences between consecutive terms are increasing powers of 2, giving the recurrence relation $a_n = a_{n-1} + 2^{n-2}$ for $n \geq 2$.

THE FIBONACCI SEQUENCE

In some recursively defined sequences, the recurrence relation (the second instruction) involves more than just the previous term a_{n-1}. One example of such a sequence is the celebrated sequence of Leonardo Fibonacci (1170-1250), also known as Leonardo of Pisa, who described the sequence in 1202 in a problem about the growth of a rabbit population (see Exploratory Exercise 3). If you have not seen this sequence before, try to guess the pattern before you read the solution.

Example 2
Describe the pattern of the Fibonacci sequence: 1, 1, 2, 3, 5, 8, 13, 21, 34, 55, ... and define it recursively.
Solution
To get the next term of the sequence, you add the previous two terms. Notice that this means that you have to know the first two terms in order to get started. The initial conditions are $a_1 = 1$ and $a_2 = 1$, and the recurrence relation is $a_n = a_{n-1} + a_{n-2}$ for $n \geq 3$.

The Fibonacci sequence has fascinated mathematicians and others for over 800 years. In addition to its connection with the population growth of rabbits, the Fibonacci sequence can be found in the spiral pattern, or phyllotaxis, of pinecones, sunflowers, pineapples, and many other plants. The Fibonacci sequence frequently appears in the field of computer science; one such example is investigated in Exploratory Exercise 4. Indeed, there is an entire scholarly journal, *The Fibonacci Quarterly*, devoted to this sequence. Some financial analysts claim that the Fibonacci sequence can be used to predict fluctuations in the stock markets.

Notice that, in addition to the recurrence relation, we need to specify the first *two* terms of the Fibonacci sequence to get started. The Lucas sequence, named after the French mathematician Anatole Lucas (1842-1891), has the same recurrence relation as the Fibonacci sequence, and the same first term, but is quite different from the Fibonacci sequence, as you can see in the next example.

Example 3
Write down the first ten terms of the Lucas sequence, defined by $a_1 = 1, a_2 = 3$, and $a_n = a_{n-1} + a_{n-2}$ for $n \geq 3$.
Solution
The Lucas sequence begins 1, 3, 4, 7, 11, 18, 29, 47, 76, 123, ...

Example 4
Write down the first ten terms of the following sequences.
(a) $a_1 = 1, a_2 = 1$, and $a_n = a_{n-1} + 2a_{n-2}$ for $n \geq 3$
(b) $a_1 = 1, a_2 = 1, a_3 = 1$, and $a_n = a_{n-1} + a_{n-2} + a_{n-3}$ for $n \geq 4$
Solution
In each case, the first ten terms of the sequence are listed.
(a) 1, 1, 3, 5, 11, 21, 43, 85, 171, 341, ...
(b) 1, 1, 1, 3, 5, 9, 17, 31, 57, 105, ...

RECURSIVE AND NONRECURSIVE FORMULAS

There are advantages and disadvantages to defining a sequence recursively. Recall that in the previous section we learned how to determine a formula for the nth term of any arithmetic or geometric sequence. In Example 2 of the previous section, we found that $a_n = 3n - 2$ is a formula for the general term of the arithmetic sequence 1, 4, 7, 10, 13, ... Notice that we do not need to know any of the previous terms of the sequence to use this formula. If we want to know the 100th term of this sequence, we can quickly and easily plug in 100 for n in the formula and compute that $a_{100} = 3(100) - 2 = 298$.

If we had to rely on the recursive formula for this sequence, $a_1 = 1$ and $a_n = a_{n-1} + 3$ for $n \geq 2$, then to find the 100th term of this sequence, we would first have to compute the 2nd term, 3rd term, 4th term, and so on up to the 99th term, before we could finally determine the 100th term. Suppose we would like to find the 100th term in the Fibonacci sequence. Since we have not yet learned a nonrecursive formula for the general term of the Fibonacci sequence, we would have to compute all of the first 100 terms of the Fibonacci sequence to get our answer. (The answer, by the way, is a rather large number: 354,224,848,179,261,915,075.)

As you can see, it is preferable to have a nonrecursive formula for the nth term of a sequence. The only problem is that it is often difficult to find one. In the exercises, you will investigate nonrecursive formulas for some of the sequences in Example 4. There is in fact a nonrecursive formula for the nth term of the Fibonacci sequence, but it is surprisingly complicated, given the simplicity of the recurrence relation. In general, recursive formulas are easier to determine, once the pattern of the sequence has been detected, but nonrecursive formulas are easier to use.

RECURSIVE SEQUENCES ON A GRAPHING CALCULATOR

A graphing calculator can be a useful tool for investigating sequences defined recursively. We will use the features on the TI-83™, although other calculators have similar features. Simple sequences in which the recurrence relation involves only the previous term a_{n-1} can be displayed on the home screen using the Last answer key ([2nd] [ANS]).

Example 5

Display the sequence in Example 1(d) defined by $a_1 = 2$ and $a_n = 2a_{n-1} - 1$.

Solution

Type in 2 [ENTER] to enter the initial value. Then type in 2 × [2nd] [ANS] – 1 (or simply × 2 – 1) to get the next term. Now simply hit [ENTER] to get each new term.

To see a graph of this sequence, or to display the terms of more complex sequences, such as the Fibonacci sequence, we can use the calculator's Sequence mode (**Seq**).

Example 6
Display the Fibonacci sequence in a table and as a graph.

Solution

After selecting the **Seq** mode, press the [Y =] button. Type in the recurrence relation on the second line after u(n) = (TI models use u(n) to represent a_n and u(n – 1) to represent a_{n-1}). It should look like $u(n) = u(n-1) + u(n-2)$, as in Figure 1, below. The u can be found above the 7 key; use the [X,T, θ, n] key for n. Type in the initial values {1, 1} on the third line after u(nMin) = . Enclose the initial values in braces (braces are found above the parentheses keys). To display the sequence as a table, press [2nd] [TABLE]; to display it as a graph, press [GRAPH]. To adjust the window size, press [WINDOW] or use one of the options in the [ZOOM] menu. See Figure 2 for the display of the table.

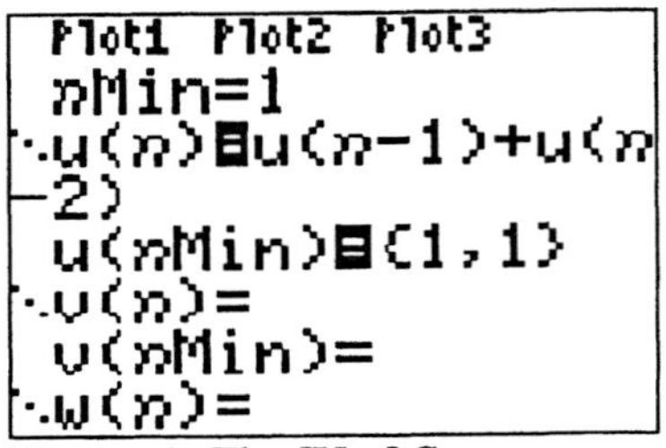

```
Plot1  Plot2  Plot3
 nMin=1
·.u(n)Bu(n-1)+u(n
-2)
 u(nMin)B(1,1)
·.v(n)=
 v(nMin)=
·.w(n)=
```
Figure 1: The [Y =] Screen

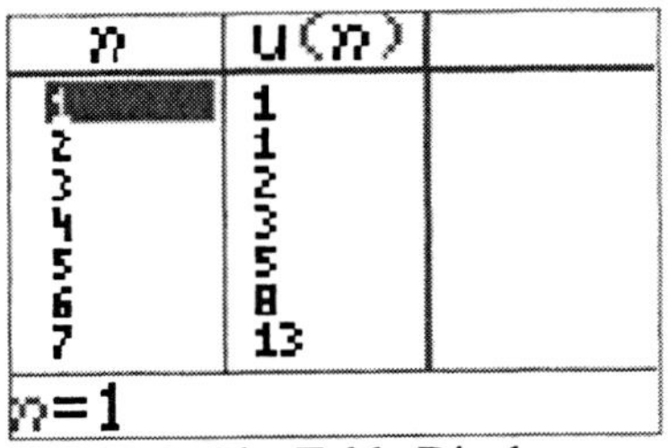

n	u(n)	
1	1	
2	1	
3	2	
4	3	
5	5	
6	8	
7	13	

n=1

Figure 2: The Table Display

WEB PLOTS

A web plot graphs a_n as a function of the previous term a_{n-1} and is particularly useful in examining the long-term behavior of a sequence. The initial value of the sequence is sometimes called the **seed**, and the following terms of the sequence comprise the **orbit** for that particular seed. Let us return to the sequence of Example 5 to demonstrate how a web plot of a sequence can illustrate the behavior of a sequence.

Example 7
Create a web plot of the sequence from Example 5.

Solution

On the [Y =] screen, type in the recurrence relation $u(n) = 2u(n-1) - 1$ on the second line and the initial condition $u(nMin) = 2$ on the third line. Choose the ZStandard option from the [ZOOM]. To display a web plot, press [2nd] [FORMAT] above the [Zoom] key, and choose Web from the top row of the screen. Now press [GRAPH]. You should see two lines: $y = 2x - 1$ (can you see the connection with the recurrence relation?) and $y = x$. Press the [TRACE] button and then the right cursor button repeatedly to see a visual representation of what happens to the terms of the sequence. You should see the web plot zigzag its way farther and farther up and to the right, as in the figure that follows. You may press [2nd] [TABLE] to see the terms of the sequence in tabular form.

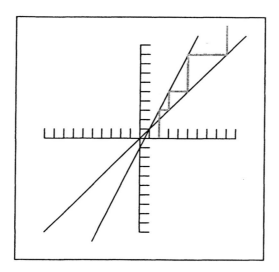

Figure 3: A web plot of the sequence from Example 5

In Example 7, we saw that when our initial value (the seed) is 2, the terms get larger and larger without bound. In this circumstance, we say that the orbit goes to infinity. Let us modify our sequence slightly by changing the initial value, but keeping the recurrence relation unchanged, and determine what happens.

Example 8

Create a web plot for the sequences with the same recurrence relation as in the previous example, but with the following seeds. Describe the web plot for each seed. Press [2nd] [TABLE] to verify your description numerically.

(a) $a_1 = 1$ (b) $a_1 = 0$ (c) $a_1 = 0.75$ (d) $a_1 = 1.1$

Solution

(a) The web plot stays fixed at the single value 1.

(b) The web plot zigzags down and to the left. The terms of the sequence are decreasing toward negative infinity: $0, -1, -3, -7, -15, -31, \dots$.

(c) The web plot starts out a bit more slowly but once again zigzags down and to the left. Once again, the terms of the sequence are decreasing toward negative infinity: $0.75, 0.5, 0, -1, -3, -7, \dots$.

(d) The web plot starts out slowly but eventually zigzags up and to the right. The terms increase toward infinity: $1.1, 1.2, 1.4, 1.8, 2.6, 4.2, \dots$.

Notice that when the seed is 1, the terms of the sequence remain fixed at 1. We call such a value a **fixed point** of the recurrence relation. Notice as well that when the seed is not equal to 1, the orbit of the sequence moves away from 1. This happens even with seeds that are quite close to 1, as in Examples (c) and (d). For this reason, we say that 1 is a **repelling** fixed point. Different types of behavior are illustrated in Exploratory Exercises 5 through 9. The behavior of such relations is studied in the branch of mathematics called dynamical systems.

RECURSIVE SEQUENCES AND SPREADSHEETS

In addition to a graphing calculator, a spreadsheet can provide a natural means of exploring recursive sequences. Spreadsheets are particularly well suited for illustrating practical applications involving compound interest rates or loan payments.

Example 9

Juan makes a deposit of $18,000 in a savings account with an annual interest rate of 6%, compounded monthly.

(a) Determine the interest earned and the balance in Juan's account over the first 3 months.

(b) Define a sequence $\{a_n\}$ recursively, where a_n gives the balance in Juan's account after n months.

(c) Use a spreadsheet to illustrate the interest earned and the balance in Juan's account over the first 6 months.

(d) Use the fact that the sequence defined in (b) is geometric to derive a nonrecursive formula for a_n.

(e) Calculate the balance in Juan's account using the formula you derived in (d). Compare your answer to the value in the spreadsheet in (c).

Solution

(a) With an annual interest rate of 6%, the monthly interest rate is 0.5%, or 0.005. After 1 month, Juan will have earned $(18,000)(0.005) = $90 interest, so the balance in his account will be $18,090. During the second month, the interest earned is 0.5% of $18,090, or $(18,090)(0.005) = $90.45, so the balance after 2 months will be $18,090 + $90.45 = $18,180.45. During the third month, the interest earned is $(18,180.45)(0.005) = $90.90 (rounded to the nearest penny), and the balance after 3 months will be $18,180.45 + $90.90 = $18,271.35.

(b) In part (a) we determined that the balance after 1 month will be $18,090, so that $a_1 = 18,090$. The interest earned during the nth month will be 0.5% of the balance after $n - 1$ months, or $(0.005)a_{n-1}$, and the balance after n months will be $a_{n-1} + (0.005)a_{n-1} = (1+0.005)a_{n-1} = (1.005)a_{n-1}$. Thus we can define the sequence by giving the initial condition $a_1 = 18,090$ and the recurrence relation $a_n = (1.005)a_{n-1}$.

(c) Table 1 shows the commands to generate the interest earned and the balance in Juan's account over the first 6 months.

Table 1: A spreadsheet for Juan's savings account (Example 9)				
	A	B	C	D	E
	A	B	C	D	E
1	Annual interest rate	Principal	Month	Interest	Balance
2	0.06	$18,000	0		=B2
3			1	=A$5*E2	=E2+D3
4	Monthly interest rate		2	=A$5*E3	=E3+D4
5	=A2/12		3	=A$5*E4	=E4+D5
6			4	=A$5*E5	=E5+D6
7			5	=A$5*E6	=E6+D7
8			6	=A$5*E7	=E7+D8

Example 9 (continued)

Note: Entering the formulas in Columns D and E can be greatly simplified by using the Fill Down command available in Excel™ and other spreadsheets. For example, after entering the formula for Cell D3, click on the lower right-hand corner of that cell and drag down to Cell D8. The entries in Columns C and E may be generated in a similar way.

(d) The first term is $a_1 = 18,090$, and the common ratio is $r = 1.005$, so the formula is $a_n = 18090 \cdot (1.005)^{n-1}$. Alternatively, since $18,090 = 18,000 \cdot (1.005)$, we could write $a_n = 18,000 \cdot (1.005)^n$.

(e) Using the formula in (d), $a_6 = 18,000 \cdot (1.005)^6 \approx 18,546.795$, which would give $18,546.80 when rounded to the nearest cent. This should agree with the amount in cell E8 in the spreadsheet in (c).

Example 10

Yolanda wants to start a college fund for her daughter Chekesha. She opens a savings account by depositing $100 and decides to deposit an additional $100 into the account every month. If the account has an APR of 3%, compounded monthly, use a spreadsheet to determine how much money will be in the account after

(a) 6 months, (b) 1 year, and (c) 10 years.

Solution

This is similar to the previous example, except that in addition to adding interest each month, we add $100 as well. We recommend that, after typing in the formulas for the first few months, you highlight the entries in Columns C, D, and E and use the Fill Down command to generate the remaining rows, as in part (c) of Example 9. The commands are shown in Table 2, and the actual numerical values are shown in Table 3, from which we see that the answers are

(a) $705.26, (b) $1,319.68, and (c) $14,109.08.

	A	B	C	D	E
	Table 2: A spreadsheet for Chekesha's college fund (Example 10)				
1	Annual interest rate	Month	Interest	Deposit	Balance
2	0.03	0		100	=D2
3		1	=A$5*E2	100	=E2+C3+D3
4	Monthly interest rate	2	=A$5*E3	100	=E3+C4+D4
5	=A2/12	3	=A$5*E4	100	=E4+C5+D5
6		4	=A$5*E5	100	=E5+C6+D6
7		5	=A$5*E6	100	=E6+C7+D7
8		6	=A$5*E7	100	=E7+C8+D8

	A	B	C	D	E
	Table 3: Numerical spreadsheet values for Example 10				
1	Annual interest rate	Month	Interest	Deposit	Balance
2	0.03	0		$100	$100
3		1	$0.25	$100	$200.25
4	Monthly interest rate	2	$0.50	$100	$300.75
5	0.0025	3	$0.75	$100	$401.50
6		4	$1.00	$100	$502.50
7		5	$1.26	$100	$603.76
8		6	$1.51	$100	$705.27
14		12	$3.04	$100	$1,319.68
122		120	$34.94	$100	$14,109.08

Exploratory Exercise Set 2.5

1.   Divide up into pairs, with Student A and Student B in each pair.
   (a) Have Student A write down the first five terms of an **arithmetic** sequence.
      (1) Without showing the sequence to Student B, Student A tells Student B the common difference d for the arithmetic sequence.
      (2) Student B writes down a recurrence relation for the sequence. What more information does Student B need to determine the sequence?
      (3) Student A tells Student B the first term of the sequence. Student B writes down the first five terms of the sequence and compares with the original sequence Student A wrote down.
   (b) Have Student B write down the first five terms of a **geometric** sequence.
      (1) Without showing the sequence to Student A, Student B tells Student A the common ratio r for the geometric sequence.
      (2) Student A writes down a recurrence relation for the sequence. What more information does Student A need to determine the sequence?
      (3) Student B tells Student A the first term of the sequence. Student A writes down the first five terms of the sequence and compares with the original sequence Student B wrote down.

2.   On the TI-73 ™ calculator, the CONSTANT function can be used to illustrate recursive sequences. Consider the sequence 3, 5, 9, 17, 33, ... with initial condition $a_1 = 3$ and recurrence relation
   $a_n = 2a_{n-1} - 1$ for $n \geq 2$.
   (a) Describe in words how to obtain a particular term from the previous term.
   (b) To set the constant function on the TI-73 ™, carry out the following steps:
      (1) Enter [2nd] [SET].
      (2) Select **Single** mode, rather than **Multiple**.
      (3) To the right of $C_1 =$, type [×]2[−]1.
      (4) Enter [2nd] [QUIT].
   (c) Now enter 2 [CONST]. On the left side of the screen you should see 2*2-1. Record what you see on the right side of the screen.
   (d) Now press [CONST] several more times and describe what you see on the screen.

3.   This exercise explores the rabbit problem that Fibonacci posed in his book *Liber Abaci*, written in 1202.

   Suppose a pair of rabbits, one female and one male, are born on January 15. After taking 2 months to mature, each pair gives birth to a new pair every month. Assuming no rabbits die, how many pairs of rabbits will there be after n months?

   (a) When will the original pair of rabbits first give birth to a new pair of rabbits?
   (b) How many pairs of rabbits will there be at the end of January? February? March? April? May?
   (c) How many *mature* pairs of rabbits will there be at the end of January? February? March? April? May?
   (d) How many *newborn* pairs of rabbits will there be at the end of January? February? March? April? May?
   (e) Record your answers from parts (a) through (c) in the table on the following page, and continue your calculations for 2 more months.

Month	Number of Pairs (f_n)	Number of Mature Pairs	Number of Newborn Pairs
January			
February			
March			
April			
May			
June			
July			

(f) Explain why the numbers in the last two columns agree after the first month.

(g) Explain why the number of mature pairs of rabbits at the end of the nth month is the same as the total number of pairs of rabbits alive 2 months ago.

(h) Explain why the number of pairs of rabbits in any given month is the sum of the number of newborn pairs in that month and the total number of pairs in the previous month.

(i) Let f_n represent the number of pairs of rabbits at the end of the nth month. Explain why

$$f_n = f_{n-1} + f_{n-2}.$$

(j) What can you conclude about the sequence $f_1, f_2, f_3, \ldots$?

4. In computer science, a *bit string* is a sequence of zeros and ones. For example, there are four bit strings of length two: 00, 01, 10, and 11. (Notice that a bit string may begin with a zero.)
  (a) List the bit strings of length one.
  (b) List the bit strings of length two that contain no two consecutive 1's.
  (c) List the bit strings of length three that contain no two consecutive 1's.
  (d) List the bit strings of length four that contain no two consecutive 1's.
  (e) What do you notice about the number of bit strings in your answers to parts (a) through (d)? Answer by completing the following conjecture (guess), which we will revisit and prove in Chapter 6.
  Conjecture: The number of bit strings of length n that contain no two consecutive 1's is

  _____ , where f_n denotes the nth term of the Fibonacci sequence:

  $1, 1, 2, 3, 5, 8, \ldots$.

5. In this exercise, we will use a spreadsheet to explore recursive sequences. You may want to review Examples 2 and 3 in this section.
  (a) In Column A of a spreadsheet, enter the number 1 in Rows 1 and 2. Next, enter the command =A1+A2 in Cell A3. (Don't forget to click on the check mark when you're done.) Now, click on Cell A3 and drag down the lower right-hand corner to "fill down" the next ten rows or so. Which familiar sequence appears in Column A?
  (b) Copy and paste the numbers from Column A into Column B. Now type the number 3 into Cell B2 and hit Enter. Which familiar sequence appears in Column B?
  (c) In Cell C1, type in the formula =AVERAGE(A1:B1) to compute the average of the two numbers in cells A1 and B1. Now "fill down" Cell C1 for as many rows as you have in Columns A and B. What familiar sequence appears in Column C?
  (d) On the basis of your observations in parts (a), (b), and (c), complete the following conjecture.
  Conjecture: If f_n represents the nth term of the Fibonacci sequence, and L_n represents the nth term of the Lucas sequence, then the average of f_n and L_n is _____ .

Exploratory Exercises 6 through 10 use a graphing calculator such as the TI-83 to explore recursive sequences with web plots. You may wish to review Examples 7 and 8 in this section before working these exercises.

6. Consider the recurrence relation $a_n = \frac{1}{2}a_{n-1} - 1$ for $n \geq 2$.
    (a) Draw web plots, as in Example 7, for the sequences with this recurrence relation and each of the following seeds: $a_1 = 1, 3, 0, -2, -3$. Use the ZDecimal option from the [ZOOM] menu.
    (b) Which of these seeds is a fixed point for this recurrence relation?
    (c) Explain why this fixed point is called an **attracting** fixed point.

7. Consider the recurrence relation $a_n = -a_{n-1} + 1$ for $n \geq 2$.
    (a) Draw web plots for the sequences with this recurrence relation and each of the following seeds: $a_1 = -1, 0, 1, 2$. Use the ZDecimal option from the [ZOOM] menu. Describe the orbit for each seed.
    (b) List the first six terms of the sequence when the seed is $a_1 = 3$. Explain why this point is called a **periodic point** rather than a fixed point. In this case the period is 2.
    (c) Does this recurrence relation have a fixed point? What is it? Test your guess by drawing a web plot for this seed.
    (d) Draw web plots for this recurrence relation using the following seeds: 0.3, 0.4, 0.6. Describe the orbit for each seed.
    (e) Explain why the fixed point you found in part (c) is neither attracting nor repelling.

*The recurrence relations in Example 7 of this section and in Exploratory Exercises 6 and 7 all have the form $a_n = ma_{n-1} + b$, where m and b are constants. Consequently, the graph drawn in the web plot, in addition to the line $y = x$, is the line with equation $y = mx + b$. As you might guess, such recurrence relations are called **linear recurrence relations**. On the basis of your experience with these linear recurrence relations, answer the following questions.*

8. (a) How many fixed points does a linear recurrence relation generally have?
    (b) There is one value of the slope m for which a linear recurrence relation has no fixed points. What is it?
    (c) Describe how you can determine the value of the fixed point(s) graphically by examining the web plot.
    (d) Find the value of the fixed point(s) of a linear recurrence relation in terms of m and b by solving the following system of equations:
    $$\begin{cases} y = mx + b \\ y = x \end{cases}$$

9. Suppose $a_n = ma_{n-1} + b$ is a linear recurrence relation.
    (a) If the linear recurrence relation has a repelling fixed point (see Example 7 in this section), what can you say about the value of the slope m? Consider positive and negative values of m.
    (b) If the linear recurrence relation has an attracting fixed point (see Exploratory Exercise 6), what can you say about the value of the slope m? Consider positive and negative values of m.
    (c) If the linear recurrence relation has a periodic point (see Exploratory Exercise 7), what can you say about the value of the slope m?
    (d) If the linear recurrence relation has no fixed point, what can you say about the value of the slope m?

10. In this exercise we will explore the quadratic recurrence relation $a_n = (a_{n-1})^2 - 2$ for $n \geq 2$. It should give you an idea of the level of complexity when the recurrence relation is no longer linear.
    (a) Draw the web plot and describe the orbit for each of the following seeds: $-3, -2, -1, 0, 1, 2, 3$. Which, if any, of these seeds are fixed points?
    (b) Draw the web plot for each of the following seeds and describe the behavior of the orbits: -1.9, $0.1, 1.1, 1.414, 1.732$.
    (c) For which initial values (seeds) does the orbit seem to jump around randomly in a fixed range without ever settling in on any particular value?

11. Ouida decides to deduct some money from her monthly paychecks to go into a retirement savings account. She starts out by putting $10 into the account, which earns 7.5% annual interest, compounded monthly. The next month she puts in $11; then $12 the next month, and so forth. Use a spreadsheet to determine how much money will she have in her account after (a) 6 months; (b) 1 year; (c) 20 years.

12. (*Writing Exercise*) You are chosen by your classmates to explain the advantages and disadvantages of defining a sequence recursively. Write a paragraph that describes what you will say.

Exercise Set 2.5

1. Write down the first five terms of each of the following recursive sequences:
    (a) $a_1 = 12$ and $a_n = a_{n-1} - 2$ for $n \geq 2$
    (b) $a_1 = 5$ and $a_n = 2a_{n-1}$ for $n \geq 2$
    (c) $a_1 = 2$ and $a_n = 3a_{n-1} - 1$ for $n \geq 2$
    (d) $a_1 = 3$ and $a_n = a_{n-1} + n$ for $n \geq 2$
    (e) $a_1 = 4$ and $a_n = a_{n-1} \div (-2)$ for $n \geq 2$

2. (a) Which of the sequences in Exercise 1 are arithmetic? Give the common difference in each case.
    (b) Which of the sequences in Exercise 1 are geometric? Give the common ratio in each case.
    (c) Which of the sequences in Exercise 1 are neither arithmetic nor geometric?

3. Write down a recurrence relation for each of the following sequences:
    (a) An arithmetic sequence with common difference $d = 5$
    (b) A geometric sequence with common ratio $r = \frac{1}{2}$
    (c) An arithmetic sequence with common difference $d = -3$
    (d) A geometric sequence with common ratio $r = -2$

4. Write down the first five terms of each of the sequences in Exercise 3 if the first term is $a_1 = 4$.

5. Write down the first five terms of each of the following sequences:
    (a) $a_1 = 1$ and $a_n = na_{n-1}$ for $n \geq 2$
    (b) $a_1 = 1$ and $a_n = na_{n-1} + 1$ for $n \geq 2$
    (c) $a_1 = 1$ and $a_n = na_{n-1} - 1$ for $n \geq 2$
    (d) $a_1 = 1$ and $a_n = na_{n-1} + (-1)^n$ for $n \geq 2$
    (e) $a_1 = 0$ and $a_n = na_{n-1} + (-1)^n$ for $n \geq 2$

6. For each of the following recurrence relations, determine whether the resulting sequence will be arithmetic, geometric, or neither; then give the common difference for each arithmetic sequence and the common ratio for each geometric sequence.
   (a) $a_n = a_{n-1} + 2$
   (b) $a_n = -3a_{n-1}$
   (c) $a_n = a_{n-1} - \frac{3}{2}$
   (d) $a_n = a_{n-1} \div 4$
   (e) $a_n = 2a_{n-1} + 1$

7. Write down an apparent recurrence relation for each of the following sequences:
   (a) $1, 6, 11, 16, 21, \ldots$
   (b) $1, 3, 9, 27, 81, \ldots$
   (c) $2, 4, 16, 256, 65536, \ldots$
   (d) $3, 4, 6, 9, 13, 18, \ldots$
   (e) $4, 7, 13, 25, 49, \ldots$

8. Suppose a sequence has the recurrence relation $a_n = a_{n-1} + a_{n-2}$ for $n \geq 3$. Write down the first ten terms of the sequence if the initial terms are
   (a) $a_1 = 1$ and $a_2 = 3$
   (b) $a_1 = 3$ and $a_2 = 1$
   (c) $a_1 = 2$ and $a_2 = -1$
   (d) $a_1 = 5$ and $a_2 = -3$

9. (a) Write down the first five terms of the sequence defined by $a_1 = 1, a_2 = 2$, and
    $a_n = (n-1)(a_{n-1} + a_{n-2})$ for $n \geq 3$.
   (b) Write down the first five terms of the sequence defined by $a_1 = 1$ and $a_n = na_{n-1}$ for $n \geq 2$.
   (c) What do you notice about the two sequences in parts (a) and (b)?

10. (a) Write down the first five terms of the sequence defined by $a_1 = 0, a_2 = 1$, and
     $a_n = (n-1)(a_{n-1} + a_{n-2})$ for $n \geq 3$.
    (b) Write down the first five terms of the sequence defined by $a_1 = 0$ and $a_n = na_{n-1} + (-1)^n$ for $n \geq 2$.
    (c) What do you notice about the two sequences in parts (a) and (b)?

11. Write down the first ten terms of the sequences defined by
    (a) $a_1 = 1, a_2 = 3$ and $a_n = 2a_{n-1} - a_{n-2}$ for $n \geq 3$
    (b) $a_1 = 1, a_2 = 1, a_3 = 1$ and $a_n = a_{n-1} + a_{n-3}$ for $n \geq 4$

12. Consider the sequence defined recursively by $a_1 = 1$, $a_2 = 3$, and $a_n = 2a_{n-1} - a_{n-2}$ for $n \geq 3$.
    (a) List the first five terms of the sequence.
    (b) Is the sequence arithmetic? If so, find the common difference d.
    (c) Find a nonrecursive formula for a_n (that is, a formula that doesn't depend on any previous terms).

13. (a) Write down the first five terms of the sequence defined recursively by $a_1 = 1$, $a_2 = 5$, and
     $a_n = a_{n-1} + 2a_{n-2}$ for $n \geq 3$.
    (b) Write down the first five terms of the sequence defined nonrecursively by $a_n = 2^n + (-1)^n$.
    (c) What do you notice about the two sequences in parts (a) and (b)?

14. (a) Write down the first five terms of the sequence defined recursively by $a_1 = 1$ and $a_n = \left(\dfrac{n+1}{n-1}\right)a_{n-1}$

for $n \geq 2$.

(b) Write down the first five terms of the sequence defined recursively by $b_1 = 1$ and $b_n = \left(\dfrac{n+2}{n-1}\right)b_{n-1}$

for $n \geq 2$.

(c) Write down the first five terms of the sequence defined recursively by $c_1 = 1$ and $c_n = \left(\dfrac{n+3}{n-1}\right)c_{n-1}$

for $n \geq 2$.

*If we add up the first few terms of a sequence, the resulting sum is called a **partial sum**. In particular, the sum of the first n terms of a sequence is called the nth partial sum of the sequence. In Exercises 15 through 20, we will investigate the partial sums of various sequences.*

15. Referring to the sequence defined in part (a) of Exercise 14, compute the following partial sums of the sequence.
   (a) a_1
   (b) $a_1 + a_2$
   (c) $a_1 + a_2 + a_3$
   (d) $a_1 + a_2 + a_3 + a_4$
   (e) $a_1 + a_2 + a_3 + a_4 + a_5$

16. Referring to the sequence defined in part (b) of Exercise 14, compute the following partial sums of the sequence.
   (a) b_1
   (b) $b_1 + b_2$
   (c) $b_1 + b_2 + b_3$
   (d) $b_1 + b_2 + b_3 + b_4$
   (e) $b_1 + b_2 + b_3 + b_4 + b_5$

17. (a) The partial sums in Exercise 15 should resemble one of the sequences in Exercise 14. Which one?
   (b) The partial sums in Exercise 16 should resemble one of the sequences in Exercise 14. Which one?

18. Find the first five partial sums of each of the following sequences.
   (a) $1, 1, 1, 1, 1, \ldots$
   (b) $1, 2, 3, 4, 5, \ldots$
   (c) $1, 2, 4, 8, 16, \ldots$
   (d) $1, 3, 5, 7, 9, \ldots$

19. Recall the Fibonacci sequence: $1, 1, 2, 3, 5, 8, 13, 21, 34, 55, \ldots$.
   (a) Write down the first ten partial sums of the Fibonacci sequence.
   (b) Find a formula for the nth partial sum of the Fibonacci sequence. (*Hint*: What do you get when you add 1 to each of the partial sums you found in part (a)?)

20. Consider the sequence defined recursively by $a_1 = 1$ and $a_n = a_{n-1} + 6(n-1)$ for $n \geq 2$. (The numbers in this sequence are known as *centered hexagonal numbers* and correspond to the number of circles in the hexagonal arrangements in Figure 4.)
   (a) Write down the first five terms of this sequence.
   (b) Write down the first five partial sums of this sequence.

(c) On the basis of your answers to part (b), find a nonrecursive formula for the nth partial sum of the sequence.

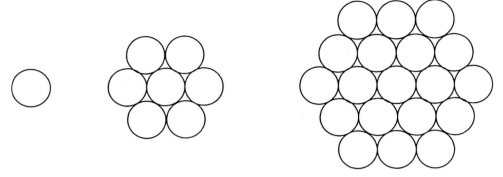

Figure 4: The Centered Hexagonal Numbers

An Introduction to Mathematical Proof

Chapter 3

3.1 Introduction to Mathematical Reasoning

In the process of reviewing the concepts of even and odd integers, Sylvia idly begins to add pairs of even integers:

$$2 + 4 = 6 \qquad 4 + 12 = 16 \qquad 12 + 18 = 30 \qquad 106 + 108 = 214$$

Suddenly she notices something. Each of the sums is an even integer also. She tries again:

$$-12 + -8 = -20 \qquad 12 + 68 = 80 \qquad -8 + 64 = 56 \qquad 12 + 2 = 14$$

Sylvia says, "I believe that I see a pattern here." She jots down the pattern a couple of different ways:

The sum of two even integers is even.

If x and y are even integers, then $x + y$ is an even integer.

Because the mathematics we study in textbooks is presented to us in finished, polished form, we often lose sight of the fact that mathematics is a dynamic, growing discipline, full of exciting discoveries and unsolved problems. Indeed, students often fail to understand the very simple dynamic by which mathematics grows. *Mathematics grows as someone observes a pattern and then finds a way to express that pattern so that it can be used and communicated to others.*

Inspired by Sylvia and possessing a spare idle moment, we connect two points on a circle with a line segment as in Figure 1(a) and observe that the interior of the circle is divided into two nonoverlapping regions. Continuing our reveries, we might connect three points on the circle as in Figure 1(b) and observe that the interior of the circle is divided into four nonoverlapping regions. As in Figure 1(c), we connect four points on the circle and observe that the interior of the circle is divided into eight regions. If we are alert, our pulse might quicken a bit at this point.

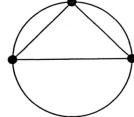
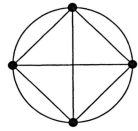

| Figure 1(a) | Figure 1(b) | Figure 1(c) | Figure 1(d) |

Quickly we place five points on the circle and get the result in Figure 1(d). Table 1 displays our results. Do we see a pattern? Can we communicate it to someone else?

Table 1				
Number of Points	2	3	4	5
Number of Regions	2	4	8	16

How about:

When n distinct points on a circle are joined by line segments, the interior of the circle is divided into 2^{n-1} nonoverlapping regions.

Well, this sounds very good. We are almost ready to impress our mathematics teacher with our expertise. For good measure, however, let us place six points on the circle and verify our pattern once more. Oh my! As we count the regions in Figure 2, we find that there are only 31 regions, not the 2^{6-1} or $2^5 = 32$ regions that we expected. Unfortunately, we have found a value of n for which our pattern does not work. We have found a counterexample.

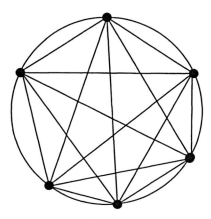

Figure 2

In view of these investigations, we need to revise our previous statement. Mathematics grows as someone observes a pattern, finds a way to express that pattern, develops a careful argument or proof that the pattern is always true, and then communicates the pattern to others. Indeed, the commitment to verify all new discoveries by proof is a distinguishing characteristic of mathematics as a discipline.

In Sylvia's investigations of the sums of even integers, she tried eight examples, and in each case, the sum was also an even integer. However, there must always lurk the suspicion that somewhere there is an example that she has not yet tried. We need to prove this result. As is often the case, the proof that her observation is always true depends very much on the definition of the objects about which she made her observations. In this case we need to remember the definitions of even and odd integers.

Definitions:

An integer n is **even** provided that there is some integer k with $n = 2k$.

An integer n is **odd** provided that there is some integer k with $n = 2k+1$.

Example 1
      6 is even since 3 is an integer with $6 = 2 \cdot 3$.
      17 is odd since 8 is an integer with $17 = 2 \cdot 8 + 1$.

Now we are ready to prove that our conjecture is true for all possible cases. When we have proved the conjecture, it is called a theorem.

Theorem 1: If x and y are even integers, then $x + y$ is even.

> *Preparation for Theorem 1*: As we have seen, it is not enough to look at specific examples to prove that this theorem is true. We must find a way to discuss all even integers at one time. To accomplish this, we will represent x by $2k$ and y by $2j$, where k and j are arbitrary integers. Since

we do not specify k and j, any result we obtain from our discussion of $2k$ and $2j$ will be true for all integers.

Proof of Theorem 1: Since x is even, there is an integer k such that $x = 2k$. Since y is also even, there is an integer j with $y = 2j$. Then $x + y = 2k + 2j = 2(j + k)$. Since $j + k$ is an integer, $x + y$ is two times an integer, and hence $x + y$ is even.■

At this point we need to reflect on the two explorations that we have just completed.
- When, after observing a pattern for several examples, we formulate a statement to describe our observations, that statement is called a **conjecture**.
- When investigating conjectures as potential theorems, we are always attentive to examples for which the conjecture does not work. An example for which the conjecture fails to hold is called a **counterexample**, and the existence of one counterexample proves the conjecture false.
- Proofs of theorems depend very much on the definitions of the objects being discussed. It is good practice to review all relevant definitions carefully before attempting a proof.
- The proofs we write are seldom complete but are, rather, only outlines of the complete argument. Typically, proofs are very dependent on knowledge gained from previously proved theorems, often theorems learned earlier in our mathematics education. For instance, in the proof of Theorem 1 we used the fact that if k and j are integers, then $k + j$ is an integer. This fact was learned in earlier courses. (Stated formally, we are using the fact that the system of integers is closed under addition.)
- Our proofs are very dependent on the patterns of valid reasoning discussed in Chapter 1. As you read the proof of Theorem 1, did you think in this way?

    If $A = 2 \cdot B$, where B is an integer, then A is an even integer. (Definition)
    $x + y = 2(k + j)$ where $k + j$ is an integer. (From the proof)

    Therefore, $x + y$ is even.

    Did you realize you were using the Rule of Detachment discussed in Chapter 1?

Because mathematical proofs are only outlines of arguments and are very dependent on information that you have learned previously, the level of detail of arguments is determined by the audience for whom one writes. In our examples and discussion, your instructor and the authors will help you become aware of the level of detail expected in the proofs that you write. For example, we will allow you to use freely the properties of operations on integers, natural numbers, and real numbers, as well as the elementary algebra that results from those properties (factoring, manipulation of algebraic expressions, and so on).

Example 2
    Investigate products of pairs consisting of an even integer and an odd integer. Form a conjecture. If, after looking at a variety of examples, you cannot find a counterexample to your conjecture, prove your conjecture.
Solution
    After multiplying several pairs such as $3 \cdot 6 = 18$ and $-7 \cdot 4 = -28$ and $41 \cdot 126 = 5166$, we might make the following conjecture:
        The product of an odd integer and an even integer is an even integer.
            or
        If x is an odd integer and y is an even integer, then the product xy is an even integer.
    After a bit more experimentation, we might attempt to prove our conjecture and thus establish a theorem.

Theorem 2: If x is an odd integer and y is an even integer, then the product xy is an even integer.

Preparation for Theorem 2: As in the proof of Theorem 1, we need to represent odd number x and even number y in their most general forms. Thus we represent x as $2k + 1$ and y as $2j$, where k and j are arbitrary integers.

Proof of Theorem 2: Since x is odd, there is an integer k with $x = 2k + 1$. Since y is even, there is an integer j with $y = 2j$. Then $xy = (2k + 1)(2j)$. Using the properties of addition and multiplication of integers, we can rewrite this as

$$xy = 4kj + 2j = 2(2kj + j)$$

Since $2kj + j$ is an integer, xy is an even integer. ∎

As noted in Section 5 of Chapter 2, *every integer is either even or odd, but never is the same integer both even and odd.* Although we will not prove this intuitively obvious fact, we will use it from time to time in the discussion in later sections.

DIVISIBILITY

The following two definitions will provide us raw material for a number of investigations in the exercise set.

Definitions: For integers a and b, a **divides** b provided that there is an integer k with $b = ak$. In this circumstance, we say that a is a **divisor** or **factor** of b. We also say that b is **divisible** by a. We denote this relation as $a \mid b$, where the vertical bar is read "divides."

Example 3

> 6 divides 42 because 7 is an integer with $42 = (6)(7)$.
> $11 \mid 121$ because 11 is an integer with $121 = (11)(11)$.
> $3 \mid 0$ because 0 is an integer with $0 = (3)(0)$.

Definition: An integer p that is greater than 1 is **prime** provided that the only positive divisors of p are 1 and p.

Example 4

> 7 is prime because its only positive divisors are 1 and 7.
> 6 is not prime because its positive divisors are 1, 2, 3, and 6.

In the exercises of this chapter, you will have the opportunity to practice the basic dynamics of mathematical reasoning: observing a pattern, expressing the pattern as a conjecture so that it can be communicated to others, and proving (or disproving) the conjecture. Be forewarned, however, that there is as much art as craft in the process of proving theorems. Proofs that seem obvious to one student will require quite a bit of perspiration by the next. However, in the process of wrestling with problems involving proofs, you will build analytical abilities that will serve you well in many different contexts, and you will be in a better position to understand the useful and intriguing mathematics presented in the remainder of this book.

The following theorems are among the results you will investigate in the exercise sets. We record them here not for careful study at this time but so that you will have them available for later reference.

Theorem 3: If x is an odd integer and y is an odd integer, then $x + y$ is an even integer.

Theorem 4: If x is an even integer and y is an odd integer, then $x + y$ is an odd integer.
Theorem 5: If x is an odd integer and y is an odd integer, then xy is an odd integer.
Theorem 6: If x is an even integer and y is an integer, then xy is an even integer.
Theorem 7: If $a \mid b$ and $a \mid c$, then $a \mid (b + c)$.
Theorem 8: If $a \mid b$ and $a \mid c$ and $a \mid d$, then $a \mid (b + c + d)$.
Theorem 9: If $a \mid b$, then $a \mid bn$ for any integer n.

Exploratory Exercise Set 3.1

1. (a) In this section, find the definition of an odd integer and copy it onto your paper. Use the definition to explain why 27 is an odd integer.
    (b) In this section, find the definition of an even integer and copy it onto your paper. Use the definition to explain why 202 is an even integer.
    (c) Add the pair of odd integers $3 + 5$.
    (d) Add five more pairs of odd integers.
    (e) One conjecture that we might make after completing parts (c) and (d) is the following statement.

    If x is an odd integer and y is an odd integer, then $x + y$ is an even integer.

    Complete the following proof of this statement.

    Proof: Since x is an odd integer, there is an integer m with $x = 2m + 1$. Since y is an odd integer, there is an integer n with $y = 2n+1$. (Now, complete the proof!)
    (f) In expressing what it means for x to be an odd integer, we used the letter m and wrote $x = 2m + 1$. In expressing what it means for y to be an odd integer, we used the letter n and wrote $y = 2n + 1$. Why was it important to use different letters in expressing x and y?
    (g) Return to the list of theorems that precedes these exploratory exercises. Which of these theorems have you now proved?

2. Investigate sums of pairs of numbers consisting of an even integer and an odd integer, form a conjecture, and prove your conjecture. You will find the outline of Exploratory Exercise 1 helpful. [Return to the list of theorems. Which of these theorems have you now proved?]

3. Explorations with the function $f(n) = n^2 - n + 41$ suggest that this function has interesting properties.
    (a) Complete the following table by computing $f(n)$ for $n = 1, 2, \ldots, 9$ and the positive integer factors of the functional values. For example, $f(1) = 1^2 - 1 + 41 = 41$ and the positive integer factors of 41 are 1 and 41. Since the only positive integer factors of $f(1)$ are 1 and the integer itself, $f(1)$ is a prime number.

n	1	2	3	4	5	6	7	8	9
$f(n)$	41								
Factors	1, 41								
Is $f(1)$ prime?	Yes								

    (b) Can you make a conjecture on the basis of the information from the table?
    (c) Now check your value by computing $f(n)$ with $n = 41$. Is your conjecture always true?
    (d) What does the result from (c) say about the truth of the following conjecture?
        For all natural numbers n, $f(n)$ is a prime number.
    (e) What does (d) say about the importance of constructing proofs of conjectures?

4. In this exercise we will investigate a special property of integers of the form
$$2^p - 1$$
    where p is a prime number.

(a) Complete the following list of the first ten prime numbers: 2, 3, 5, 7, ...

(b) Complete the following table for numbers of the form $2^p - 1$.

p	$2^p - 1$	Positive divisors of $2^p - 1$	Is it a prime?
2	$2^2 - 1 = 3$	1, 3	
3			
5			
7			

(c) On the basis of these explorations, what conjecture might you make?

(d) Continue your investigations of this conjecture with the prime 11. (*Hint:* What is 23·89?)

5. (*Writing Exercise*) A friend observes that 4 divides 100 and that 4 divides 1000 and that 4 divides 10,000. He then claims that this "proves" that 4 divides all positive powers of 10. You clearly need to educate him. First explain in general terms why it is not sufficient just to look at examples to determine whether a conjecture is true or false. Then find a counterexample to his conjecture.

Exercise Set 3.1

1. Explain why
    (a) 3 divides 27.
    (b) 12 | 0.
    (c) 4 | 16.
    (d) –2 | 12.
    (e) $x \mid 6xy$ for integers x and y.
    (f) For integers r, s, and t, $r \mid rst$.

2. Find all positive divisors of these integers.
    (a) 16   (b) 24   (c) 42

3. Explain why
    (a) 0 does not divide 7.
    (b) 6 divides –42.
    (c) 1 is not a prime number.
    (d) 17 is a prime number.
    (e) there is only one even prime number.

4. Rewrite these statements using the word *factor*.
    (a) 7 divides 28.
    (b) 5 does not divide 36.

5. Rewrite the statements of Exercise 4 using the word *divisible*.

6. Find the first twenty prime numbers.

7. (a) Observe that 3 divides 12 and 3 divides 24. Does 3 divide (12 + 24)? Explain.
    (b) Observe that 5 | 25 and 5 | 15. Does 5 | (25+15)?
    (c) Find two other examples that would support the following conjecture:
        If $a \mid b$ and $a \mid c$, then $a \mid (b + c)$.
    (d) Complete the following proof of this conjecture.
        Since $a \mid b$, there is an integer k with _____.
        Since $a \mid c$, there is an integer m with _____.
        $(b + c) = ak + am = a(\underline{\hspace{1cm}})$.
        Since $k + m$ is an _____, $a \mid (b + c)$.

(e) Return to the list of theorems just before the exploratory exercises. Which of these theorems have you now proved?

8. Follow carefully the steps of Exercise 7 to prove Theorem 8: "If $a \mid b$ and $a \mid c$ and $a \mid d$, then $a \mid (b + c + d)$."

9. (a) Observe that $4 \mid 12$. Does $4 \mid (12 \cdot 5)$? Explain.
   (b) Observe that $3 \mid 6$. Does $3 \mid (6 \cdot 11)$? Explain.
   (c) Find two other examples that would support the following conjecture:
       If $a \mid b$, then $a \mid bn$ for any integer n.

   (d) Complete the following proof of this conjecture.
       Since $a \mid b$, there is an integer k with _____.
       $(bn) = (ak)n = a(\underline{})$.
       Since kn is an integer, _____.
   (e) Return to the list of theorems just before the exploratory exercises. Which of these theorems have you now proved?

10. Consider the transitive property of divisibility:
       If $a \mid b$ and $b \mid c$, then $a \mid c$.
    (a) Verify this property with $a = 3$, $b = 6$, and $c = 18$.
    (b) Verify this property with several other values for a, b, and c.
    (c) Write a proof that this statement is true for all integers a, b, and c.

11. Review Exploratory Exercise 1. Investigate products of pairs of odd integers.
    (a) Try several examples.
    (b) Form a conjecture.
    (c) Prove your conjecture.
    (d) Return to the list of theorems just before the exploratory exercises. Which of these theorems have you now proved?

12. Review Exploratory Exercise 1. Investigate products of pairs of integers in which at least one integer is even.
    (a) Try several examples.
    (b) Form a conjecture.
    (c) Prove your conjecture.
    (d) Return to the list of theorems just before the exploratory exercises. Which of these theorems have you now proved?

Divisibility Rules: The following divisibility rules enable us to determine easily whether a large integer is divisible by integers such as 2, 3, 4, and 5.

Rule for 2: 2 divides the integer z if 2 divides the last (rightmost) digit of z.
   Example: $2 \mid 156$ since $2 \mid 6$.
Rule for 3: 3 divides the integer z if 3 divides the sum of the digits of z.
   Example: $3 \mid 156$ since $3 \mid (1+5+6)$.
Rule for 4: 4 divides the integer z if 4 divides the integer represented by the last two digits of z.
   Example: $4 \mid 124$ since $4 \mid 24$.
Rule for 5: 5 divides the integer z if the last (rightmost) digit is 0 or 5.
   Example: $5 \mid 175$ since the rightmost digit is 5.

These rules are true for all integers z and can easily be proved using Theorem 8 and 9. In Exercises 13 through 16 we will prove these rules. To avoid notational issues, we will prove these statements for integers z with $0 < z < 1000$. Observe that each such integer can be written in expanded form as

$z = a(100)+b(10) + c$, where a, b, and c are from the set {0, 1, 2, ..., 9}

13. Review the italicized text preceding this exercise and then
    (a) Write the integer 744 in expanded form as $a(100) + b(10) + c$.
    (b) Explain why the "Rule for 2" ensures that $2 \mid 744$.
    (c) Check that $2 \mid 744$ by finding an integer k such that $744 = 2k$.
    (d) Complete the following proof of the "Rule for 2."
        > If 2 divides the last digit of z, then 2 divides z.
        > Proof:
        > z can be written in expanded notation as $z = a(100) + b(10) + c$.
        > $2 \mid 100$ since $100 = 2(50)$.
        > Therefore, $2 \mid [a(100)]$ by Theorem 9.
        > $2 \mid 10$ since _____.
        > Therefore, $2 \mid [b(10)]$ by _____.
        > "$2 \mid c$" is the hypothesis of the "Rule for 2."
        > Since $2 \mid [a(100)]$ and $2 \mid [b(10)]$ and $2 \mid c$, then by _____, $2 \mid z$.

14. Review the "Rule for 3" discussed in the italicized text immediately preceding Exercise 13.
    (a) Write the integer 654 in expanded form as $a(100) + b(10) + c$.
    (b) Explain why the "Rule for 3" tells us that $3 \mid 654$.
    (c) Check that $3 \mid 654$ by finding an integer k such that $654 = 3k$.
    (d) Complete the following proof of the "Rule for 3."
        > If 3 divides the sum of the digits of z, then 3 divides z.
        > Proof: z can be written in expanded notation as $z = a(100) + b(10) + c$.
        > $100 = 99+1$; $10 = 9+1$.
        > Therefore, $z = a(99+1) + b(9+1) + c = a(99) + b(9) + (a + b + c)$. Explain.
        > $3 \mid 99$ since _____.
        > Therefore, $3 \mid [a(99)]$ by Theorem ____.
        > $3 \mid 9$ since _____.
        > Therefore, $3 \mid [b(9)]$ by Theorem ____.
        > $3 \mid (a + b + c)$ is the hypothesis of the "Rule for 3."
        > Now, complete the proof of the "Rule for 3."

15. Write a proof of the "Rule for 4" for three-digit integers z. Pay careful attention to the argument in Exercise 13.

16. Write a proof of the "Rule for 5."

17. The divisibility rule for 9 is very similar to the divisibility rule for 3.
    > "Rule for 9": $9 \mid z$ if 9 divides the sum of the digits of z.
    > Prove the "Rule for 9."

18. The divisibility rule for 11 for three-digit integers can be stated as follows:
    > Rule for 11: If $z = a(100) + b(10) + c$, then $11 \mid z$ if $11 \mid (a - b + c)$.
    > Prove the "Rule for 11."

19. Each of the arguments in Exercises 13-18 were completed for integers z between 0 and 1000, that is, for integers with one, two, or three digits.
    (a) How would you modify the arguments to prove the theorems for positive integers with four digits?
    (b) The result of Theorem 8 would need to be expanded to complete this argument. State the expansion of Theorem 8 that is needed.
    (c) Prove your statement from (b).
    (d) Rewrite the argument in Exercise 13 so that it would include positive integers that involve four digits when written in expanded form.

20. The relation $a \mid b$ (a "divides" b) on the set of positive integers has two of the important properties discussed in Chapter 2. Review the definitions of reflexive, symmetric, and transitive relations.
    (a) Prove that "divides" is reflexive.
    (b) Explain why Exercise 10 shows that "divides" is transitive.
    (c) Show that "divides" is not symmetric.
    (d) Prove that if $a \mid b$ and $b \mid a$, then $a = b$.

3.2 Proving Theorems

In this section we will continue to think about how to write careful proofs that will convince both your classmates and your instructor that a conjecture is true and hence can be called a theorem. In some of our discussion and in exercise sets, we will give attention to concepts and theorems that we introduced in Chapter 2. For example, suppose we wish to prove that the subset relation on a collection of sets is transitive. That is, for sets A, B, and C, suppose we wish to prove

$$\text{If } A \subseteq B \text{ and } B \subseteq C, \text{ then } A \subseteq C.$$

Where would we begin in the formulation of a proof of this statement?

DIRECT PROOFS

In Chapter 1 we saw that a statement of the form "If p, then q" is false only when p is true and q is false. Hence we can prove a theorem of the form "If p, then q" by starting with the assumption that p is true and then demonstrating that this means that q must also be true. A proof in which we begin with the assumption that the **hypothesis** (p) is true and then use the hypothesis, definitions, previously proved theorems, and mathematical reasoning to argue that the **conclusion** (q) must be true is called a **direct proof**. Each of the proofs we constructed in Section 3.1 was a direct proof.

Before attempting to write a direct proof or any other proof, we need to spend some time in preparation. We must look at enough examples to form a conjecture, or, if the conjecture is already provided, we must look at enough examples to make certain we understand the conjecture. Because many of the statements we will prove are direct consequences of the definitions we are studying, we will need to remind ourselves of the relevant definitions. As we begin to accumulate several theorems that we have proved, we will need to review those theorems in the hope they will provide help in our present task. Finally, we must reflect on this information as we try to determine a strategy for completing our proof. Because preparation prior to writing a proof is so important, and because the temptation is to begin writing a proof without adequate preparation, we will introduce many of the proofs in this text with a brief discussion titled "Preparation." Now let us return to the theorem with which we began this section.

Theorem 10: Suppose A, B, and C are sets. If $A \subseteq B$ and $B \subseteq C$, then $A \subseteq C$.

Preparation for Theorem 10:
Example: Let $A = \{p, q\}$, $B = \{p, q, r\}$, and $C = \{p, q, r, s, t\}$. Is $A \subseteq B$? Is $B \subseteq C$? Is $A \subseteq C$? In such circumstances, must it always follow that $A \subseteq C$? How do you know? Look carefully at the individual elements. Is $p \in A$? Since $A \subseteq B$, what does this say about p? Since B is also a subset of C, what then does this say about p?
Important Definition: Since this is a theorem about the relation "is a subset of," we must give careful attention to this definition.

Definition: $X \subseteq Y$ if and only if each element of X is an element of Y.
Reflection: We are trying to show that when $A \subseteq B$ and $B \subseteq C$ then $A \subseteq C$ by showing that any element of A is also an element of C. Following our definition, if we let a be an arbitrary element of A and argue that it must also be an element of C, we will be done. Hence the first sentence in

our proof will be "Let a be an arbitrary element of A." We will then look carefully at our hypotheses, "$A \subseteq B$" and "$B \subseteq C$," to see whether we can show that a must be an element of C.

Proof of Theorem 10: Let a be an arbitrary element of A. Since by hypothesis $A \subseteq B$, the definition of "is a subset of" requires that $a \in B$. Similarly, since by hypothesis $B \subseteq C$, then $a \in C$. Thus $A \subseteq C$.■

Goodness, did you note that the proof was much shorter than the preparation? That will be true of most of the theorems we will prove. After we have thought a bit about the theorem and reviewed the appropriate definitions, the proof will often be very straightforward.

Important Note: Observe that in writing proofs, it is always important to choose the most arbitrary objects for the argument. In proving Theorem 10, we chose an arbitrary element of A. In our proof of Theorem 1 (even integer + even integer = even integer), we did not discuss specific even integers like 4 and 10 but, rather, discussed arbitrary even integers $2k$ and $2j$. If in another context we were proving theorems about isosceles triangles, we would not choose an isosceles triangle with an included right angle to make our argument; rather, we would discuss an arbitrary isosceles triangle.

SHOWING SETS ARE EQUAL

In Chapter 2 we defined sets A and B to be equal provided that $A \subseteq B$ and $B \subseteq A$. Hence, in proving a theorem that asserts that two sets are equal, we often make our argument in two parts; we demonstrate that the first set is a subset of the second and that the second set is a subset of the first. We will practice this strategy in proving the DeMorgan's Law for the complement of the union of sets A and B.

Theorem 11: If A and B are sets, then $\overline{A \cup B} = \overline{A} \cap \overline{B}$.

Preparation for Theorem 11:
Examples: Let $U = \{p, q, r, s, t\}$, $A = \{p, q\}$, and $B = \{q, r\}$. Now verify Theorem 11 for these sets. Does this give you a clearer sense of what the theorem asserts always happens?
Definitions: The crucial definitions are the definitions for intersection, union, and complement.
   Definition: $A \cap B = \{x \mid x \in A \text{ and } x \in B\}$
   Definition: $A \cup B = \{x \mid x \in A \text{ or } x \in B\}$
   Definition: $\overline{C} = \{x \mid x \in U \text{ and } x \notin C\}$
Previous Theorems: We will also need to remember DeMorgan's Laws for forming the negation of a disjunction of statements. Remember from Chapter 1 that the negation of "p or q" is "$\sim p$ and $\sim q$."
Reflections: Since we are proving that two sets are equal, we will first demonstrate that $\overline{A \cup B} \subseteq \overline{A} \cap \overline{B}$ and then demonstrate that $\overline{A} \cap \overline{B} \subseteq \overline{A \cup B}$.
This requires that we review the definition of "is a subset of" discussed in the preparation for the proof of Theorem 10. Quite often in proving theorems about sets, we focus on the elements of the set as we move from statement to statement. Hence, our initial strategy will be to let x be any element of $\overline{A \cup B}$ and show that x must also be an element of $\overline{A} \cap \overline{B}$. Then we use a similar argument to show containment in the other direction.

Proof of Theorem 11:
Show that $\overline{A \cup B} \subseteq \overline{A} \cap \overline{B}$. Let $x \in \overline{A \cup B}$. Then $x \notin A \cup B$. Said another way, "It is not true that $x \in A$ or $x \in B$." We can rewrite this statement using DeMorgan's Laws for the negation of disjunction as "$x \notin A$ and $x \notin B$." Thus $x \in \overline{A}$ and $x \in \overline{B}$. By definition of intersection, $x \in \overline{A} \cap \overline{B}$. Therefore, $\overline{A \cup B} \subseteq \overline{A} \cap \overline{B}$.

Show that $\overline{A} \cap \overline{B} \subseteq \overline{A \cup B}$. Let $x \in \overline{A} \cap \overline{B}$. Then $x \in \overline{A}$ and $x \in \overline{B}$. Said in words, "x is not an element of A, and x is not an element of B." By DeMorgan's Law for negation of a disjunction, this statement can be rewritten as "It is not true that x is an element of A or x is an element of B." $x \notin A \cup B$. $x \in \overline{A \cup B}$. Therefore, $\overline{A} \cap \overline{B} \subseteq \overline{A \cup B}$.

Since we have now shown set containment in both directions, we have shown that the two sets are equal. $\overline{A \cup B} = \overline{A} \cap \overline{B}$. ∎

PROOF BY CONTRADICTION

In Chapter 1 we observed that the negation of "If p, then q" is equivalent to the statement "p and not q." This fact is the basis of a very important proof technique known as **proof by contradiction**. In a proof by contradiction of a theorem of the form "If p then q," we first assume that the negation of the theorem "p and not q" is true and then show that this leads to a contradiction. Thus the negation of the theorem must be false, which means that the theorem is true. Let us use a proof by contradiction to prove the following theorem.

Theorem 12: For an integer n, if n^2 is even, then n is even.

Preparation for Theorem 12:
Examples: Observe that 16, 64, and 144 are square numbers that are even. Further, $16 = 4^2$, $64 = 8^2$, and $144 = 12^2$. Since 4, 8, and 12 are all even, these examples give evidence for this conjecture.
Previous Theorems: We need to give careful attention to one of the theorems from Section 3.1.
      Theorem 5: If x and y are odd, then xy is odd.
Reflections: Since we will prove this theorem by contradiction, we will start by assuming "p and not q." Our first statement in the proof will be
      Suppose the integer n^2 is even and n is odd.
Do you see that we will quickly arrive at a contradiction? (Notice that we choose an arbitrary integer n for our argument, rather than one of the specific integers we used in our preparatory work.)

Proof of Theorem 12: Suppose the integer n^2 is even and n is odd. By Theorem 5, since n is odd, $n \cdot n$ or n^2 is odd. This is a contradiction of the fact that n^2 is even. Hence our assumption that n^2 is even and n is odd is false, making our theorem true. ∎

For a second example of a proof by contradiction, we will turn to a classic theorem proved by the mathematicians of ancient Greece. We will prove that the number $\sqrt{2}$ is irrational, that is, the number $\sqrt{2}$ cannot be expressed as the ratio of two integers. (It is interesting to note that when mathematicians of the Pythagorean religious cult proved this result in the fifth century BCE, it shattered a basic assumption of the cult and led to its dissolution.)

Theorem 13: If x is the number $\sqrt{2}$, then x is irrational.

Preparation for Theorem 13:
Definitions: Critical to understanding and proving this theorem are the definitions of rational number, irrational number, and even integer.

      Definition: x is a **rational number** provided that x can be represented as a quotient $\dfrac{m}{n}$, where m and n are integers with $n \neq 0$.
      Definition: x is an **irrational number** provided that x is not rational.
      Definition: x is an even number provided there is an integer k with $x = 2k$.
Previous Theorems: We will use Theorem 12 in a critical way in this argument. In addition, we will need a fact from our earlier study of fractions. We will need to remember that every fraction

$\dfrac{m}{n}$ can be written in reduced form such that integers m and n have no common positive divisors other than 1.

Reflections: Since we will prove this theorem by contradiction, we will start by assuming "p and not q." Our first statement in the proof will be

Suppose $x = \sqrt{2}$ and x is not irrational … or, better yet,

Suppose $x = \sqrt{2}$ and x is rational.

Proof of Theorem 13:

Suppose $x = \sqrt{2}$ and x is rational.   Then there are integers m and n with

$$x = \sqrt{2} = \frac{m}{n}$$

such that m and n have no common factors other than 1. Since $\sqrt{2} = \dfrac{m}{n}$, we can square both sides of the equation to learn that

$$2 = \frac{m^2}{n^2} \quad \text{or} \quad 2n^2 = m^2$$

Since m^2 is the product of 2 and the integer n^2, m^2 is even. By Theorem 12, this means that m is even, so there is an integer k with $m = 2k$. Thus $m^2 = (2k)^2 = 4k^2$. Hence

$$2n^2 = 4k^2 \text{ and dividing by 2 reveals that } n^2 = 2k^2.$$

Using Theorem 12 once again indicates that since n^2 is even, n is even. Now we have shown that both m and n are even, that is, 2 is a divisor (or factor) of both m and n. This contradicts our assumption that m and n share no common positive divisors other than 1. Since we have found a contradiction, we have proved the theorem. ■

ONE LAST THEOREM

We next state a theorem that we will help you prove in the exercise set.

Theorem 14: Suppose a, b, d, q, and r are integers with $a = bq + r$.
  Part 1: If $d \mid b$ and $d \mid r$, then $d \mid a$.
  Part 2: If $d \mid a$ and $d \mid b$, then $d \mid r$.

The importance of Theorem 14 is not immediately clear. For now, it will simply give us an opportunity to practice constructing a direct proof. In Section 3.4, however, this theorem will play an important role in helping us understand the Euclidean Algorithm.

In closing this section, we want to emphasize that even after we have learned two different strategies for proving theorems, direct proof and proof by contradiction, the task of proving theorems does not become mechanical. It still requires a great deal of thought. However, with careful preparation that includes looking at examples, collecting relevant definitions and theorems, and reflecting on strategies, we can devise proofs of many conjectures with surprising success.

Exploratory Exercise Set 3.2

1. In Chapter 2 we used Venn Diagrams to suggest the truth of the Distributive Property of Intersection over Union. In this exercise we will prove this theorem.

> The Distributive Property of Intersection over Union: If A, B, and C are sets, then
> $A \cap (B \cup C) = (A \cap B) \cup (A \cap C)$

   (a) *Example*: Use the sets $A=\{r, s, t\}$, $B=\{t, u\}$, and $C=\{s, t, v\}$ and verify that in this example, the statement is true.

   (b) *Definitions*: Look up and record the definitions of union and intersection.

   (c) *Reflections*: We are trying to prove that the set $A \cap (B \cup C)$ is equal to the set $(A \cap B) \cup (A \cap C)$.

   In order to show that these sets are equal we need to show that

   $$A \cap (B \cup C) \subseteq (A \cap B) \cup (A \cap C)$$

   and that _____.

   In the proof that $A \cap (B \cup C) \subseteq (A \cap B) \cup (A \cap C)$, the first statement will be _____.

   In the proof of containment in the other direction, the first statement will be _____.

   (d) Give the reasons in the following proof that $A \cap (B \cup C) \subseteq (A \cap B) \cup (A \cap C)$.
   Let $a \in A \cap (B \cup C)$. Then $a \in A$ and $a \in B \cup C$. *Why?*
   Since $a \in B \cup C$, then $a \in B$ or $a \in C$. *Why?*
   Thus $a \in (A \cap B)$ or $a \in (A \cap C)$. *Why?*
   Thus $a \in (A \cap B) \cup (A \cap C)$. *Why?*
   Explain carefully why you have now shown $A \cap (B \cup C) \subseteq (A \cap B) \cup (A \cap C)$.

   (e) Complete a similar argument showing that $(A \cap B) \cup (A \cap C) \subseteq A \cap (B \cup C)$.

   (f) Explain carefully why you have now shown that $A \cap (B \cup C) = (A \cap B) \cup (A \cap C)$.

2. Prove this theorem by contradiction:
   If x is a rational number and y is an irrational number, then $x + y$ is an irrational number.

   (a) *Examples:* Suppose x is 2 and $y = \pi$. What is $x + y$? Find three other sums of rational and irrational numbers.

   (b) *Definitions:* Find and rewrite the definitions of a rational number and an irrational number.

   (c) *Reflection:* In order to prove a theorem of the form $p \to q$ by contradiction, we start with the negation $p \wedge \sim q$. Complete the first line of this proof.
   Suppose x is a rational number and y is an irrational number and ....
   Notice that we will reach a contradiction if we can show that the irrational number y can be written as a quotient of integers. Why?

   (d) Complete the following proof by contradiction.
   Proof: Suppose x is a rational number, y is an irrational number, and $x + y$ is a rational number.
   Since x is rational, there are integers m and n ($n \neq 0$) with $x =$ _____.
   Since we are supposing $x + y$ is rational, there are integers k and l ($l \neq 0$) with
   $x + y =$ _____.
   Thus _____ $+ y = \dfrac{k}{l}$.

Show that when we solve for y, we find that $y = \dfrac{nk - ml}{nl}$.

Why is this a contradiction?

(e)   Explain carefully why we have proved that the sum of a rational number and an irrational number is irrational.

3.   On the set of integers define a relation $\equiv$ as follows:

Definition: a is related to b ($a \equiv b$) provided that $3 \mid (a - b)$.

Show that this relation is an equivalence relation.

(a)   Preparation:

(1)   *Examples*:

1.   Explain why $1 \equiv 4$ and $4 \equiv 1$.
2.   Explain why $5 \equiv 2$ and $2 \equiv 5$.
3.   Explain why $0 \equiv 6$ and $6 \equiv 6$.

(2)   *Definitions*: Write the definition of $3 \mid n$ where n is a natural number. Review Chapter 2 and then write the definitions of a reflexive relation, a symmetric relation, a transitive relation, and an equivalence relation.

(3)   *Reflection*: We can show that $\equiv$ is an equivalence relation by showing that $\equiv$ is _____, _____, and _____.

(b)   The proof: Complete the following proof.

(1)   Show that $\equiv$ is reflexive: Let a be an integer.   Then $a \equiv a$ because _____.

(2)   Show that $\equiv$ is symmetric: Let a and b be integers with $a \equiv b$.

(*Note:* We need to show that _____)

Since $a \equiv b$, _____.

Since $3 \mid (a - b)$, there is an integer k with _____.

Think carefully; then complete the proof by showing that $b \equiv a$.

(3)   Show that $\equiv$ is transitive. Let a, b, and c be integers with $a \equiv b$ and $b \equiv c$.

(Note: We need to show that _____)

Since $a \equiv b$, $3 \mid (a - b)$. This means there is an integer k with _____

Since $b \equiv c$, _____. This means there is an integer l with _____

Think carefully; then complete the proof by showing that $a \equiv c$.

Note: This equivalence relation is very important in mathematics. The sophisticated mathematics terminology for "a is related to b" for this relation is "a is congruent to b modulo 3." The usual notation for $a \equiv b$ is $a \equiv b(\bmod 3)$.

Exercise Set 3.2

1.   Let $A = \{v, w, x, y, z\}$ and let $B = \{r, s, t, u, v, w\}$

(a)   Compute $A \cap B$.

(b)   Recall and rewrite the definition of intersection and the relation "is a subset of."

(c)   Explain carefully why, in the case above, $A \cap B \subseteq A$.

2.   After reviewing your work from Exercise 1, prove the following theorem: If A and B are sets, then $A \cap B \subseteq A$.

(a)   *Reflection:* Review the *Reflections* section of the proof of Theorem 10, and then write the first line of the proof of this theorem and a sentence or two describing the strategy you will use in proving this theorem.

(b)   Complete the proof of the theorem:

(1)   Let $x \in A \cap B$. Then $x \in$ ___ and $x \in$ ___

(2)   Therefore, $A \cap B \subseteq$ ___.

3. Prove the following theorem: If A and B are sets, then $A \subseteq A \cup B$.
   (a) *Example:* Suppose $A = \{1, 2, 3\}$ and $B = \{3, 5\}$. Verify that this conjecture is true for this example.
   (b) *Definitions:* Recall and rewrite the definitions of union and the relation "is a subset of."
   (c) *Reflection:* Review the *Reflections* section of the proof of Theorem 10, and then write the first line of the proof of this theorem and a sentence or two describing the strategy you will use in proving this theorem.
   (d) Write a proof of the theorem.

In the problems that follow, you will be asked to prove a theorem. In preparation for constructing each proof, look at appropriate examples, collect related definitions as well as facts and theorems from previous study, and reflectively determine a strategy before proving the theorem.

4. After reviewing the proof of Theorem 11 in the text, prove DeMorgan's Law for the complement of an intersection of two sets:
   $$\text{If } A \text{ and } B \text{ are sets, then } \overline{A \cap B} = \overline{A} \cup \overline{B}.$$

5. After reviewing your work on Exploratory Exercise 1, prove the Distributive Property of Union over Intersection:
   $$\text{If } A, B, \text{ and } C \text{ are sets, then } A \cup (B \cap C) = (A \cup B) \cap (A \cup C).$$

6. After reviewing Exploratory Exercise 2, prove this theorem:
   If x is a non-zero rational number and y is an irrational number, then xy is an irrational number.

7. For each of these conjectures, either prove the conjecture or disprove it by providing a counterexample.
   (a) If x is a rational number and y is a rational number, then $x + y$ is a rational number.
   (b) If x is a rational number and y is a rational number, then xy is a rational number.
   (c) If x is an irrational number and y is an irrational number, then $x + y$ is an irrational number.
   (d) If x is an irrational number and y is an irrational number, then xy is an irrational number.

In Exercises 8 through 9 you will complete the proof of Theorem 14 from this section. Suppose a, b, d, q, and r are integers with $a = bq + r$. In preparation, review Theorems 8 and 9 from Section 3.1.

8. Theorem 14, Part 1: Show that if $d \mid b$ and $d \mid r$, then $d \mid (bq + r)$.

9. Theorem 14, Part 2: Show that if $d \mid a$ and $d \mid b$, then $d \mid r$.
   (a) Use Theorems 8 and 9 to prove this result: If $d \mid x$ and $d \mid y$, then $d \mid (x - y)$.
   (b) With the observation from (a), complete the proof of Part 2 of Theorem 14.

10. Theorem 12 is most often proved using the fact that a conditional statement and its contrapositive are equivalent.
    (a) Write the contrapositive of Theorem 12.
    (b) Construct a direct proof of the contrapositive of Theorem 12.

In Exploratory Exercise 3 we showed that the relation on the integers defined by "a is related to b if $3 \mid (a - b)$" is an equivalence relation. In Chapter 2 we learned that an equivalence relation on a set partitions the set into disjoint subsets. With a little careful thought (that is not included here), we can see that the relation of Exploratory Exercise 3 partitions the integers into the three disjoint sets that are described fancifully below as smooth numbers, rough numbers, and abrasive numbers.
> *Definition: An integer n is smooth if there is an integer k such that $n = 3k$.*
> *Definition: An integer n is rough if there is an integer k such that $n = 3k + 1$.*
> *Definition: An integer n is abrasive if there is an integer k such that $n = 3k + 2$.*
In Exercises 11 through 14 we will investigate some of the properties of smooth, rough and abrasive integers. In Exercise 15 we will use these properties to show that the square root of 3 is irrational.

11. (a) Find several examples of pairs of smooth integers and compute their products.
    (b) Make a conjecture about the product of two smooth integers.
    (c) Prove your conjecture.

12. (a) Find several examples of pairs of rough integers and compute their products.
    (b) Make a conjecture about the product of two rough integers.
    (c) Prove your conjecture.

13. (a) Find several examples of pairs of abrasive integers and compute their products.
    (b) Make a conjecture about the product of two abrasive integers.
    (c) Prove your conjecture.

14. After reviewing the proof of Theorem 12, prove this theorem:

    If a^2 is a smooth integer, then a is a smooth integer.

15. Review the proof of Theorem 13 in the text and the results from Exercises 11 through 14, and prove this theorem:

    If x is the number $\sqrt{3}$, then x is irrational.

16. We say that two integers a and b are "congruent modulo 5" and denote this relation by $a \equiv b(\mathrm{mod}5)$ provided that $5 \mid (a - b)$. After reviewing your work on Exploratory Exercise 3, prove that this is an equivalence relation on the set of integers.

17. Prove: If n^2 is even, then n^2 is divisible by 4.

3.3 Proof by Mathematical Induction

Children often enjoy arranging large numbers of dominos in patterns, knocking down the first domino, and then watching all the dominos come tumbling down. The arrangements of dominos can be quite complex, and the fact that all dominos always fall seems to amaze those who observe. On the other hand, the process of creating such an array of dominos is relatively simple. There are two critical issues to which we must attend as we arrange the dominos. We must make certain that we can topple the first domino, and we must make sure that if a domino falls, the domino that follows it falls. Said more mathematically, the two critical issues would read like this:

    A. The first domino must fall.
    B. For each value of k, if the kth domino falls, the $(k + 1)^{st}$ domino must fall.

If we attend to these two issues as we arrange our dominos in a pattern, we can be sure that all the dominos will fall.

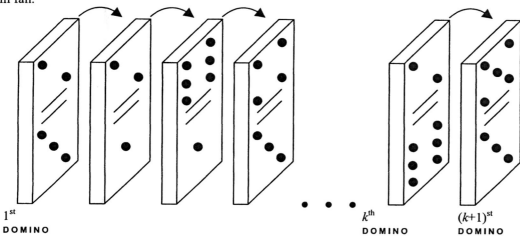

1st DOMINO k^{th} DOMINO $(k+1)^{st}$ DOMINO

Similar patterns of thinking are the basis of a very powerful theorem-proving technique that can be used to prove a certain class of theorems. Suppose that after trying several examples, we make this conjecture:

$$\text{For all natural numbers } n, \; 1 + 2 + 3 + \ldots + n = \frac{n(n+1)}{2}.$$

In this instance, we wish to prove that the open sentence $1 + 2 + 3 + \ldots + n = \dfrac{n(n+1)}{2}$ is true whenever a specific number is used in place of n in the sentence. This is an example of a conjecture that can be written in this general form:

        For all natural numbers n, $P(n)$.
             or
        If n is a natural number, then $P(n)$.

In such cases, $P(n)$ is a sentence involving the natural number n whose truth value can be determined when n is given a value. [$P(n)$ is an *open sentence* as described in Section 1.1.] For conjectures of this form, we may wish to prove the theorem using the Principle of Induction, a statement that we will accept as an axiom of the natural numbers (positive integers).

Principle of Induction: If $P(n)$ is an open sentence, and if

1. $P(1)$ is true and
2.   for every $k \geq 1$, whenever $P(k)$ is true, $P(k + 1)$ is true,

then $P(n)$ is true for all natural numbers n.

Example 1

Which of the following statements might we hope to prove using the Principle of Induction?

For all integers $n \geq 1$, $1 + 2^1 + 2^2 + \ldots + 2^n = 2^{n+1} - 1$.

If n is a positive integer, 5 divides $8^n - 3^n$.

If a set S has n elements for a natural number $n \geq 1$, then S has 2^n subsets.

Solution

Each of these statements is an example of a statement that we might hope to prove using an induction argument. In each case we have an open sentence $P(n)$ that we wish to show true for all natural numbers.

$P(n)$ is the assertion "$1 + 2^1 + 2^2 + \ldots + 2^n = 2^{n+1} - 1$."

$P(n)$ is the assertion "5 divides $8^n - 3^n$."

$P(n)$ is the assertion "if a set S has n elements, then S has 2^n subsets."

We will now use a proof by induction to establish a formula for the sum of the first n natural numbers.

Theorem 15: If n is a natural number, then $1 + 2 + 3 + \ldots + n = \dfrac{n(n+1)}{2}$.

Preparation for Theorem 15:

Examples: If we prove this theorem, we are actually establishing the truth of an infinite number of statements such as

$P(1)$: $1 = \dfrac{1(1+1)}{2}$

$P(2)$: $1 + 2 = \dfrac{2(2+1)}{2}$

$P(3)$: $1 + 2 + 3 = \dfrac{3(3+1)}{2}$

$P(4)$: $1 + 2 + 3 + 4 = \dfrac{4(4+1)}{2}$

We could determine the truth of any one of the statements just by checking the arithmetic. For instance, we can verify the truth of $P(4)$ by observing that

$1 + 2 + 3 + 4 = 10$ and $\dfrac{4(4+1)}{2} = 10$. The purpose of our proof will be to create a single

argument that will show that each of the infinite number of statements is true.

Reflection: All proofs by induction have the same outline; each proof consists of two steps, the Basis Step and the Inductive Step.

The Basis Step: Show that $P(1)$ is true.

The Inductive Step: For all natural numbers k, show that if $P(k)$ is true, then $P(k + 1)$ is true.

In terms of our domino analogy, in the Basis Step we show that the first domino will fall over. In the Inductive Step we show that the dominos are arranged so that if the kth domino falls, the $k + 1^{st}$ domino falls.   As a prelude to completing the inductive step, we recommend writing carefully $P(k)$ and $P(k+1)$. For instance, in this theorem,

$P(k)$ is $1 + 2 + 3 + \ldots + k = \dfrac{k(k+1)}{2}$.

$$P(k+1) \text{ is } 1 + 2 + 3 + \ldots + k + (k+1) = \frac{(k+1)((k+1)+1)}{2}.$$

Proof of Theorem 15:

Basis Step: Verify that $P(1)$ is true.

$P(1)$ is the statement that $1 = \frac{1(1+1)}{2}$. Evaluating $\frac{1(1+1)}{2}$ yields $\frac{1(2)}{2} = 1$.

Thus $P(1)$ is true.

Inductive Step: We wish to prove the statement "If $P(k)$, then $P(k + 1)$." Our hypothesis, then, is the statement $P(k)$:

$$1 + 2 + 3 + \ldots + k = \frac{k(k+1)}{2}.$$

We must find some way to connect the hypothesis with what we wish to prove.

Observe that if we add $(k + 1)$ to both sides of the equation given in the induction hypothesis, the left-hand side of the resulting equation is clearly the same as the left-hand side of $P(k + 1)$. We then perform algebraic manipulation on the right-hand side of the resulting equation with the following results.

$$1 + 2 + 3 + \ldots + k + (k+1) = \frac{k(k+1)}{2} + (k+1)$$

Find a common denominator for the expression on the right.

$$1 + 2 + 3 + \ldots + k + (k+1) = \frac{k(k+1) + 2(k+1)}{2}$$

Factor out the common factor $(k + 1)$.

$$1 + 2 + 3 + \ldots + k + (k+1) = \frac{(k+1)(k+2)}{2}$$

Since this last statement is $P(k + 1)$, we have shown that when $P(k)$ is true, $P(k + 1)$ is true. ∎

Although every proof by induction involves these same two steps, proofs by induction often require a great deal of ingenuity to complete. For the examples of proof by induction found in this text, the Basis Step will usually be quite easy to verify. More work will usually be necessary to determine how to show that when $P(k)$ (called the **inductive hypothesis**) is true, then $P(k + 1)$ is true. In some cases a successful strategy will involve thinking hard about $P(k)$ and determining how to modify it to arrive at $P(k + 1)$. This is the strategy used in the previous proof and in the proof of Theorem 16.

Theorem 16: For all positive integers n, $1 + 2^1 + 2^2 + \ldots + 2^n = 2^{n+1} - 1$.

Preparation for Theorem 16:
Examples:

When $n = 1$, $1 + 2^1 = 2^{1+1} - 1$ since $3 = 4 - 1$.
When $n = 2$, $1 + 2^1 + 2^2 = 2^{2+1} - 1$ since $7 = 8 - 1$.
When $n = 3$, $1 + 2^1 + 2^2 + 2^3 = 2^{3+1} - 1$ since $15 = 16 - 1$.

Such exploration helps us be certain we understand the theorem being proved, sometimes yields ideas about how the proof should be constructed, and certainly illustrates how to complete the Basis Step of the proof.

Reflection: In preparation for completing the Induction Step, it is useful to write $P(k)$ and $P(k + 1)$ carefully and look for connections between the two. In this case,

$P(k)$: $1 + 2^1 + 2^2 + \ldots + 2^k = 2^{k+1} - 1$
$P(k + 1)$: $1 + 2^1 + 2^2 + \ldots + 2^k + 2^{k+1} = 2^{(k+1)+1} - 1$

Observe that if we add 2^{k+1} to both sides of $P(k)$, we get a sentence closely related to $P(k + 1)$. We will try this strategy in the proof.

Proof of Theorem 16:

Basis Step: $P(1)$ is the statement $1 + 2^1 = 2^{1+1} - 1$, which is true since

$$1 + 2^1 = 3 \text{ and } 2^{1+1} - 1 = 2^2 - 1 = 4 - 1 = 3.$$

Inductive Step: Our inductive hypothesis is the statement
$$1 + 2^1 + 2^2 + \ldots + 2^k = 2^{k+1} - 1.$$
Adding 2^{k+1} to both sides yields
$$\begin{aligned}
1 + 2^1 + 2^2 + \ldots + 2^k + 2^{k+1} &= 2^{k+1} - 1 + 2^{k+1} \\
&= (2^{k+1} + 2^{k+1}) - 1 \\
&= 2 \cdot 2^{k+1} - 1 \\
&= 2^{k+2} - 1 \\
&= 2^{(k+1)+1} - 1
\end{aligned}$$

Thus $1 + 2^1 + 2^2 + \ldots + 2^k + 2^{k+1} = 2^{(k+1)+1} - 1$. We have completed the Inductive Step. ∎

In the proofs of Theorems 15 and 16, we found the connection between the inductive hypothesis, $P(k)$, and $P(k + 1)$ by thinking hard about $P(k)$ and determining how to modify it to arrive at $P(k + 1)$. An alternative strategy is to look hard at $P(k + 1)$ and try to modify part of $P(k + 1)$ in such a way that $P(k + 1)$ can be connected to $P(k)$. This is the strategy used in the proof of Theorem 17.

Theorem 17: For all natural numbers n, 5 divides $8^n - 3^n$.

Preparation for Theorem 17:
Examples:
> When $n = 1$, 5 certainly divides $8^1 - 3^1 = 5$.
> When $n = 2$, 5 certainly divides $8^2 - 3^2 = 55$.
> When $n = 3$, 5 certainly divides $8^3 - 3^3 = 485$.

Theorems: Review the theorems on divisibility from Section 3.1. In particular, review Theorems 7 and 9.

Reflection: In preparation for completing the Inductive Step, we write $P(k)$ and $P(k + 1)$ carefully and look for connections between the two.
> $P(k)$: 5 divides $8^k - 3^k$
> $P(k + 1)$: 5 divides $8^{k+1} - 3^{k+1}$

Connections between these statements are not immediately obvious. Further, it is not clear how to connect 5 with the expression $8^{k+1} - 3^{k+1}$. Perhaps if we write 8 as $(5 + 3)$ and write the 8^{k+1} in $P(k + 1)$ as $8 \cdot 8^k = (5 + 3)8^k$, we might find the connections we need. We will try this strategy in the proof.

Proof of Theorem 17:

Basis Step: As we saw above, when $n = 1$, 5 certainly divides $8^1 - 3^1 = 5$.

Inductive Step: Assume that k is a natural number and that 5 divides $8^k - 3^k$. Show that this ensures that 5 divides $8^{k+1} - 3^{k+1}$. We observe that

$$\begin{aligned}
8^{k+1} - 3^{k+1} &= 8 \cdot 8^k - 3^{k+1} \\
&= (5 + 3) \cdot 8^k - 3^{k+1} \\
&= 5 \cdot 8^k + 3 \cdot 8^k - 3^{k+1} \\
&= 5 \cdot 8^k + 3 \cdot 8^k - 3 \cdot 3^k \\
&= 5 \cdot 8^k + 3(8^k - 3^k)
\end{aligned}$$

Now we have the connection with $P(k)$ that we needed. By $P(k)$, the inductive hypothesis, 5 divides $8^k - 3^k$. Theorems 7 and 9 ensure that 5 divides $8^{k+1} - 3^{k+1}$. ∎

The technique of proof by induction is not very useful for discovering relationships but is quite useful in proving relationships uncovered by other means. In an exploratory exercise from Section 2.1, we looked at several examples and then conjectured that if a set has n elements, then the set has 2^n subsets. We will now prove that this is true for all natural numbers n.

Theorem 18: If S is a nonempty set with n elements, then S has 2^n subsets.

Preparation for Theorem 18:

Examples: Suppose $S_1 = \{x\}$ is a set with a single element. Then the subsets of S_1 are $\emptyset$ and $\{x\}$. Thus S has $2 = 2^1$ subsets.

Suppose $S_2 = \{x, y\}$ is a set with two elements. Then the subsets of S_2 are $\emptyset$, $\{x\}$, $\{y\}$, and $\{x, y\}$. Thus S has $4 = 2^2$ subsets.

Suppose $S_3 = \{x, y, z\}$ is a set with three elements. The subsets of S_3 are $\emptyset$, $\{x\}$, $\{y\}$, $\{z\}$, $\{x, y\}$, $\{x, z\}$, $\{y, z\}$, and (x, y, z). Thus S has $8 = 2^3$ subsets.

Reflection: In completing the Inductive Step, we will need to find a connection between

$P(k)$: A set with k elements has 2^k subsets.

$P(k + 1)$: A set with $k + 1$ elements has 2^{k+1} subsets.

A useful connection might be seen by comparing the subsets of $S_2 = \{x, y\}$ and $S_3 = \{x, y, z\}$. Note that the subsets of S_3 can be partitioned into two collections of subsets:

Collection 1: $\emptyset$, $\{x\}$, $\{y\}$, $\{x, y\}$
Collection 2: $\{z\}$, $\{x, z\}$, $\{y, z\}$, $\{x, y, z\}$

Note that Collection 1 is just the subsets of S_2 whereas Collection 2 consists of sets formed by forming the union of subsets of S_2 with the set consisting of the element z. Thus each subset of S_2 corresponds to exactly two subsets of S_3. This connection may prove useful in our proof.

Proof of Theorem 18:

Basis Step: Suppose a set S contains a single element. That is, suppose that $S = \{x\}$. Then its subsets are S and $\emptyset$. Thus set S has 2 subsets.

Inductive Step: Assume that if k is a natural number and a set has k elements, then the set has 2^k subsets. Let T be a set with $k + 1$ elements. We must show that T has 2^{k+1} subsets. Let $z \in T$ and let $S = T - \{z\}$. Then S has k elements and, by inductive hypothesis, has 2^k subsets. The subsets of T can be obtained from the subsets of S in the following way. For each subset A of S there are two associated subsets of T: A and $A \cup \{z\}$. These constitute all the subsets of T, and all are distinct. Since S has 2^k subsets, T has $2 \cdot 2^k = 2^1 \cdot 2^k = 2^{k+1}$ subsets. ∎

It is interesting to note that the only subset of the empty set is the empty set itself. Hence it has $2^0 = 1$ subset. Thus the number of subsets of a set with n elements is 2^n for all sets. It is also interesting to note that Proof by Induction can be used with starting points other than $k = 1$. See the discussion of the Extended Principle of Induction that follows Exercise 14. (Using the Extended Principle of Induction, we could have begun the proof of Theorem 18 by starting with $k = 0$.)

Exploratory Exercise Set 3.3

1. The method of proof by induction is an excellent way to establish that a formula is valid for all positive integers. However, it is not a useful mechanism for discovering such formulas. Often one observes a pattern in unrelated investigations and then verifies it with a proof by induction. With this in mind, let us explore the sum of the first n odd natural numbers.

   $n = 1$: $1 = 1$
   $n = 2$: $1 + 3 = 4$
   $n = 3$: $1 + 3 + 5 = 9$

   (a) Compute the sum of the first n odd numbers when $n = 4, 5,$ and 6.

$$n = 4: \quad 1 + 3 + 5 + 7 = ?$$
$$n = 5: \quad 1 + 3 + 5 + 7 + 9 = ?$$
$$n = 6: \quad 1 + 3 + 5 + 7 + 9 + 11 = ?$$

(b) Observe that the nth odd number is $2n - 1$. Make a conjecture about the sum of the first n odd numbers:

$$1 + 3 + 5 + \ldots + (2n - 1) = ?$$

2. In Exploratory Exercise 1, we conjectured that the sum of the first n odd numbers is n^2. Let us prove this conjecture using proof by induction.

(a) We wish to prove the following:

If n is a positive integer, then $1 + 3 + 5 + \ldots + (2n - 1) = n^2$.

In this section we suggested that we could regard this as an open sentence $P(n)$ about an arbitrary positive integer . What is $P(n)$ for this conjecture?

(b) *Examples:* In the Basis Step of the proof we will try to establish that $P(1)$ is true. What is $P(1)$ for this theorem? What is $P(2)$? What is $P(3)$? Verify that each of these three sentences is true.

(c) *Reflection:* In the Inductive Step we will need to prove that if $P(k)$ is true, then $P(k + 1)$ is true. State $P(k)$ and $P(k + 1)$.

(d) Complete the Basis Step of the proof.

(e) We have started the Inductive Step below. Complete the Inductive Step of this proof.

Our inductive hypothesis is $1 + 3 + 5 + \ldots + (2k - 1) = k^2$. Observe that the $(k + 1)^{st}$ odd integer is $2(k + 1) - 1 = 2k + 1$. Adding $2k + 1$ to both sides of the equation given by our inductive hypothesis yields

$$1 + 3 + 5 + \ldots + (2k - 1) + (2k + 1) = k^2 + (2k + 1) = \underline{\hspace{2cm}}?$$

(f) Explain why your work in (d) and (e) proves that for all positive integers,

$$1 + 3 + 5 + \ldots + (2n - 1) = n^2.$$

3. Let us prove the following theorem by induction:

4 divides $6^n - 2^n$ for all positive integers n.

(a) What is the open sentence $P(n)$ that we wish to show is true for all positive integers n?

(b) *Definitions and Theorems:* Success with proving this theorem will clearly require careful thought about the notion of divisibility. Return to Section 3.1 and record the definition of "a divides b." Also record Theorems 7 and 9 of that section.

(c) *Examples:* State $P(1)$, $P(2)$, and $P(3)$ and explain why each is true.

(d) *Reflection:* State $P(k)$ and $P(k + 1)$. In order to complete the Inductive Step, we must show that when 4 divides $6^k - 2^k$, then 4 divides $6^{k+1} - 2^{k+1}$. This requires that we find a connection between $6^k - 2^k$ and $6^{k+1} - 2^{k+1}$ while keeping 4 in the picture. Perhaps the key comes from the facts that 6^{k+1} can be written as $6 \cdot 6^k$ and $6 = (4 + 2)$. Explain why
$$6^{k+1} - 2^{k+1} = 4 \cdot 6^k + 2(6^k - 2^k).$$

(e) Complete the Basis Step of this proof.

(f) The Inductive Step of the proof is started below.
Assume 4 divides $6^k - 2^k$.

$$6^{k+1} - 2^{k+1} = 6 \cdot 6^k - 2^{k+1} = 4 \cdot 6^k + 2(6^k - 2^k) \quad \text{(why?)}$$

Now, use the theorems from (b) and the inductive hypothesis to complete this argument.

4. Induction proofs are not the only way to establish a theorem of the form "For all positive integers n, $P(n)$." Sometimes there is a direct proof that reveals more about why the theorem is true than the induction proof. Let us discover a direct proof of Theorem 15.

For all natural numbers n, $1 + 2 + 3 + \ldots + n = \dfrac{n(n+1)}{2}$.

(a) We will get our idea by experimenting with an unusual way to compute the sum $S = 1 + 2 + 3 + \ldots + 10$. Observe that S also equals $10 + 9 + 8 + \ldots + 3 + 2 + 1$. Now add both representations of S, as shown below.

$$S = 1 + 2 + 3 + \ldots + 8 + 9 + 10$$

$$\underline{S = 10 + 9 + 8 + \ldots + 3 + 2 + 1}$$

$$2S = 11 + 11 + 11 + \ldots + 11 + 11 + 11$$

Since the right side of the equation is the sum of 10 copies of 11,

$$2S = 10(11).$$

What is the sum S?

(b) Use this strategy to compute the sum $S = 1 + 2 + 3 + \ldots + 100$.

(c) Use this strategy to compute the sum $S = 1 + 2 + 3 + \ldots + n$, thereby proving this result:

$$1 + 2 + 3 + \ldots + n = \frac{n(n+1)}{2} \text{ for all positive integers } n.$$

5. (*Writing Exercise*) Read this "proof by induction" carefully. Since the statement being proved is clearly not true, there is an error somewhere in the argument. Write a careful, line-by-line analysis of the argument, attempting to determine which parts of the argument are valid and to find the one part of the argument where the error is made.

Non-Theorem: For all positive integers n, if a parking lot contains n cars, then all the cars in the lot have the same color.

Non-Proof:
    Basis Step: Let $n = 1$; that is, suppose a parking lot contains a single car. Clearly, all the cars in the lot have the same color.

    Inductive Step: Assume that if a parking lot contains k cars, then all the cars have the same color. Consider a parking lot with $k + 1$ cars. Let us number these cars $1, 2, 3, \ldots, k, k + 1$. Suppose Car $k + 1$ drives away from the parking lot. Then only k cars remain in the parking lot, so they all have the same color. Suppose Car $k + 1$ returns and Car 1 departs. Again, only k cars remain in the parking lot, so they all have the same color. Since the two collections of k cars overlap, all $k + 1$ cars must have the same color.

Exercise Set 3.3

Prove the results in Exercises 1 through 8 by induction.

1. For all natural numbers n, $3 + 6 + 9 + \ldots + 3n = \dfrac{3n(n+1)}{2}$.

    Hint: Review the proof of Theorem 15 and Exploratory Exercise 2 of this section.

2. For all positive integers n, $1^2 + 2^2 + 3^2 + \cdots + n^2 = \dfrac{n(n+1)(2n+1)}{6}$.

    Hint: Review the proof of Theorem 15 and Exploratory Exercise 2 of this section.

3. For all natural numbers n, $1^3 + 2^3 + 3^3 + \cdots + n^3 = \left[\dfrac{n(n+1)}{2}\right]^2$.

4. For all positive integers n, $1 + 4 + 7 + \ldots + (3n - 2) = \dfrac{n(3n-1)}{2}$.

5. For all positive integers n, $n < 2^n$.

6. For all positive integers n, 5 divides $9^n - 4^n$.
   Hint: Review the proof of Theorem 17 and Exploratory Exercise 3 of this section.

7. For all positive integers n, 2 divides $3^n - 1$.

8. For all positive integers n, 3 divides $n^3 - n$.

9. (a) Experiment with several small values of n and make a conjecture about a formula for this sum.

$$\frac{1}{1\cdot 2} + \frac{1}{2\cdot 3} + \frac{1}{3\cdot 4} + \ldots + \frac{1}{n(n+1)}$$

   (b) Prove your conjecture.

10. Prove the result in Exercise 8 without using a proof by induction. (*Hint*: Factor the expression $n^3 - n$ completely.)

11. In Exercise 7 of Section 3.1, we proved that if $a \mid b$ and $a \mid c$, then $a \mid (b + c)$. Use a proof by induction to prove the following extension of this theorem. For all positive integers n, if $a \mid b_1, a \mid b_2, \ldots, a \mid b_n$, then $a \mid (b_1 + b_2 + \ldots + b_n)$.

Review the divisibility rules found before Exercise 13 of Section 3.1. In the exercises of Section 3.1 we proved divisibility rules we called the "Rule for 2," the "Rule for 3," the "Rule for 4," and so on. However, we proved these rules only for positive integers with no more than three digits. In Exercises 12 through 14 we invite you to prove some of these divisibility rules for positive integers with n digits using a proof by induction. Remember that a positive integer z with n digits can be written as

$$z = a(10)^{n-1} + (an\ integer\ with\ n - 1\ digits)$$

12. Prove: For all positive integers n, the "Rule for 2" is true.

13. Prove: For all positive integers n, the "Rule for 4" is true.

14. Prove: For all positive integers n, the "Rule for 5" is true.

A more general version of the Principle of Induction reads like this:

Extended Principle of Induction: *Suppose P(n) is an open sentence and r is an integer. If*
   (a) *P(r) is true and*
   (b) *whenever k ≥ r and P(k) is true, P(k + 1) is true,*

   then P(n) is true for all integers n ≥ r.

15. Use the domino analogy to describe the meaning of the Extended Principle of Induction.

In Exercises 16 through 18, use the Extended Principle of Induction to prove the stated result.

16. For integers $n \geq 4$, $2n + 1 \leq 2^n$.

17. For integers $n \geq 5$, $n^2 < 2^n$.

18. Complete this induction proof of this theorem: With an unlimited supply of 3-cent and 5-cent stamps, one can form postage for any amount of postage of 8 cents or more.
    (a) *Basis Step:* Explain how to form postage for an amount of 8 cents with 3-cent and 5-cent stamps.
    (b) *Induction Step:* Suppose that you can form postage for an amount of k cents with 3-cent and 5-cent stamps.
        (1) Suppose that you used at least one 5-cent stamp in forming the postage for k cents. Explain how you could modify that postage to obtain postage for an amount of $k + 1$ cents.
        (2) Suppose that you used only 3-cent stamps in forming the postage for k cents. Explain why you must have used at least three 3-cent stamps. Explain how you could modify that postage (using 5-cent stamps) to obtain postage for an amount of $k + 1$ cents.

19. Review Exercise 18. Show that any amount of postage of 12 cents or more can be formed with an unlimited supply of 4-cent and 5-cent stamps.

3.4 From Mathematical Proof to Number Theory

Flipping through her daughter's sixth-grade mathematics book, Wendy found this problem:

Find the greatest common factor of 24 and 36.

Since Wendy could not quite remember the definition of greatest common factor from her days in school, she reasoned that it had something to do with factors. Hence she made a list of the positive factors of the two numbers.

Positive factors of 24: 1, 2, 3, 4, 6, 8, 12, 24
Positive factors of 36: 1, 2, 3, 4, 6, 9, 12, 18, 36

Then Wendy observed that the two numbers had the following common factors.
Common Factors: 1, 2, 3, 4, 6, 12

Wendy then concluded that the greatest of these common factors was 12.

Indeed, without benefit of the definition, Wendy was able to find the greatest common factor (often called the greatest common divisor) of two numbers. The formal definition follows.

Definition: The **greatest common divisor (factor)** of natural numbers a and b is the greatest natural number that divides both a and b. This number is denoted **gcd(a, b)** or **gcf(a, b)**.

Further, Wendy found a perfectly legitimate way to compute the greatest common divisor of two numbers. Let's try it again.

Example 1
Find
(a) gcd(18, 27)
(b) gcd(8, 15)
Solution
(a) The positive divisors of 18 are 1, 2, 3, 6, 9, 18, and the positive divisors of 27 are 1, 3, 9, 27. Clearly the common divisors are 1, 3, 9, and the greatest of these is 9.
(b) The positive divisors of 8 are 1, 2, 4, 8, and the positive divisors of 15 are 1, 3, 5, 15. The only positive integer that divides both 8 and 15 is 1, so gcd(8, 15) = 1.

In a circumstance such as found in Example 1(b), where gcd(a, b) = 1, we say that a and b are **relatively prime**. Thus 8 and 15 are relatively prime; 3 and 5 are relatively prime; and 28 and 33 are relatively prime.

Although Wendy found a technique that will find the greatest common divisor of pairs of small integers quite easily, her technique would be quite cumbersome for working with pairs of large integers. (Try enumerating the lists of divisors for 4576 and 3360). For that reason, we would like to examine a recursive process called the Euclidean algorithm, which was first developed by Greek mathematicians in the third century BCE. To understand the Euclidean algorithm, we will need to remember the technique for dividing one integer by another that we first learned in about the fourth grade. After completing the division of 28 by 11, we had a result like this:

$$11\overline{)28} \;\; \genfrac{}{}{0pt}{}{2}{}$$

$$\begin{array}{r} 2 \\ 11\overline{)28} \\ \underline{22} \\ 6 \end{array}$$

In this circumstance, we call 2 the quotient and 6 the remainder and observe that the dividend 28 can be written as

$$\text{Dividend} = (\text{Divisor})(\text{Quotient}) + (\text{Remainder})$$
$$28 = 11(2) + 6$$

Note that the remainder 6 is less than the divisor 11. Indeed, by a somewhat subtle proof that we shall not include here, we could prove the following theorem, which asserts that this kind of result is always possible:

Theorem 19: The Division Theorem   If a and b are integers with $b > 0$, then there is an integer q (the quotient) and an integer r (the remainder) such that

$$a = bq + r \text{ where } 0 \le r < b$$

> **Example 2**
>
> Find the quotient and remainder when
> (a) $a = 32$ is divided by $b = 14$
> (b) $a = 42$ is divided by $b = 12$
>
> **Solution**
> (a) $q = 2$ and $r = 4$, so $32 = 14(2) + 4$
> (b) $q = 3$ and $r = 6$, so $42 = 12(3) + 6$

THE EUCLIDEAN ALGORITHM

The success of the Euclidean Algorithm rests on the assertions of Theorem 14 in Section 3.2. Theorem 14 asserts that if integers a, b, q, and r are related by the equation

$$a = bq + r$$

then the set of common divisors of a and b is equal to the set of common divisors of b and r. It follows, then, that $\gcd(a, b) = \gcd(b, r)$. The Euclidean Algorithm is a recursive process that uses this observation to replace the task of finding $\gcd(a, b)$ where a and b are large numbers by the task of finding the greatest common divisor in a very simple case. [Observe that $\gcd(A, 0)$ is easily computed to be A. (See Exercise 15.)]

There are many ways to present the Euclidean Algorithm. Since recursion is one of the important themes of this text, we will present the algorithm in a way that emphasizes its recursive nature.

The Euclidean Algorithm for finding the greatest common divisor of a and b:
1.   Let $A = a$ and $B = b$.
2.   Find Q and R such that $A = BQ + R$ with $0 \le R < B$.
     Replace A by B and B by R.
3.   If $R > 0$, repeat Step (2).
     When $R = 0$, the current value of B (the last non-zero value of R) is the greatest common divisor of a and b.

Example 3

    (a) Find the greatest common divisor of 36 and 14 using the Euclidean Algorithm.

    (b) Use the observation from Theorem 14 (when A, B, Q, and R are related by $A = BQ + R$ then $\gcd(A, B) = \gcd(B, R)$) to show that the Euclidean Algorithm did indeed produce the greatest common divisor of 36 and 14.

Solution

    (a) We organize the work of the Euclidean Algorithm in a table to facilitate later investigations. Consider Table 2. We start with $A = 36$ and $B = 14$. In each row of Table 2 we record the equation that represents A in terms of B and R for each of the successive values of A, B, and R.

Table 2			
Equation	A	B	R
(1) $36 = 14(2) + 8$	36	14	8
(2) $14 = 8(1) + 6$	14	8	6
(3) $8 = 6(1) + 2$	8	6	2
(4) $6 = 2(3) + 0$	6	2	0

Example 3 *(continued)*

    The current value of B (the last non-zero value of R) is the greatest common divisor of 36 and 14; $\gcd(36,14) = 2$.

    (b) We can use the equations in column 1 together with our insight from Theorem 14 to conclude that $\gcd(36, 14) = \gcd(14, 8) = \gcd(8, 6) = \gcd(6,2) = \gcd(2,0) = 2$.

In the process of completing the Euclidean Algorithm, we obtain a surprising side effect. We generate the information we need to write $\gcd(a, b)$ as a sum of integer multiples of a and b. The fact that this occurs is the content of a very important theorem in number theory.

Example 4

    (a) Find the greatest common divisor of 252 and 180 using the Euclidean Algorithm.

    (b) Use your work from (a) to write $\gcd(252, 180)$ as a sum of integer multiples of 252 and 180.

Solution

    (a) $a = 252$ and $b = 180$. We can use the Division Theorem to write $252 = 180(1) + 72$. The rest of the work is found in Table 3.

Table 3			
Equation	A	B	R
(1) $252 = 180(1) + 72$	252	180	72
(2) $180 = 72(2) + 36$	180	72	36
(3) $72 = 36(2) + 0$	72	36	0

Example 4 *(continued)*

    The last value of B (the last non-zero remainder) is 36; $\gcd(252,180) = 36$.

    (b) By solving Equations (1) and (2) for their remainders and then substituting the result from Equation (1) in Equation (2), we can write 36 as a sum of integer multiples of 252 and 180. (See Table 4.)

Table 4		
Equation	Solve for remainders	Substitute the remainder from (1) into *
(1) $252 = 180(1) + 72$	$252 - 1(180) = 72$	
(2) $180 = 72(2) + 36$	$*180 - 2(72) = 36$	$180 - 2[252 - 1(180)] = 36$ $3(180) + (-2)(252) = 36$

Example 4 *(continued)*

Thus gcd(252, 180) = 3(180) + (–2) 252.

The reasoning employed in Example 4 can be used to prove Theorem 20.

Theorem 20: For positive integers a and b, we can find integers M and N such that gcd$(a, b) = Ma + Nb$.

Although we will omit the technical details of the proof of Theorem 20, the underlying reasoning is described in Example 4. Let follow this reasoning again in Example 5.

Example 5

In Example 3 we learned that gcd(36, 14) = 2. Use the work from Example 3 to show that we can write gcd(36, 14) as a sum of integer multiples of 36 and 14.

Solution

In column 1 of Table 5 we find the work from Example 3. We solve for each of the remainders in the equations, and then we substitute until we can write gcd(36, 14) = 2 in terms of 36 and 14.

Table 5			
Equation	Solve for remainders	Substitute the remainder from (2) into *.	Substitute the remainder from (1) into **.
(1) $36 = 14(2) + 8$	$36 - 2(14) = 8$		
(2) $14 = 8(1) + 6$	$14 - 1(8) = 6$		
(3) $8 = 6 (1) + 2$	$* 8 - 1(6) = 2$	$8 - 1[14 - 1(8)] = 2$ $**2(8) + (-1)(14) = 2$	$2[36 - 2(14)] + (-1)14 = 2$ $(2)(36) + (-5)(14) = 2$

Example 5 *(continued)*

Hence gcd(36, 14) = 2(36) +(–5) (14).

Theorem 20 leads to a very important property of prime numbers.

Theorem 21: If p is a prime number, and p divides the product ab, then either $p \mid a$ or $p \mid b$.

Preparation for the Proof of Theorem 21:
Examples: Note that $6 \mid (10 \cdot 15)$ but that 6 does not divide 10 and 6 does not divide 15, so this theorem is definitely not true for numbers that are not prime. However, note that $3 \mid 6 \cdot 7$ and $3 \mid 6$. We hope to show that this property is true for all prime numbers.
Definition: An integer $p > 1$ is prime provided that the only divisors of p are 1 and itself.

Theorems: We have available all the theorems on divisibility and Theorem 20. Read Theorem 20 carefully.

Reflection: If p is a prime, its only positive divisors are 1 and p. Hence if p does not divide a, the only common divisor of p and a is 1. Hence $\gcd(p, a) = 1$. We should be able to use this fact together with Theorem 20 in this proof. However, even with the help of Theorem 20, we will need a bright idea to complete this argument.

Proof of Theorem 21: By hypothesis, p is a prime number. We will consider two cases.

Case 1: Suppose $p \mid a$. Then we are done. We have satisfied the conclusion, p divides a or p divides b.

Case 2: Suppose p does not divide a. Then $\gcd(p, a) = 1$. By Theorem 20 we can find integers M and N with

$$1 = Mp + Na$$

At this point we need a bright idea, some way to connect this result $1 = Mp + Na$ to the number b. Perhaps we might try multiplying both sides of the equation by b. This yields

$$b = Mpb + Nab$$

Note that p obviously divides Mpb. By hypothesis, $p \mid ab$ and thus $p \mid Nab$. Finally, by Theorem 7, p divides the sum of Mpb and Nab, or $p \mid b$. ∎

Exploratory Exercise Set 3.4

In these exercises we will explore a somewhat fanciful problem whose solution will shed light on some of the reasoning of this section. Imagine, if you will, that we want to weigh potatoes using a balance scale and an unlimited supply of 3-ounce weights and 5-ounce weights. In Figure 1 you see "weighings" of an 11- ounce potato and a 1- ounce potato.

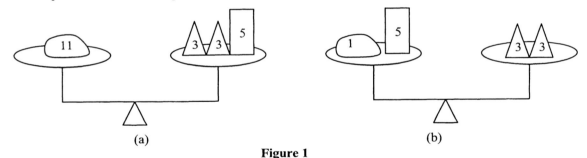

Figure 1

1. Below you are instructed to represent a method of weighing several potatoes. Work with a partner to complete these representations.
   (a) Draw a diagram that shows how to weigh the following potatoes if you have a balance scale and an unlimited supply of 3-ounce and 5-ounce weights.
      (1) 9-ounce potato
      (2) 4-ounce potato
      (3) 13-ounce potato
      (4) 7-ounce potato
   (b) If we represent the number of weights that are on the balance pan with the potato with a negative coefficient and the number of weights that are on the balance pan opposite the potato with a positive coefficient, the sum of the products, (coefficient) · (weight), is a mathematical expression for the weight of the potato. For example, the "weighing" in Figure 1(a) can be represented as $2(3) + 1(5) = 11$, and the "weighing" in Figure 1(b) can be represented as $2(3) + (-1)(5) = 1$. Write a mathematical expression that represents each of the "weighings" you drew in part (a).
   (c) Suppose all potatoes weigh an integer number of ounces. Make a guess about which potatoes you can weigh with 3-ounce and 5-ounce potatoes.

2. Work with a partner to investigate which potatoes can be weighed with an unlimited number of 6-ounce and 9-ounce weights. For example, in Figure 2 you will find a diagram of a balance pan weighing a 3-ounce potato. You will also find an arithmetic expression that represents this "weighing."

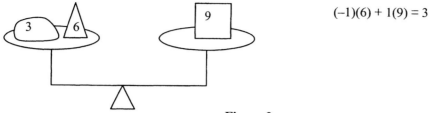

$$(-1)(6) + 1(9) = 3$$

Figure 2

(a) Represent with a diagram a method to weigh these potatoes with 6-ounce and 9-ounce weights if the weighing is possible.
    (1) 12-ounce potato
    (2) 21-ounce potato
    (3) 11-ounce potato
    (4) 6-ounce potato
    (5) 33-ounce potato
    (6) 2-ounce potato
(b) For each of the possible "weighings" of (a), represent the weighing with an arithmetic expression.
(c) Suppose all potatoes weigh an integer number of ounces. Make a guess about which potatoes you can weigh with 6-ounce and 9-ounce weights.

3. Work with a partner to answer the following questions about weighing potatoes using 5-ounce and 7-ounce weights.
    (a) Draw a diagram representing each of these arithmetic expressions as a method of weighing a potato using 5-ounce and 7-ounce weights.
        (1) $17 = 2(5) + 1(7)$
        (2) $3 = 2(5) + (-1)(7)$
        (3) $1 = 3(5) + (-2)(7)$
    (b) Explain how you could use the weighing described in (3) of (a) to determine weighings of the following potatoes.
        (1) 2-ounce potato [*Hint*: What do you learn when you multiply both sides of the equation in (3) of (a) by 2?]
        (2) 4-ounce potato
        (3) 9-ounce potato
        (4) *n*-ounce potato

4. (*Writing Exercise*) Suppose we are trying to weigh potatoes using *a*-ounce weights and *b*-ounce weights and that $d = \gcd(a, b)$. With a partner, carefully revisit Theorem 20 in the context of this potato-weighing problem, and discuss each of the three issues found below. After discussion, each of you should independently write a short paper summarizing your conclusions, giving examples as needed in your explanations.
    •   Explain why Theorem 20 guarantees that we can always weigh a potato that weighs *d* ounces.
    •   Explain why Theorem 20 guarantees that we can always weigh any multiple of *d* ounces. [*Hint*: See Exploratory Exercise 3(b).]
    •   Suppose all potatoes weigh an integer number of ounces. Explain why Theorem 20 guarantees that if $\gcd(a, b) = 1$, then we can weigh all potatoes with an unlimited number of weights of *a*-ounces and *b*-ounces. [Hint: See Exploratory Exercise 3(b).]

Exercise Set 3.4

---

1. For each of the following pairs of integers:
    (1) List all positive divisors of each number in the pair.
    (2) List all common divisors of each number in the pair.
    (3) Find the greatest common divisor of the pair.
    (a) 6, 15   (b) 12, 54   (c) 21, 63   (d) 28, 54   (e) 17, 36

2. Answer the following questions:
    (a) An elementary art teacher has classes with 21 and 35 students, respectively. She desires to order equipment that can be used in equal-sized groups in each class. What is the largest number of students that can be in each group so that all groups have the same number of students?
    (b) We must distribute 18 pears and 24 apples into baskets so that each basket contains the same number of pears and each basket contains the same number of apples. What is the greatest number of baskets that we can fill and meet this requirement?
    (c) What do the problems in (a) and (b) have to do with the concept of greatest common divisor?

3. Review Example 3.
    (a) Use the following table to find the greatest common divisor of 14 and 62.

Equation	A	B	R
(1) $62 = 4(14) + 6$	62	14	6
(2) $14 = 2(6) + 2$			
(3) $6 = 3(2) + 0$			

    (b) Use the fact that when $A = BQ + R$, $\gcd(A, B) = \gcd(B, R)$ to show that this algorithm did produce the greatest common divisor of 14 and 62.

4. Review Example 3.
    (a) Use the following table to find the greatest common divisor of 82 and 36.

Equation	A	B	R
	82	36	

    (b) Use the fact that when $A = BQ + R$, $\gcd(A, B) = \gcd(B, R)$ to show that this algorithm did produce the greatest common divisor of 82 and 36.

5. Use the Euclidean Algorithm to find the greatest common divisor of each pair in Exercise 1.

6. Use the Euclidean Algorithm to find the greatest common divisor of each of the following pairs.
    (a) 180, 144   (b) 315, 220   (c) 4576, 3360

7. Review Example 4 and your work in Exercise 3. Use this table to write gcd(14, 62) as a sum of integer multiples of 14 and 62.

Equation	Solve for remainders	Substitute the remainder from (1) into *
(1) $62 = 4(14) + 6$	$62 - 4(14) = 6$	
(2) $14 = 2(6) + 2$	*	

8. Review Example 5 and your work in Exercise 4. Use this table to write gcd(82, 36) as a sum of integer multiples of 82 and 36.

Equation	Solve for remainders	Substitute the remainder from (3) into ...	Substitute the remainder from (2) into ...	Substitute the remainder from (1) into ...
(1) 82 = 2(36) + 10	82 − 2(36) = 10			
(2)				
(3)				
(4)				

9. Use your work from Exercise 5 to write gcd(a, b) as a sum of integer multiples of a and b for each pair of natural numbers a and b.

10. Find four different pairs of integers M and N so that gcd(2, 3) = $M \cdot 2 + N \cdot 3$.

11. The **least common multiple** of a and b is the smallest positive integer that is a multiple of both a and b; it is denoted **lcm(a, b)**. Devise a procedure to find lcm(a, b) analogous to the one developed by Wendy on the first page of this section, and use it to find the least common multiple of the following pairs of numbers.
    (a) 8, 12   (b) 6, 15   (c) 18, 27   (d) 16, 42

12. Complete the following table, and use the results to form a conjecture about a connection between gcd(a, b) and lcm(a, b).

a	b	gcd(a, b)	lcm(a, b)	$a \cdot b$
8	12			
6	15			
18	27			
16	42			

13. Use the connection described in Exercise 12 to find the least common multiple of the pairs in Exercise 1.

14. Answer the following questions:
    (a) A cyclist and a walker travel around a track in the same direction. The cyclist completes the loop every 4 minutes, and the walker completes the loop every 10 minutes. If they start at the same time and place on the loop, how long will it be before they are back at that place together?
    (b) Before checking with the caterer, the cook cuts one cake into 24 equal pieces and an identical cake into 36 equal pieces. The caterer insists that both cakes must be cut exactly alike. Into how many pieces will both cakes be cut after the additional cuts are made?
    (c) What do the problems in (a) and (b) have to do with the concept of least common multiple?

15. Explain carefully why gcd(A, 0) = A for any integer A.

16. (*Proof Exercise*) Let A and B be positive integers. Suppose there are integers M and N so that $1 = MA + NB$.
    (a) What can you say about gcd(A, B)?
    (b) Prove your answer.

17. (*Proof Exercise*) Let A and B be positive integers. Suppose there are integers M and N so that $9 = MA + NB$.
    (a) What can you say about gcd(A, B)?
    (b) Prove your answer.

18. (*Proof Exercise*) Use a proof by induction to prove this extension of Theorem 21: If p is a prime number, and p divides the product $a_1 \cdot a_2 \cdot a_3 \cdot \ldots \cdot a_s$, then p divides one of the a_i's.

19. (*Proof Exercise*) Complete the proof of this theorem: If p and q are two distinct primes and both p and q divide N, then pq divides N.
    (a) Since p divides N, there is an integer k with $N = \underline{\hspace{1cm}}$.
    (b) Since q divides N, what can we learn from Theorem 21?
    (c) Complete the proof of the theorem.

20. (*Proof Exercise*) Use a proof by induction to prove this extension of the theorem in Exercise 19:
    If $p_1, p_2, \ldots, p_r$ are distinct primes, each of which divides N, then $p_1 \cdot p_2 \cdot \ldots \cdot p_r$ divides N.

Theorem 20 was very important in proving Theorem 21. Use Theorem 20 in establishing the result of Exercise 21.

21. (*Proof Exercise*) If D divides both A and B, then D divides $\gcd(A, B)$.

3.5 What Is So Special About Primes?

In previous sections we have referred to the notion of a prime number, an integer greater than 1 whose only positive factors are 1 and itself. Also of interest is the concept of a **composite number**, an integer greater than 1 that has a positive factor other than 1 and itself. (Since the factors of 12 are 1, 2, 3, 4, 6, 12, then 12 is a composite number.) In the last five centuries BCE, the years when Greek mathematicians were hammering out our first understanding of mathematical proofs, they found the relationship between prime numbers and composite numbers to be quite fascinating. When they attempted to write composite numbers such as 21, 12, 168, and 76 as products of smaller integers, they observed results like this:

$$21 = 3 \cdot 7 \qquad 12 = 2 \cdot 2 \cdot 3 \qquad 168 = 2 \cdot 2 \cdot 2 \cdot 3 \cdot 7 \qquad 76 = 2 \cdot 2 \cdot 19$$

When they realized that these factors were primes, and when they had checked many more examples, they conjectured that

Theorem 22: If n is a composite integer, then n can be written as product of primes.

> *Preparation for Theorem 22 (Reflection):*
> If n is a composite integer, then it has a factor a with $1 < a < n$. Since $a \mid n$, there is a positive integer b with $n = ab$. Why must it also be true that $1 < b < n$?

Proof of Theorem 22: Since n is a composite number, there are integers a and b such that $n = ab$ with $1 < a < n$ and $1 < b < n$. If both a and b are prime numbers, then we have written n as a product of primes. Otherwise, either a or b is composite and can be written as a product of two smaller natural numbers. Continuing in this way, we ultimately can write n as a product of primes. ∎

You might be somewhat suspicious of a "proof" that contains the phrase "continuing in this way." Be assured that it is possible to construct a proof by induction that confirms this fact. However, the argument given above describes well a process for finding a prime factorization for a particular integer.

> **Example 1**
> Find a prime factorization for (a) 136 and (b) 702.
> **Solution**
> (a) 136 is a composite number because it can be written as $4 \cdot 34$. Neither 4 nor 34 is a prime number, so we in turn factor each of them. In the diagram in Figure 1(a), we find factors for 4 and 34, all of which are primes. Therefore $136 = 2 \cdot 2 \cdot 2 \cdot 17$. [In Chapter 5 we will learn that the diagrams in Figures 1(a) and 1(b) are examples of diagrams that are called trees. Hence we will call these diagrams **factor trees**.]
> (b) From the factor tree in Figure 1(b), we see that $702 = (9)(78) = (3 \cdot 3)(2 \cdot 39) = (3 \cdot 3)(2)(3 \cdot 13)$. Thus, if we write the prime factors in increasing order, we see that $702 = 2 \cdot 3 \cdot 3 \cdot 3 \cdot 13$.

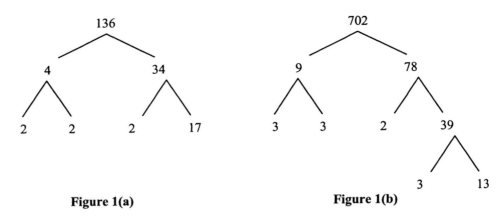

Figure 1(a)　　　　　　　　　**Figure 1(b)**

Theorem 22 shows that prime numbers play a very special role; in some sense they are the building blocks of all other integers. Another statement of Theorem 22 reads, "Every integer > 1 *is either a prime number or a product of prime numbers.*" However, Greek mathematicians found that this special property of prime numbers could be extended; they proved that each composite number can be written as a product of primes *in exactly one way.* This result is known as the Fundamental Theorem of Arithmetic.

Theorem 23: The Fundamental Theorem of Arithmetic: Every positive integer greater than 1 is either prime or can be written uniquely as a product of primes.

We need to state carefully what we mean by "written uniquely." After all, 6 can be written as a product of primes in two ways, $6 = 2 \cdot 3$ and $6 = 3 \cdot 2$. However, for present purposes, we shall regard these as the same prime factorization. Observe that these products differ only in the order in which the primes are listed. We need to be a bit more precise in the statement of Theorem 23.

Theorem 23, A Restatement: Suppose $N > 1$ is an integer that can be written in two ways as a product of primes. Suppose $N = p_1 \cdot p_2 \cdot \ldots \cdot p_r$ where the p's are primes and N also is equal to $q_1 \cdot q_2 \cdot \ldots \cdot q_s$ where the q's are primes. Then $r = s$ and the q's can be reordered so that for each integer i from 1 to r, $p_i = q_i$.

Preparation for Theorem 23:
Theorem: We shall find Theorem 21 from the previous section very useful: If p is a prime number, and p divides the product ab, then either $p \mid a$ or $p \mid b$.
Reflection on a Special Case: We will ultimately prove this theorem by induction on the number of primes used in a prime factorization of the number. We can understand the heart of the argument, however, by examining a special case. Suppose that a number n can be written as the product of two primes in two different ways:
$$n = p_1 \cdot p_2 = q_1 \cdot q_2 \qquad (**)$$
Since p_1 divides $q_1 \cdot q_2$, then, by Theorem 21, $p_1 \mid q_1$ or $p_1 \mid q_2$. Suppose $p_1 \mid q_1$. Since p_1 and q_1 are primes, the fact that $p_1 \mid q_1$ implies that $p_1 = q_1$. Dividing both sides of (**) by p_1 reveals that $p_2 = q_2$, which is the promised result. What happens in the case that $p_1 \mid q_2$? Can you devise a similar argument that shows that $p_1 = q_2$ and $p_2 = q_1$? By renumbering q_2 as q_1 and q_1 as q_2, we have the desired result.
One More Theorem: For the general case we will need an extension of Theorem 21 that reads: If p is a prime number, and p divides the product $a_1 \cdot a_2 \cdot a_3 \cdot \ldots \cdot a_s$, then p divides one of the a's. (See Exercise 18 from Exercise Set 3.4.)

Proof of Theorem 23: We will actually prove this equivalent statement of the theorem.

For all natural numbers n, if $p_1 \cdot p_2 \cdot \ldots \cdot p_n = q_1 \cdot q_2 \cdot \ldots \cdot q_s$, where the p's and q's are primes, then $n = s$ and the q's can be reordered so that $p_i = q_i$ for all integers i from 1 to n.

Basis Step: Suppose $p_1 = q_1 \cdot q_2 \cdot \ldots \cdot q_s$. Since p_1 is a prime, s must be 1. Thus $p_1 = q_1$.

Induction Step: Assume that the following is true for positive integer k.

If $p_1 \cdot p_2 \cdot \ldots \cdot p_k = q_1 \cdot q_2 \cdot \ldots \cdot q_s$, where the p's and q's are primes, then $k = s$ and the q's can be reordered so that $p_i = q_i$ for all integers i from 1 to k.

We must show that this assumption implies the following:

If $p_1 \cdot p_2 \cdot \ldots \cdot p_k \cdot p_{k+1} = q_1 \cdot q_2 \cdot \ldots \cdot q_r$ (***), where the p's and q's are primes, then $k + 1 = r$ and the q's can be reordered so that $p_i = q_i$ for all integers i from 1 to $k +1$.

We will start by examining the equation $p_1 \cdot p_2 \cdot \ldots \cdot p_k \cdot p_{k+1} = q_1 \cdot q_2 \cdot \ldots \cdot q_r$. Clearly $p_{k+1} \mid q_1 \cdot q_2 \cdot \ldots \cdot q_r$, so by the Extension of Theorem 21, there is an integer j such that $p_{k+1} \mid q_j$, and hence $p_{k+1} = q_j$. Dividing both sides of (***) by p_{k+1} reveals that

$$p_1 \cdot p_2 \cdot \ldots \cdot p_k = (\text{product of } r - 1 \text{ of the } q\text{'s})$$

By the induction hypothesis, $k = r - 1$ and the q's can be reordered so that, factor by factor, the p's = q's. Thus $k + 1 = r$ and q_j can be renumbered as the $(k + 1)^{st}$ q. Thus $p_{k+1} = q_{k+1}$. We have completed the proof. ∎

DIVISORS AND THE FUNDAMENTAL THEOREM OF ARITHMETIC

The Fundamental Theorem of Arithmetic can tell us much information about the divisors of numbers whose prime factorizations we know. With this information, we can easily compute greatest common divisors.

Example 2

Consider the integer $N = 2^3 5^2$.

(a) Suppose $d > 1$ is a divisor of N. Use the Fundamental Theorem of Arithmetic to learn as much as possible about the prime factors of d.

(b) Write the prime factorizations of all the divisors of N that are greater than 1.

Solution

(a) Let $p_1 \cdot p_2 \cdot \ldots \cdot p_r$ be the prime factorization of d. Since d divides N, there is a positive integer b with $N = d \cdot b$ or $N = p_1 \cdot p_2 \cdot \ldots \cdot p_r \cdot b$. Either $b = 1$, or b has a prime factorization.

Thus $N = 2^3 5^2 = 2 \cdot 2 \cdot 2 \cdot 5 \cdot 5 = p_1 \cdot p_2 \cdot p_3 \cdot \ldots \cdot p_r \cdot [(\text{prime factorization of } b) \text{ or } 1]$

In either case, because the Fundamental Theorem of Arithmetic declares that the prime factorization of N is unique, we know that each of the primes $p_1, \ldots, p_r$ is either a 2 or a 5. Further, there can be at most three 2's and two 5's in this list of primes.

(b) From our thoughts in (a), the positive divisors of N must be 1, 2, 2^2, 2^3, 5, 5^2, $2 \cdot 5$, $2 \cdot 5^2$, $2^2 \cdot 5$, $2^2 \cdot 5^2$, $2^3 \cdot 5$, $2^3 \cdot 5^2$.

Example 3

Find the greatest common divisor of $5^2 7^3$ and $2^4 5^3 7^2$.

Solution

Let d be a common divisor of $5^2 7^3$ and $2^4 5^3 7^2$. By the reasoning of the solution to Example 2(a), the fact that d divides $5^2 7^3$ requires that the prime factors of d be 5's and 7's. Further, there are no more than two 5's and three 7's. Similarly, since d divides $2^4 5^3 7^2$, the prime divisors of d can be only 2's, 5's, and 7's, and there can be no more than four 2's, three 5's, and two 7's. Putting these two observations together, we learn that d must look like $5^k 7^r$, where $k \le 2$ and $r \le 2$. Since $5^2 7^2$ is the greatest of these common divisors, $\gcd(5^2 7^3, 2^4 5^3 7^2) = 5^2 7^2$.

Because prime numbers have special properties, they have been much studied from the fifth century BCE until the present. We will examine some of the properties that have been discovered, but first we will look at a scheme the ancient Greeks developed for finding primes.

THE SIEVE OF ERATOSTHENES

Table 1 lists the integers from 1 to 100. The following iterative algorithm has been applied to the integers in the table.

Algorithm for Sieve of Eratosthenes:
1. Strike out 1.
2. Find the next number x that has not been circled or struck out, and circle it. Strike out all numbers that are multiples of x.
3. If some number less than 100 is still uncircled and unstruck, return to Step 2.
4. Record the numbers that have been circled.

Observe that after 1 is crossed out, 2 is the next integer that is circled. We then cross out all multiples of 2. The next integer that is not yet crossed out is 3. We circle 3 and cross out all multiples of 3. As we proceed through the steps of this iteration, we circle precisely those integers that have no positive divisors (other than 1) that are smaller than themselves. Hence, by recording the numbers that have been circled, we find all the primes between 1 and 100.

Table 1: The Primes Between 1 and 100

~~1~~	②	③	~~4~~	⑤	~~6~~	⑦	~~8~~	~~9~~	~~10~~
⑪	~~12~~	⑬	~~14~~	~~15~~	~~16~~	⑰	~~18~~	⑲	~~20~~
~~21~~	~~22~~	㉓	~~24~~	~~25~~	~~26~~	~~27~~	~~28~~	㉙	~~30~~
㉛	~~32~~	~~33~~	~~34~~	~~35~~	~~36~~	㊲	~~38~~	~~39~~	~~40~~
㊶	~~42~~	㊸	~~44~~	~~45~~	~~46~~	㊷	~~48~~	~~49~~	~~50~~
~~51~~	~~52~~	㊼	~~54~~	~~55~~	~~56~~	~~57~~	~~58~~	㊾	~~60~~
㊹	~~62~~	~~63~~	~~64~~	~~65~~	~~66~~	㊻	~~68~~	~~69~~	~~70~~
㉛	~~72~~	㉞	~~74~~	~~75~~	~~76~~	~~77~~	~~78~~	㊻	~~80~~
~~81~~	~~82~~	㉟	~~84~~	~~85~~	~~86~~	~~87~~	~~88~~	㊾	~~90~~
~~91~~	~~92~~	~~93~~	~~94~~	~~95~~	~~96~~	㊾	~~98~~	~~99~~	~~100~~

When we look at the list of primes between 1 and 100, several questions might occur to us.

Question 1: Between 70 and 80 there are only three primes, between 80 and 90 there are only two primes, and only one prime is found between 90 and 100. As we look at larger numbers, the primes appear to be getting more scarce. Is it possible that we might "run out" of primes? To put it another way,

Is there an end to the primes? (Are there only a finite number of primes?)

Answer: In Exploratory Exercise 2 we will follow the reasoning of ancient Greek mathematicians to learn that in fact there are an infinite number of primes.

Question 2: We note that primes like 3 and 5, 11 and 13, and 17 and 19 differ by 2. Pairs of primes that differ by 2 are called **twin primes**. We might ask,

Are there an infinite number of twin prime pairs?

Answer: This is an old, old question to which no one knows the answer. It is a question that is currently being investigated by mathematicians.

Question 3: In our discussion of Question 1, we observed that the primes seem to be getting further apart, the bigger they get. Looking back at Table 1, we see that between 83 and 89 there are five integers, none of which is prime. Between 89 and 97 there are seven integers, none of which is prime. A natural question is this:

As we look at larger and larger numbers, can we find arbitrarily long lists of sequential integers, none of which is prime?

For example, can we find a list of eight sequential integers, none of which is prime?

Answer: The answer to both questions is yes. Let $n = 9 \cdot 8 \cdot 7 \cdot 6 \cdot 5 \cdot 4 \cdot 3 \cdot 2 \cdot 1$. Explain why

$$n + 2, n + 3, n + 4, n + 5, n + 6, n + 7, n + 8, n + 9$$

is a list of eight sequential numbers, none of which is prime. In Exploratory Exercise 3, we will learn how to find a list of R sequential integers (for any large integer R), none of which is prime.

Exploratory Exercise Set 3.5

1. Work in pairs to explore the Prime Number Machine, which, when provided with a list of primes, spews out a new list of primes.

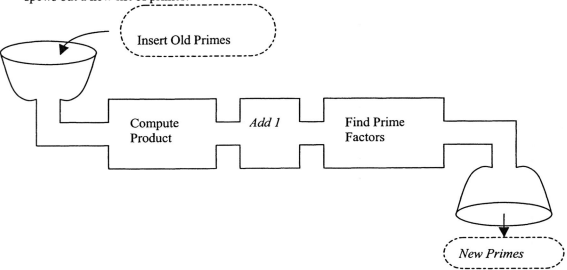

(a) Complete the following steps to use the Prime Number Machine. We will provide the machine with the numbers 2, 5, 11.
   (1) Compute the product $2 \cdot 5 \cdot 11$.
   (2) Compute the sum $2 \cdot 5 \cdot 11 + 1$.
   (3) Find all the prime factors of the sum from (2).
(b) Using the following lists of primes, use the Prime Number Machine. In each case, record the new list of primes that are produced.
   (1) 2, 3, 5
   (2) 3, 7, 13
   (3) 5, 11, 17
(c) In each of the four cases you examined in (a) and (b), compare the list of primes you placed in the machine with the new list of primes that were produced by the machine. Make a conjecture.
(d) Give the reasons in the following proof of this statement: If p_1, p_2, and p_3 are primes and p is a prime that divides $(p_1 \cdot p_2 \cdot p_3 + 1)$, then p is not equal to any one of p_1, p_2, or p_3.
   Preparation: We will do a proof by contradiction. Let us assume that p is equal to p_1, p_2, or p_3. Without loss of generality, we can assume $p = p_1$.
   (1) Explain why p divides $p_1 \cdot p_2 \cdot p_3$.
   (2) By hypothesis, p divides $p_1 \cdot p_2 \cdot p_3 + 1$. Explain why the result from (a) means that p must divide 1.
   (3) Why is this a contradiction?

The result obtained in Exploratory Exercise 1 generalizes completely. Suppose p_1, p_2, ..., p_n are primes and p is a prime that divides $(p_1 \cdot p_2 \cdot ... \cdot p_n + 1)$. Then p is a prime not equal to any of the p_i's. [The primes that come out of the Prime Number Machine are different from those that go in.]

2. (*Proof Exercise*) Greek mathematicians used the result about the Prime Number Machine to prove that there are an infinite number of primes. Work with a partner to complete the following proof by contradiction.
   (a) Write the negation of the statement "There are an infinite number of primes."

(b) Suppose there are a finite number of primes; indeed, suppose there are s primes. Then we can find a complete list of them and write them as $p_1, p_2, \ldots, p_s$. Now, let us put this list of primes into the Prime Number Machine (that is, let us form the sum $p_1 \cdot p_2 \cdot \ldots \cdot p_s + 1$ and find the prime factors of this sum). Let p be one of the primes spewed out by the Prime Number Machine.
   (1) What do we know about this prime?
   (2) Explain why this is a contradiction.

3. Work together in pairs to examine the question of whether we can find long sequences of consecutive integers, none of which is a prime number.
   (a) Consider the number $N = 5 \cdot 4 \cdot 3 \cdot 2 \cdot 1$. (This number is called 5 factorial and is denoted 5!)
      (1) Explain how you know 2 divides $N + 2$ without computing N.
      (2) Explain how you know 3 divides $N + 3$ without computing N.
      (3) Explain how you know 4 divides $N + 4$ without computing N.
      (4) Explain how you know 5 divides $N + 5$ without computing N.
      (5) Explain how you know that the four numbers $N + 2$, $N + 3$, $N + 4$, and $N + 5$ are all composite numbers.
   (b) Let $K = 10 \cdot 9 \cdot 8 \cdot \ldots \cdot 3 \cdot 2 \cdot 1 = 10!$ Explain how you know that $K + 2$, $K + 3$, $K + 4$, $K + 5$, …, $K + 10$ are all composite numbers.
   (c) Find a sequence of 20 consecutive positive integers, none of which are prime numbers.
   (d) Let R be a positive integer greater than 1. Explain how you can find R consecutive integers, none of which are prime numbers.

4. (*Writing Exercise*) The study of prime numbers is part of a larger field of study called number theory. Number theory is rich in easily stated conjectures that have still not been proved or disproved after centuries of effort. Using a Web search engine such as Google™, search using the following keywords. Write a paragraph on each of the following conjectures. In your paragraphs, explain carefully what each conjecture states, give examples, and summarize other information found during the search.
   - Twin primes conjecture
   - Goldbach conjecture

Exercise Set 3.5

1. Complete the factor trees to find the prime factorizations for 36 and 1050.

(a)   36   →   9   4

(b)   1050   →   25   42

2. Use a factor tree to find the prime factorization for the following positive integers.
   (a) 124   (b) 225   (c) 440   (d) 1184

3. Use a Sieve of Eratosthenes to find all prime numbers between 1 and 200.

4. Using your work from Exercise 3, determine how many primes are less than
   (a) 50   (b) 100   (c) 150   (d) 200

5. Using your work from Exercise 3, determine how many composite numbers are less than
   (a) 50   (b) 100   (c) 150   (d) 200

6. Find all pairs of twin primes between 1 and 100.

7. (a) Follow the reasoning of Example 2 and use the Fundamental Theorem of Arithmetic to find out as much as possible about the divisors of $2^2 3^3$. Write a sentence about what you learn.
   (b) Find the prime factorizations of all the divisors of $2^2 3^3$.

8. (a) Follow the reasoning of Example 2 and use the Fundamental Theorem of Arithmetic to find out as much as possible about the divisors of $5^3 7^1$. Write a sentence about what you learn.
   (b) Find the prime factorizations of all the divisors of $5^3 7^1$.

9. Write the prime factorizations of all the divisors of
   (a) $3^4 7^2$
   (b) $2^1 5^2 13^2$

10. Find the greatest common divisor of $2^2 3^3$ and $2^3 3^4$. Explain your reasoning carefully, as in Example 3.

11. Find the greatest common divisor of $5^4 7^2$ and $2^3 5^3 7^3$. Explain your reasoning carefully, as in Example 3.

12. Find the greatest common divisors of
    (a) $3^6 7^4$ and $3^3 7^2$
    (b) $2^3 7^2 11$ and $2^5 7^3 11^2$

13. Find the greatest common divisor of 96 and 144 in three ways:
    (a) By finding all the common divisors of the two numbers and choosing the greatest.
    (b) By using the Euclidean Algorithm.
    (c) By finding the prime factorizations of the two numbers and using the reasoning of this section.

14. Following the outline of Exercise 13, find the greatest common divisor of 48 and 220 in three ways.

15. (*Proof Exercise*) The prime 3 is one less than a square. That is, $3 = 2^2 - 1$. Can you find other primes that can be written as $N^2 - 1$ for some positive integer N.
    (a) Make a conjecture.
    (b) Prove it.

16. The interval consisting of the six consecutive positive integers 2, 3, 4, 5, 6, 7 contains the four primes 2, 3, 5, 7. Can you find another interval of six consecutive positive integers that contains four primes? Explain.

17. (*Proof Exercise*) Use the Fundamental Theorem of Arithmetic to prove that every positive divisor of 3^{50} is a power of 3. [Remember that $1 = 3^0$ is a power of 3.]

18. (*Proof Exercise*)
    (a) Find all positive integers less than 30 that have exactly three positive divisors.
    (b) Make a conjecture about positive integers that have exactly three positive divisors.
    (c) By focusing on prime factorizations, prove your conjecture.

19. (*Proof Exercise*)
    (a) Find all positive integers less than 30 that have exactly four divisors.
    (b) Make a conjecture about positive integers that have exactly four positive divisors.
    (c) By focusing on prime factorizations, prove your conjecture.

Graph Theory

Chapter 4

4.1 The Highway Inspector and the Traveling Salesman

The Highway Inspector and the Traveling Salesman meet for breakfast at Nell's Diner and spread the map of the county on the table between them. The Highway Inspector needs a route that will allow her to travel over each section of road in the county, but in the interest of economy, she does not wish to travel over any one section of road more than once. The Traveling Salesman needs to visit every town in the county exactly once, but he has no interest in driving on each stretch of highway. Look closely at their map in Figure 1. Can you find a route that satisfies the need of the Inspector? Now find a route for the Salesman.

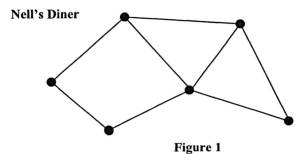

Nell's Diner

Figure 1

Example 1

Find routes for the Highway Inspector and the Traveling Salesman in the map in Figure 1.

Solution

A solution for the Salesman is shown in Figure 2(a). A solution for the Inspector is found in Figure 2(b).

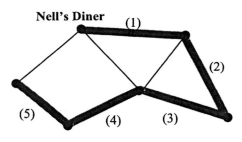

Nell's Diner (1)
(2)
(5) (4) (3)

Figure 2(a)

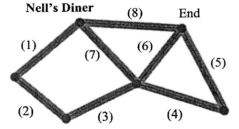

Nell's Diner (8) End
(1) (7) (6)
(5)
(2) (3) (4)

Figure 2(b)

Example 2

Try to find routes for the Highway Inspector and the Salesman for the simple maps in Figure 3.

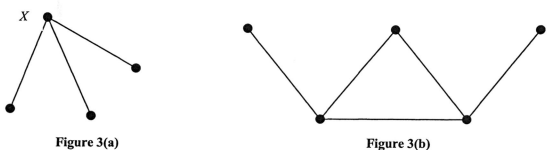

Figure 3(a)   Figure 3(b)

Solution

Neither the Inspector nor the Salesman has an economical route in the road map in Figure 3(a). If the inspector begins to inspect a segment of road at Town X, she must retrace that segment in order to travel on another segment. Since she has three segments to inspect, she will have to initiate the inspection of at least two segments at X. Similarly, the Salesman will need to visit Town X at least twice in order to visit all four towns.

The map in Figure 3(b) can be traveled happily by the Salesman, as seen in Figure 4, but there is no route for the Inspector that will not require her to travel at least one segment of the road twice.

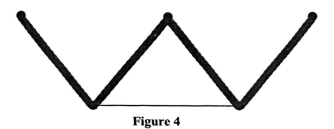

Figure 4

Let us now focus on the problem of the Highway Inspector. Clearly, we could attempt to answer this question for each new map by trial and error. However, we would like to be able to identify some simply recognized property that would enable us to answer the question easily for each new map we encounter. Consider the map given in Figure 5, and let's focus on what happens at Town A, a junction for three sections of road.

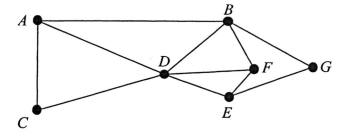

Figure 5

If Town A is not where the Inspector starts the trip, then on her first visit to Town A she will enter on one of the three sections of road and exit on another. (See Figure 6(a).) Thus there is only one section of road terminating at A that has not been inspected.

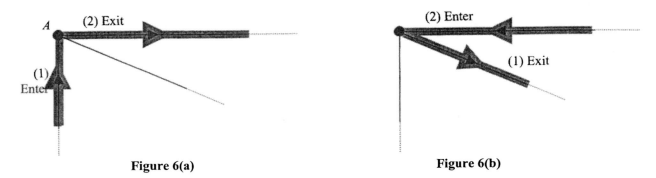

Figure 6(a)   **Figure 6(b)**

It follows that in order to complete the inspection without driving on road sections already inspected, she must ultimately end her trip at Town *A*. Similarly, if the inspector begins at Town *A*, then as she begins her trip, she inspects one of the sections of road. (See Figure 6 (b).) When later she returns, she will use a second section and hence must leave to inspect the final section of road that connects Town *A*. Thus, if an economical path for the Inspector can be found, she must either begin her trip at Town *A* or end her trip at Town *A*. Apply similar reasoning to argue that since Town *D* is a terminus for five sections of road, she must either begin or end her trip at Town *D*.

Now let us apply the same analysis to Town *B*, a connecting point for four sections of road. (See Figure 7.) If the road inspector starts at *B*, then on her first trip out, she will inspect one section of road, leaving three sections that have not been inspected.

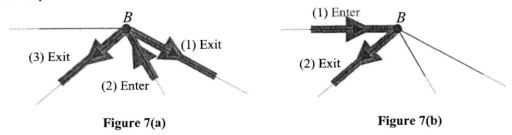

Figure 7(a)   **Figure 7(b)**

When she returns, she will inspect one road as she enters and another as she leaves, leaving one road that has not been inspected. (See Figure 7(a).) Hence, she will need to end her trip at Town *B* if she starts there. If the road inspector does not start at Town *B*, then after she enters and leaves *B*, she will have inspected two of the four sections. (See Figure 7(b).) The next time she passes through, she can both enter and leave on sections that have not yet been inspected. Further, she will never need to return to Town *B*, because all sections connecting to that town will have been inspected.

It appears that the number of road sections terminating at each town is a very important number. We will call the number of sections of road that end at a town the **degree** of that town. By the reasoning we used above, we have the following theorems.

Theorem 1: If the degree of a town is odd, then an economical trip by the Inspector must either begin or end in that town.

Theorem 2: If the degree of a town is even, then economical trips by the Inspector will have the following properties:
  (1) If the Inspector begins at that town, then she must end at that town.
  (2) If she does not begin at that town, then each time she enters the town, she has a non-inspected section on which to leave.

Looking back at Figure 5, we can make another observation. Since Towns *A*, *D*, *E*, and *F* all have odd degree, it is impossible to find an economical trip for the Inspector. That is, since a town of odd degree must be either the beginning or the end of an economical trip for the Inspector, a map that admits such a trip can have at most two towns of odd degree.

Theorem 3: If a map has more than two towns of odd degree, then the Inspector has no economical route.

Clearly, a road system that has towns of odd degree is problematical. But what about a road system whose towns all have even degree? Before we answer this question, we need to make one definition. A road system is **connected** if you can get from any town to any other town by traveling on roads in the system. The comments that follow will deal only with connected road systems. Now, try to find several paths for the Highway Inspector in the road map of Figure 8, each of whose towns has even degree.

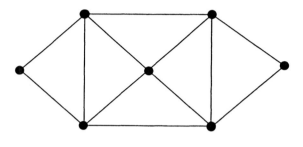

Figure 8

Were you successful? Perhaps you are beginning to suspect that the following theorem is true.

Theorem 4: If each of the towns of a connected road system has even degree, the Inspector has an economical tour. Further, the Inspector ends the tour in the town where she starts.

<u>**Proof of Theorem 4:**</u> We will indicate a proof of the theorem by describing how to construct an economical tour.
1. Start at any town on the map. Call it *S*.
2. Drive from town to town making random choices about which road section to inspect but not inspecting any section of road twice. Since there are a finite number of sections of road, you must eventually enter a town from which there are no new sections on which you can leave. Because all towns were of even degree, Theorem 2 tells us that this town at which we must stop is *S*.
3. Examine the map. If all sections of road have been inspected, we are done. Suppose there is some section that has not been inspected. Since the map is connected, there must be some town *T* that has already been visited in a previous trip but is also the terminus of a section that has not been inspected. Further, since we use road sections connected to towns two at a time (we enter on one road and exit on the other), town *T* is the terminus of an even number of road sections that have not been inspected. Pick such a town *T* and start a random side trip at *T*, making sure that you do not cover any section that has previously been inspected.
4. Again, you must eventually stop, and your stopping point will be Town *T*.
5. At this point you will have covered the whole map, or there will be some town with sections that have not been inspected, at which point you can return to step 3 for another side trip.

Because there are a finite number of sections in the map, we will cover all sections with a finite number of side trips. Finally, we can design one single economical inspection tour for the Inspector. We will start at *S* and travel the original tour, with the proviso that each time we come to a town that is the start of a side trip (like *T*), we will take the side trip beginning and ending at that town before continuing. ∎

Example 3

Use the reasoning for Theorem 4 to design an economical trip for the Inspector for the map in Figure 9.

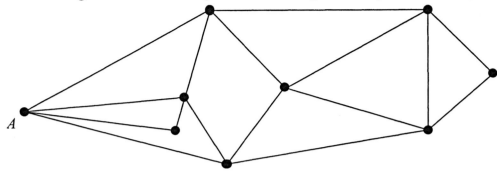

Figure 9

Solution

Let us start at Town A and let us drive on the map until we have to stop at A. Suppose our result is the Trip 1 shown in Figure 10(a). Now we have sections at Town C that have not been inspected, so let us begin Side Trip 1 and continue until we must stop. (See Figure 10(b).) We still have sections to inspect at Town P, but we can cover them with the Side Trip 2 shown in Figure 10(c). Now we can build the original trip and two side trips into an economical trip for the inspector. We will start at A, travel Trip 1 until we arrive at C, travel Side Trip 1 until we arrive at P, complete Side Trip 2, complete Side Trip 1, and finally complete Trip 1.

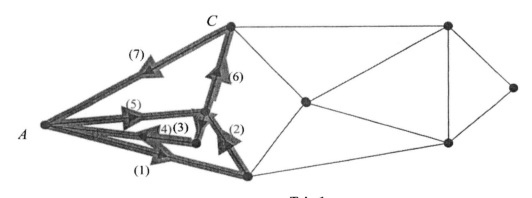

Trip 1
Figure 10(a)

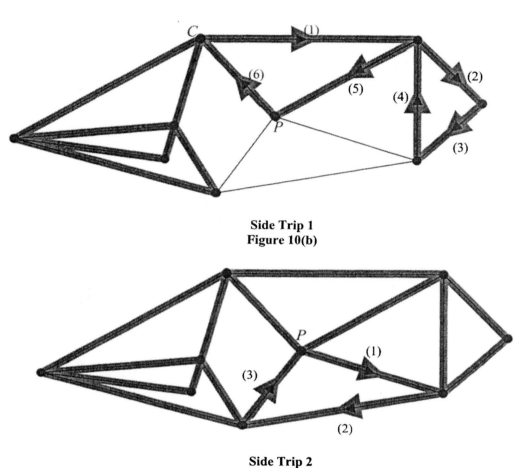

Side Trip 1
Figure 10(b)

Side Trip 2
Figure 10(c)

We now know that the Highway Inspector cannot find an economical route if there are more than two towns of odd degree, and we know that she can find an economical route if all towns are of even degree. What if a map has exactly two towns of odd degree? By an argument very similar to the argument for Theorem 4, we can prove the following theorem.

Theorem 5: If a map of a connected road system has exactly two towns of odd degree, there is an economical route for the Highway Inspector.

This, of course, leaves only one open issue. What happens in maps with exactly one town of odd degree? This issue is closed remarkably easily with Theorem 6.

Theorem 6: The sum of the degrees of all towns on a map is an even number equal to twice the number of sections of road on the map.

Proof of Theorem 6: Since each section of road on the map connects two towns, the sum of the degrees of the towns is 2 · (number of sections). ■

Consequence of Theorem 6: There are no maps with exactly one town of odd degree.

Example 4

Use Theorems 3, 4, and 5 to verify our decisions about the possible success of the Highway Inspector in Figures 1, 3(a), and 3(b).

Solution

The road map in Figure 1 has two towns of degree 3, and all other towns are of even degree. Hence, by Theorem 5, there is an economical path for the Highway Inspector that begins at one town of odd degree and ends at the other. Compare this conclusion to what we observed in Figure 2(b).

By Theorem 3, the road maps in Figures 3(a) and 3(b) have no economical route for the Inspector because each map has four towns of odd degree.

Notice how nicely we have solved the Highway Inspector problem. To determine whether an economical route exists, we need only compute the degree of each town on the map. If there is a route, the simple scheme described in the proof of Theorem 4 will provide the route. Unfortunately, we cannot do such a nice job for the Traveling Salesman; that is, the task of looking for a route that passes through all towns without duplication is less easily solved. For any particular map, we can use trial-and-error schemes until either we find a satisfactory route or we have tried all possible routes and failed. However, for large maps, the task of looking for all possible routes seems to be prohibitively difficult. Whether there is an efficient solution to the Traveling Salesman problem remains an open question in theoretical computer science.

The problems of finding routes for the Salesman and the Inspector on road maps are specific instances of problems in a very important area of mathematics called **graph theory**. In the next section, we will build on what we learned in this section to introduce this lively area more formally.

Exploratory Exercise Set 4.1

1. Working in pairs, have each person in the pair draw a map of a connected road system. Exchange maps and use the theorems of this chapter to determine whether there is an economical route for the Highway Inspector. Confirm your conclusions with your partner.

2. Each town in this road map has even degree. Work in pairs to find an economical path for the Highway Inspector by implementing the steps in the proof of Theorem 4. Have one student drive (trace the proposed path). Let the other student (while not looking at the map) navigate by randomly choosing the section of road taken at each town. [As the original trip and the side trips are traced, the driver should number the unused sections at each town and ask the navigator to choose the number of the section on which to proceed.]

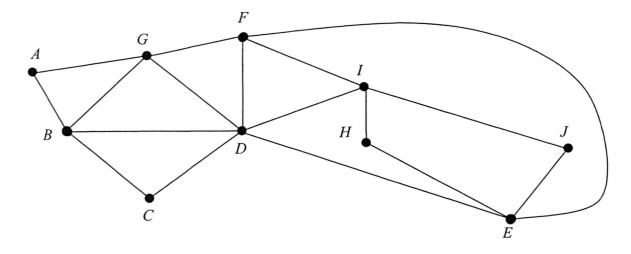

3. (*Proof Exercise*) Choose a partner and work with the partner to modify the proof in Theorem 4 to produce a proof of Theorem 5.

4. (*Writing Exercise*) Write careful explanations of the following assertions. Use pictures to illustrate your reasoning.
   - Suppose that Town *D* is a town in a road map that has degree 6.
     - If the Highway Inspector begins an economical inspection trip at *D*, then the trip must end at *D*.
     - If the Highway Inspector begins an economical inspection trip at some town other than *D*, then each time she enters Town *D*, she can leave on a section that has not been inspected.

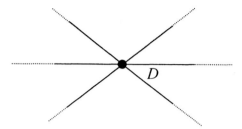

   - Suppose Town *E* is a town in a road map that has degree 5.
     - If the Highway Inspector begins an economical inspection trip at *E*, then each time she enters Town *E*, she has a non-inspected section on which to leave.
     - If the Highway Inspector begins an economical inspection trip at some town other than *E*, then eventually her trip must end at Town *E*.

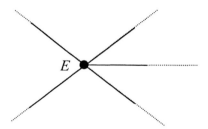

Exercise Set 4.1

1. If possible, find a route for the Salesman in this system. Find a route for the Inspector.

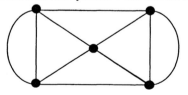

2. If possible, find a route for the Salesman in this system. Find a route for the Inspector.

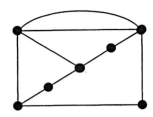

3. Complete the table for the following road maps.

Map	Number of Towns of Odd Degree	Number of Towns of Even Degree	Route for Inspector (Yes or No)
(a)			
(b)			
(c)			
(d)			
(e)			
(f)			

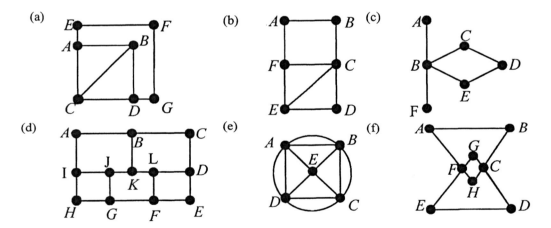

4. For each of the maps in Exercise 3, find the sum of the degrees of the towns. Compare your results with the assertion in Theorem 6.

5. The Highway Inspector problem was first solved by Leonhard Euler in 1735 when thinking about a puzzle problem that he described as follows: "In the town of Königsberg there is an island called Kneiphof, with two branches of the river Pregel flowing around it. There are seven bridges crossing the two branches. The question is whether a person can plan a walk in such a way that he will cross each of these bridges once but not more than once."

Below is a diagram of the seven bridges of Königsberg. Redraw this as a map with A, B, C, and D representing towns and the bridges representing sections of road. Can you answer Euler's puzzle?

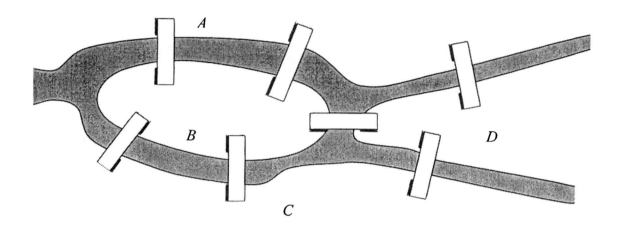

6. A city is built along both sides of a river and includes three islands A, B, C and nine bridges as shown. Is it possible to walk around the city and cross each bridge exactly once?

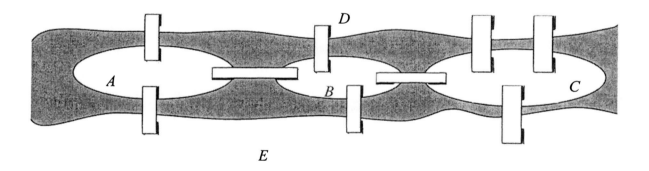

7. A person wishes to find a path through this house that will take him through each door exactly once. (a) Create a model for this problem using a road map. Represent the rooms as towns and the doors as roads between towns. (b) Explain why this model for the problem is equivalent to the Highway Inspector problem. (c) Determine a solution to the problem on your road map. (d) Find the desired path through the house.

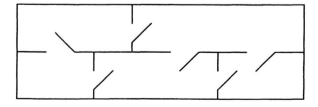

8. Can a person find a path through this house that will take him through each door exactly once?

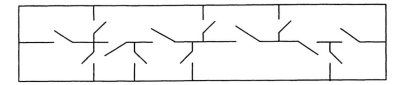

9. Can the Salesman visit each of the towns of this road system exactly once? Explain.

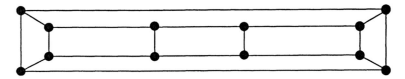

10. Look again at the Königsburg bridge problem in Exercise 5. Could the citizens of Königsberg find an acceptable route by building one more bridge? Explain.

11. Use the technique provided in the proof of Theorem 4 to find a route for the Inspector in this system.

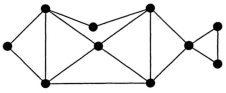

12. Explain carefully why Theorem 6 indicates that every road system must have an even number of towns of odd degree. Explain carefully why Theorem 6 indicates that there cannot be a road system with exactly one town of odd degree.

13. Although we cannot completely characterize road maps that admit an efficient trip for the Salesman, we can identify certain conditions that will ensure that the Salesman has such a trip. For example, explain why the Salesman has an efficient trip in a connected road system in which each town has degree 2.

14. Review your work from Exploratory Exercise 3. Use the technique provided in this work to find a route for the Inspector in this road map. (See also your work in Exploratory Exercise 2.)

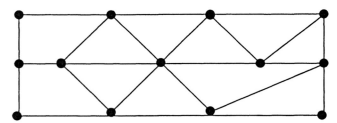

15. The following road map has four towns of odd degree. Try to find a route for the Inspector using the technique outlined in Exploratory Exercise 3. What goes wrong?

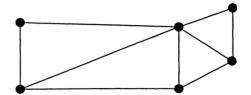

4.2 An Introduction to Graphs

In the previous section we answered some important questions about how a Highway Inspector and a Traveling Salesman could best conduct their jobs on a given road system. Without too much imagination, we can imagine similar important questions arising in many other situations, situations as diverse as computer networks, telephone systems, and gas pipeline systems. Indeed, the diagram in Figure 1 that we could have used in the previous section to represent a road system could also represent a computer network, with dots representing computers and lines representing communication links between computers. Similarly, the dots could represent telephone switching stations, and the lines could represent phone lines or microwave communication links. Because such diagrams are used to represent the relationships in many application areas, the study of the mathematics of such diagrams has been one of the most active and growing areas of mathematics in recent decades. Diagrams such as that in Figure 1 are geometric representations of mathematical objects called graphs, and in this section we will continue the study of graph theory that we began in Section 4.1.

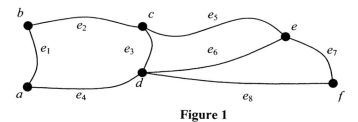

Figure 1

Note that we have points, called **vertices,** connected by pieces of curves, called **edges.** In Figure 1 the vertices are labeled a, b, c, d, e, and f. We label the edges e_1, e_2, e_3, e_4, e_5, e_6, e_7, and e_8. Although we will often think of graphs in terms of a geometric representation, we have learned that in mathematics, it is both useful and necessary to represent objects in multiple ways. The formal definition of a graph is given by describing its vertices and edges using the language of sets.

Definition: A graph $G = (V, E)$ consists of V, a nonempty, finite set of vertices, and E, a set of edges, with each edge associated with a pair of vertices. [*Note*: A graph described by such an ordered pair of sets may have many different geometric representations. (See Example 1.)]

If vertices u and v in a graph are associated with a unique edge e, we may write $e = \{u, v\}$. For instance, since each pair of vertices in Figure 1 is connected by at most one edge, we can represent each edge by an unordered pair that specifies the vertices that it connects. Thus e_1 might be named $\{a, b\}$, e_2 might be named $\{c, b\}$, and so on.

Example 1
Find two different geometric representations of the graph $G = (V, E)$, where $V = \{p, q, r, s, t\}$ and $E = \{\{p, q\}, \{p, r\}, \{p, s\}, \{r, s\}, \{r, t\}\}$.
Solution
Two different geometric representations of $G = (V, E)$ are found in Figures 2(a) and 2(b).

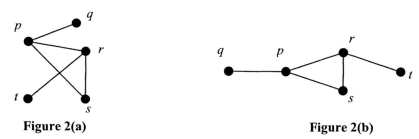

Figure 2(a)   **Figure 2(b)**

Important Note: In many pre-college texts the objects that we have called graphs are called **vertex-edge graphs**. This terminology is used to help young students avoid confusion. Young students are also studying the representations of curves on coordinate systems and calling these representations *graphs*. By using the modifier *vertex-edge,* we help such students distinguish between these two mathematical objects.

Observe that in the representation of the graph found in Figure 2(a), the edge $\{p, s\}$ intersects the edge $\{r, t\}$, but this does not indicate that there is a vertex at that point of intersection. In geometric representations of graphs, intersections of edges do not define a vertex unless the vertex is explicitly indicated in the diagram.

In the graphs in Figure 2, the edge $\{p, q\}$ connects vertices p and q. In this case we say that p and q are endpoints of the edge and that the edge is incident with p and q. In general, if edge e is associated with a vertex u, then we say that e **is incident with** u and that u is an **endpoint** of e. If vertices u and v are distinct and there is an edge e incident with both vertices u and v, we say that u and v are **adjacent vertices** and that e **connects** u and v.

Example 2

Answer the following questions for the graph in Figure 3.
  (a) Name the vertices that are endpoints of e_1.
  (b) Name the edges that are both incident with vertex x and incident with vertex y.
  (c) Are vertices z and w adjacent? Why?
  (d) Are vertices x and z adjacent? Why?

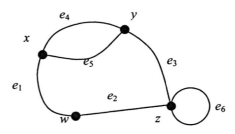

Figure 3

Solution

  (a) x and w are the endpoints of e_1.
  (b) e_4 and e_5 are incident with vertices x and y.
  (c) Vertices z and w are adjacent because e_2 is incident with vertices z and w.
  (d) Vertices x and z are not adjacent because there is no edge that is incident to both vertices.

In Figure 3 we notice that we have two different edges incident with both vertex x and vertex y. In this case it would not be appropriate to name e_4 or e_5 as $\{x, y\}$; indeed we should not use pairs of vertices to name edges in this graph. When two different edges are incident with the same set of vertices, we say the edges are **parallel**. In Figure 3, edges e_4 and e_5 are parallel. We also notice in Figure 3 that edge e_6 is incident only with the vertex z. We call such an edge a loop. A **loop** is an edge that is incident with exactly one vertex.

Because graph theory is a relatively young and very vigorous area of mathematics, there is not complete agreement on the definition of some of the standard terms in graph theory. For instance, some authors do not permit loops in their definitions of graphs. For this reason, as you read books and articles on graph theory, you must pay careful attention to the definitions of terms used in those publications.

As we take a trip in a highway system, we alternate between traveling along road sections and passing through the towns connected by the road sections. By identifying the road sections traveled and the towns visited, we could completely describe such a trip to a friend. In graph theory we call such a trip a path. The formal definition of a path follows.

> **Definition:** Let u and v be vertices in a graph and n be a positive integer. A **path from u to v of length n** is an alternating sequence of $n + 1$ vertices and n edges $(x_0, e_1, x_1, e_2, x_2, \ldots, x_n)$, where $u = x_0$, $v = x_n$, and edge e_i is incident with vertices x_{i-1} and x_i. A path from vertex u to vertex u is called a **circuit**; that is, a circuit is a path that begins and ends at the same vertex.

In Figure 3, $(x, e_1, w, e_2, z, e_3, y)$ is a path from vertex x to vertex y of length 3. Similarly, (x, e_5, y, e_4, x) is a circuit of length 2. Note that in a graph with no parallel edges and no loops, there is no ambiguity if we omit reference to the edges. For instance, the path $(a, e_1, b, e_2, c, e_3, d)$ from a to d in Figure 1 could be named as (a, b, c, d).

Example 3

In the graph of Figure 4,
(a) Find a path of length 3 from vertex x to vertex y.
(b) Find a path of length 4 from vertex x to vertex y.
(c) Find a path of length 5 from vertex x to vertex y.

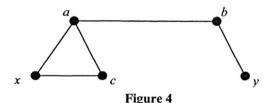

Figure 4

Solution

(a) A path of length 3: (x, a, b, y)
(b) A path of length 4: (x, c, a, b, y)
(c) A path of length 5: In this case we must think harder. We must use some of the edges more than once. Two possible answers are (x, a, c, a, b, y) and (x, a, x, a, b, y).

In Example 3 we saw that we repeat edges in some paths and circuits. In terms of the language of the Highway Inspector problem we discussed in Section 4.1, a path in which an edge (a road segment) is repeated is not an economical trip since the Highway Inspector wishes to pass over each road segment exactly once. Hence we define a **simple path** to be a path in which no edge is repeated. In the previous example, the paths discussed in (a) and (b) are simple, whereas the paths identified in (c) are not simple. Similarly, a **simple circuit** is a circuit in which no edge is repeated.

At this point we would like to restate the theorems about the Inspector that we discovered in Section 4.1 in the language of graphs. To do this, we need a few more definitions.

> **Definitions:**
>
> • An **Euler path** in a graph is a simple path that includes each edge of the graph.
> • An **Euler circuit** is a circuit that is an Euler path.
> • The **degree of a vertex** v is the number of edges that are incident with v. If there is a loop at v, then the loop is counted twice in computing the degree of v.
> • A graph is **connected** if, for each pair of vertices v and w, there is a path from v to w.

Note: Leonhard Euler was an extraordinary mathematician of the eighteenth century who did groundbreaking work in many different mathematical fields. Because of his work on the puzzle of Königsburg, he is rightly understood as the father of graph theory (see Exercise 5 of Exercise Set 4.1).

Example 4

In the graph in Figure 5,

(a) Determine whether the path $(a, e_6, c, e_4, d, e_5, a, e_6, c, e_3, b)$ is an Euler path from vertex a to vertex b. If it is not an Euler path, explain why.

(b) Determine whether the path $(c, e_3, b, e_2, b, e_1, a, e_6, c)$ is an Euler circuit. If it is not an Euler circuit, explain why.

(c) Find the degree of vertex a and vertex b.

(d) Find an Euler path in the graph in Figure 5.

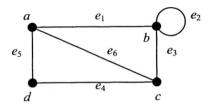

Figure 5

Solution

(a) This path fails to be an Euler path for two reasons: The edge e_6 appears twice on the path, and the path does not include the two edges e_1 and e_2.

(b) Since the path begins and ends with vertex c, it is a circuit. However, it fails to be an Euler circuit because it does not include the edges e_4 and e_5.

(c) Since three edges $\{e_1, e_5, e_6\}$ are incident with vertex a, the degree of a is 3. The edges e_1, e_2, and e_3 are incident with vertex b. Since e_2 is a loop and is counted twice, the degree of b is 4.

(d) The path $(a, e_1, b, e_2, b, e_3, c, e_4, d, e_5, a, e_6, c)$ is an Euler path that starts at a and ends at c.

With this terminology available to us, we are able to restate the theorems of Section 4.1 in the language of graph theory. For example, Theorem 1 can be restated as Theorem 7.

Theorem 7: If vertex v in a graph has odd degree, any Euler path for that graph must either begin or end at v.

Similarly, Theorems 2 through 5 can also be restated as Theorems 8 through 11.

Theorem 8: If the degree of a vertex is even, then any Euler path that begins at that vertex must end at that vertex.

Theorem 9: If a graph has more than two vertices of odd degree, then there is no Euler path for the graph.

Theorem 10: If a connected graph has more than one vertex and each of the vertices has even degree, then there is an Euler circuit for the graph.

Theorem 11: If a connected graph has exactly two vertices v and w of odd degree, then there is an Euler path from v to w.

Finally, Theorem 6 can be restated.

Theorem 12: The sum of the degrees of the vertices of a graph is an even number equal to twice the number of edges of the graph.

As a consequence, there are no graphs with an odd number of vertices of odd degree.

In Section 4.1, the Salesman wished to travel a highway system passing through each town exactly once. For efficiency, he preferred to begin and end at the same town. In order to describe the problem of the Salesman in the language of graphs, we need the following definitions.

> **Definitions:**
> - A **Hamilton path** in a graph is a path that includes each vertex of the graph exactly once.
> - A **Hamilton circuit** is a circuit that includes each vertex of a graph exactly once except for the initial vertex and the final vertex, which are the same.

Note: Lord William Rowan Hamilton (1805-1863) was the most noted Irish scientist and mathematician of his century. See Exercise 19 of this section for a graph problem that arose in a game invented by Hamilton.

The question of the Salesman becomes "Can I find a Hamilton path for the graph?" If the Salesman wishes to return to the town where he started, the question becomes "Can I find a Hamilton circuit for the graph?" We have seen that it is relatively easy to determine whether Euler paths and circuits exist and to find those paths and circuits. For graphs with small numbers of vertices and edges, one can answer the Hamilton questions with trial-and-error techniques. For many large graphs, the trial-and-error techniques require too much computation to be implemented even on high-speed computers. Researchers in many different disciplines are actively trying to determine whether there is a computationally efficient way to answer the Hamilton questions.

Exploratory Exercise Set 4.2

1. Choose a partner. Each of you should independently sketch a graph satisfying each of the conditions below (draw seven graphs each). Compare your work with the work of your partner.
    (a) A graph with five vertices, six edges, no parallel edges, and no loops
    (b) A graph with four vertices and two pairs of parallel edges
    (c) A connected graph with three vertices and a loop incident with each vertex
    (d) A graph with four vertices that is not connected
    (e) A graph that has no simple circuits
    (f) A graph with an Euler path
    (g) A graph with a Hamilton circuit

2. Choose a partner. Working independently, find, in this graph, an example of the terms that follow. Compare and discuss your work.

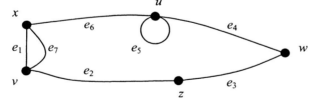

    (a) An edge incident with vertices x and u
    (b) Two vertices that are adjacent
    (c) Two vertices that are not adjacent
    (d) Two edges that are parallel
    (e) A loop
    (f) A path from x to w that is simple
    (g) A path from x to w that is not simple
    (h) A simple circuit that includes the vertex w
    (i) A circuit that includes the vertex w and is not simple

3. Graphs are used to model problems in applications as diverse as computer science, biology, chemistry, and the social sciences. However, any time that graphs are used to model a problem, we must first decide what objects in the problem will be the vertices and what relationship will determine which vertices are connected by an edge. In each of the following situations, work with a partner to determine the set to be used as vertices and the relationship that will serve to determine whether vertices are connected by an edge.

   (a) **Computer Network:** With a graph we would like to model a computer network. In the model we would like to be able to represent when two of the computers are connected directly by a cable.
   (1) What will we use as vertices in the graph?
   (2) What relationship will determine whether vertices are connected by an edge?
   (3) Suppose our network consists of computers *A, B, C, D, E,* and *F.* Suppose further that the following pairs of computers are connected directly by cable: *A* and *B, A* and *C, B* and *C, B* and *D, B* and *F, C* and *D, C* and *E, D* and *E,* and *D* and *F.* Sketch a graph that models this situation.

   (b) **Niche Overlap Graphs in Ecology:** With a graph we would like to represent the competition for food that occurs in an ecological system. In our representation, we would like to record the various species of animals in our system and whether a pair of species competes for the same food resources.
   (1) What will we use as vertices in our graph?
   (2) What relationship will determine whether vertices are connected by an edge?
   (3) Suppose our ecosystem contains raccoons, hawks, owls, squirrels, opossums, woodpeckers, and mice. Suppose further that the following pairs compete for the same food resources: raccoons and owls; raccoons and hawks; raccoons and squirrels; hawks and owls; squirrels and opossums; squirrels and woodpeckers; squirrels and mice; opossums and woodpeckers; and opossums and mice. Sketch a graph that models this situation.

4. In the proof of Theorem 4 in Section 4.1, we describe a method for finding a trip for the Highway Inspector on a road map on which all of the towns have even degree. In this exploration you will work with a partner to translate that method into the language of graph theory and, in the process, produce an indication of a proof for Theorem 10.

   (a) (*Proof Exercise*) Rewrite the proof of Theorem 4 using the language of graphs to describe a method for finding an Euler circuit for a graph all of whose vertices have even degree.
   (b) Use Theorem 10 to determine whether this graph has an Euler circuit.

   (c) Use your work from (a) to find an Euler circuit for the graph from (b).
   (d) Explain briefly how you would modify your work from (a) to describe how to find an Euler path in a connected graph with exactly two vertices of odd degree.

5. The geometric representations of graphs that enable humans to understand the connections modeled by the graph are not useful when one is preparing to investigate a graph using a computer. If a graph has no parallel edges, then one useful way to prepare the information in a graph for a computer is with an **adjacency matrix**. An adjacency matrix is a square table of numbers with one row for each vertex and one column for each vertex. Each entry of the table is either 0 or 1. If vertices v_i and v_j are adjacent (connected by an edge), we place a 1 in the ith row and j^{th} column; otherwise, we place a 0 there. Consider graph G and its associated adjacency matrix. Notice that since vertices v_1 and v_2 are adjacent, a 1 appears as the entry in the first row, second column of the matrix (and in the second row, first column of the matrix).

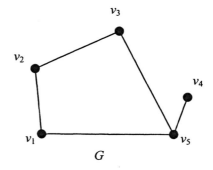

	v_1	v_2	v_3	v_4	v_5
v_1	0	1	0	0	1
v_2	1	0	1	0	0
v_3	0	1	0	0	1
v_4	0	0	0	0	1
v_5	1	0	1	1	0

G

Since vertices v_2 and v_4 are not adjacent, a 0 appears as the entry in the second row, fourth column (as well as in the fourth row, second column).

(a) Sketch a graph with vertices v_1, v_2, v_3, and v_4 that corresponds to the information in the adjacency matrix. Assume the rows and columns are labeled in the order v_1, v_2, v_3, v_4.

(1)

0	0	1	1
0	0	1	1
1	1	0	1
1	1	1	0

(2)

0	1	0	0
1	0	1	0
0	1	0	1
0	0	1	0

(b) In forming an adjacency matrix for a graph, we label each of the rows with a different vertex, and we must label the columns in the same order as the rows. Hence, if we change the order in which we label the rows, we change the matrix. Illustrate this fact by finding two different adjacency matrices for graph G using the row and column labels found below.

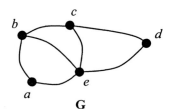

G

(1)

	a	b	c	d	e
a					
b					
c					
d					
e					

(2)

	b	d	e	c	a
b					
d					
e					
c					
a					

(c) For each of these graphs, choose an ordering for the vertices and form the adjacency matrix.

(1) 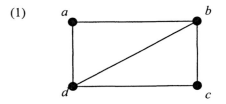   (2)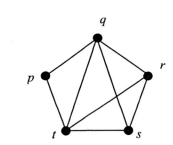

6. *(Writing Exercise)* You have been asked by a friend who teaches middle school mathematics to visit and give help to her students. The goal is to help them understand the difference between a graph of a curve or function on a coordinate system and the concept of a vertex-edge graph. Do the necessary research and write notes for this lecture, together with examples.

Exercise Set 4.2

In Exercises 1 through 4 describe each of the following graphs by determining the set of vertices V and the set of edges E. Since there are no parallel edges or loops in the graphs in Exercises 1 through 4, each edge can be described by giving the pair of endpoints.

1.

2.

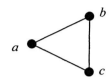

3.

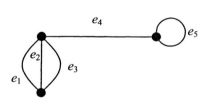

4.

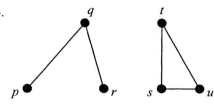

5.   Sketch a geometric representation of these graphs.
   (a) $V = \{a, b, c, d, e, f\}$; $E = \{\{a, b\}, \{a, c\}, \{b, d\}, \{b, e\}, \{d, f\}, \{e, f\}\}$
   (b) $V = \{x, y, z\}$; $E = \{\{x, y\}, \{x, z\}, \{y, z\}\}$
   (c) $V = \{P, Q, R, T\}$, $E = \{\ \}$

In Exercise 6 and 7, list the loops and parallel edges found in the graph.

6.

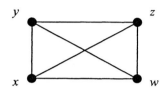

7.

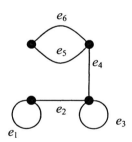

8.   In this graph, find

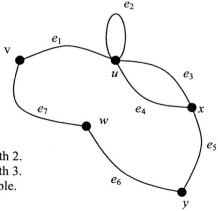

   (a) Two edges incident with vertex v.
   (b) Two vertices adjacent to vertex x.
   (c) Two edges that are parallel.
   (d) A loop.
   (e) A simple path from vertex v to vertex x of length 2.
   (f) A simple path from vertex v to vertex x of length 3.
   (g) A path from vertex v to vertex x that is not simple.
   (h) A circuit beginning and ending at vertex y.
   (i) A vertex of degree 3.
   (j) A vertex of degree 5.

9. Draw a graph with
    (a) Four vertices, each with degree 1
    (b) Four vertices, each with degree 2
    (c) Four vertices, each with degree 3

10. Sketch the following graphs that model the following relationships in which the vertex set consists of yourself, your brothers and sisters, and your parents.
    (a) The relationship "have the same color eyes"
    (b) The relationship "were born in the same state"

For each of the graphs in Exercises 11 through 14, determine whether there is an Euler path or Euler circuit by looking at the degrees of the vertices. If the graph has Euler paths or Euler circuits, find one of them.

11.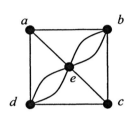

12.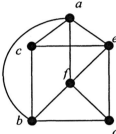

13.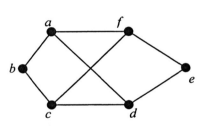

14.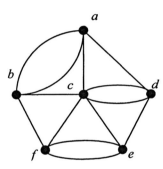

For each of the graphs in Exercises 15 through 17, find a Hamilton path or circuit or determine that no such path or circuit exists. If no Hamilton path or circuit exists, explain why.

15.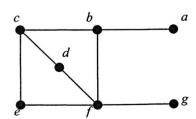

16.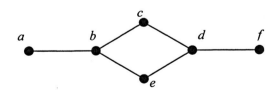

17.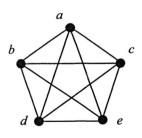

18. Give an example of a connected graph with
    (a) An Euler circuit and a Hamilton circuit.
    (b) Neither an Euler circuit nor a Hamilton circuit.
    (c) An Euler circuit but no Hamilton circuit.
    (d) A Hamilton circuit but no Euler circuit.

19. In 1857, the distinguished Irish mathematician and scientist William Rowan Hamilton invented a game that was marketed under the name "A Voyage Round the World." The game consisted of a dodecahedron with a peg at each vertex. The object was to use a string to mark a path from vertex to vertex (peg to peg) so that the path visited each vertex exactly once and then returned to the vertex from which the path started. Below is a graph that is equivalent to Hamilton's game. Find a "Voyage Round the World," a Hamilton circuit, for this graph.

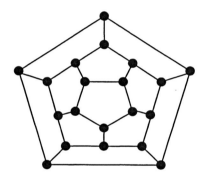

4.3 Complete Graphs, Isomorphic Graphs, and Subgraphs

Consider the following situations:

(1) Four executives meet for lunch. Each of the four persons must shake hands with the other three persons as they arrive.

(2) Five students rent a single, two-seated bicycle. Because of the limited time of the rental, it is important that each student get exactly one ride with each other student.

Can you model these problem situations with graphs?

Example 1

Create graphs that will serve as a geometric representation or model for the two situations described above.

Solution

For problem situation (1), do you get a graph similar to the graph in Figure 1(a)? In Figure 1(a) we have represented each executive as a vertex and each handshake as an edge. For problem situation (2), do you get a graph like the one in Figure 1(b)? In this figure we have represented each student as a vertex and each shared bicycle ride as an edge.

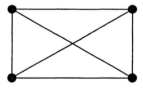

Figure 1(a)

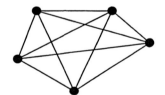

Figure 1(b)

The graph in Figure 1(a) is a geometric representation of a graph with four vertices and six edges; Figure 1(b) represents a graph with five vertices and ten edges. Both of these graphs are examples of a special kind of graph called a *complete graph*.

Before we can give a careful definition to the notion of a complete graph, we need to note that sometimes we prefer to restrict our attention to collections of graphs that do not have parallel edges or loops. A graph in which each pair of vertices is connected by at most one edge (no parallel edges) and in which we have no edge incident with a single vertex (no loops) is called a **simple graph**.

Example 2

Determine whether any of the graphs in Figure 2 are simple graphs.

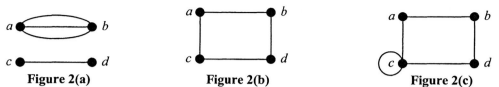

Figure 2(a)   **Figure 2(b)**   **Figure 2(c)**

Solution

The graph in Figure 2(a) is not simple since it has three edges connecting vertices a and b.
The graph in Figure 2(b) is a simple graph.
The graph in Figure 2(c) is not a simple graph since it has a loop at vertex c.

Now that we understand the notion of a simple graph, we are ready to define a complete graph.

Definition: A simple graph with n vertices in which there is exactly one edge incident with each pair of vertices is called **the complete graph on n vertices**. The complete graph on n vertices is denoted by the symbol K_n.

Geometric representations of K_1, K_2, K_3, and K_4 are found in Figure 3. A geometric representation of K_5 is found in Figure 1(b).

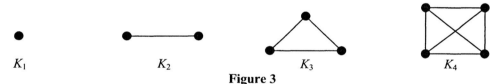

K_1 K_2 K_3 K_4

Figure 3

We need to pause and deal with one confusing aspect of the definition of complete graphs. In the definition we spoke of *the* complete graph on n vertices. However, in Figure 4 we find representations of two different simple graphs, each of which has four vertices and each of which has the property that there is exactly one edge incident with each pair of vertices. In order to understand this situation better, we need to talk about isomorphic graphs.

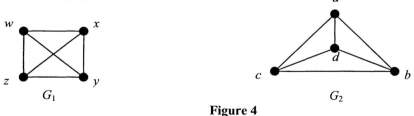

G_1 G_2

Figure 4

In mathematics we often need to identify objects that are composed of different sets of points but share all essential properties being discussed. For example, in geometry we say that different figures are congruent when they have the same size and shape. In graph theory, two graphs that share all essential properties are said to be isomorphic. In order to define "isomorphic," we need to remember three important definitions from Section 2.3.

- A function f from V_1 to V_2 is a rule that assigns to each element of V_1 (called an input) exactly one element of V_2 (called an output).
- A function f from V_1 to V_2 is one-to-one if it does not send two distinct inputs to the same output.
- A function f from V_1 to V_2 is onto if every element of V_2 is an output for some input in V_1.

Definition: Two simple graphs $G_1 = (V_1, E_1)$ and $G_2 = (V_2, E_2)$ are isomorphic provided that there is a one-to-one function f from V_1 onto V_2 so that vertices u and v of G_1 are adjacent if and only if $f(u)$ and $f(v)$ are adjacent vertices of G_2. The function f is called an **isomorphism**.

In short, an isomorphism is a one-to-one, onto function between the vertex sets so that the function preserves the adjacency relationship in both directions.

Example 3

Show that the two graphs in Figure 5 are isomorphic by finding an isomorphism from the vertices of G_1 to the vertices of G_2.

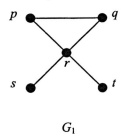

G_1

G_2

Figure 5

Solution

Consider the one-to-one function f from the vertices of G_1 onto the vertices of G_2.

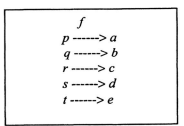

First, in Table 1 we verify that "If vertices u and v of G_1 are adjacent, then $f(u)$ and $f(v)$ are adjacent vertices of G_2."

Table 1	
p is adjacent to q and r	$a = f(p)$ is adjacent to $b = f(q)$ and $c = f(r)$
q is adjacent to p and r	$b = f(q)$ is adjacent to $a = f(p)$ and $c = f(r)$
r is adjacent to p, q, s, and t	$c = f(r)$ is adjacent to $a = f(p)$, $b = f(q)$, $d = f(s)$, and $e = f(t)$
s is adjacent to r	$d = f(s)$ is adjacent to $c = f(r)$
t is adjacent to r	$e = f(t)$ is adjacent to $c = f(r)$

Example 3 *(continued)*

Finally, we verify that "If $f(u)$ and $f(v)$ are adjacent vertices of G_2, then u and v are adjacent in G_1." For example, vertex $d = f(s)$ is adjacent to $c = f(r)$ in G_2. Certainly s is adjacent to r. Since each pair of adjacent vertices in G_2 is discussed in the right-hand column of Table 1, we know that each pair of adjacent vertices in G_2 is the output of a pair of adjacent vertices in G_1.

Thus f is an isomorphism; graphs G_1 and G_2 are isomorphic. [*Note*: f is one of four isomorphisms from G_1 to G_2. Can you find another isomorphism?]

Now we return to the notion of complete graphs.

Example 4

Show that the two complete graphs in Figure 4 are isomorphic by finding an isomorphism from the vertices of G_1 to the vertices of G_2.

Solution

Consider the function f that sends w to a, x to b, y to c, and z to d. This function is a one-to-one mapping of the vertices of G_1 onto the vertices G_2. Now let us check adjacencies. In G_1, w is adjacent to x. In G_2, $a = f(w)$ is adjacent to $b = f(x)$. Similarly, w is adjacent to y while $a = f(w)$ is adjacent to $c = f(y)$; and w is adjacent to z while $a = f(w)$ is adjacent to $d = f(z)$. However, we need not continue this case-by-case check. Since we are working with complete graphs and each vertex in either graph is adjacent to all other vertices in that graph, the adjacency requirement follows. For this reason, any one-to-one function from the vertices of G_1 onto the vertices of G_2 would be an isomorphism.

Indeed, the reasoning used in Example 4 extends to all complete graphs. If G_1 and G_2 are two complete graphs with n vertices, then any one-to-one function from the set of vertices of G_1 to the set of vertices of G_2 is an isomorphism. Hence they are essentially the same graph as far as the properties of graphs are concerned. Thus we are able to talk about *the* complete graph on n vertices.

It can be quite difficult to determine whether two simple graphs are isomorphic, particularly if the graphs have a large number of vertices. We must choose one of the many possible one-to-one functions between the vertex sets of the two graphs, and then we must show that the function preserves adjacencies in both directions. However, in some cases it is quite easy to show that pairs of simple graphs are not isomorphic; we need only show that the two graphs differ on some essential property. Properties that isomorphic simple graphs must share are called **invariant properties**. Among the invariant properties are the following:

INVARIANT PROPERTIES OF ISOMORPHIC SIMPLE GRAPHS

(1) Isomorphic simple graphs must have the same number of vertices.

(2) Isomorphic simple graphs must have the same number of edges.

(3) If d is a non-negative integer, isomorphic simple graphs must have the same number of vertices of degree d.

Invariant property (1) is a consequence of the fact that an isomorphism is a one-to-one and onto function between the vertex sets of the graphs. Invariant property (2) results from the fact that the one-to-one and onto function between the vertex sets gives rise to a one-to-one and onto function between edge sets of isomorphic simple graphs. If u is a vertex of degree k in G_1, then $f(u)$ must be a vertex of degree k in G_2 because u is adjacent to v if and only if $f(u)$ is adjacent to $f(v)$. Property (3) is a consequence of this fact.

Example 5

Explain why the simple graphs in Figure 6 are not isomorphic.

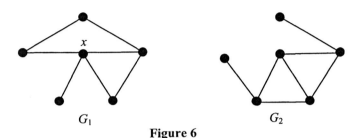

Figure 6

Solution

Observe that simple graphs G_1 and G_2 both have six vertices and seven edges. However, the degree of vertex x in G_1 is 4, whereas G_2 has no vertex of degree 4. Further, G_2 has two vertices of degree 1, whereas G_1 has only one vertex of degree 1. For either of these reasons, we know that the two simple graphs are not isomorphic.

We close this section with a brief discussion of two additional concepts that we will need as we continue our study of graph theory.

As we might expect, a subgraph of a graph is a new graph formed from subsets of the vertices and edges of the original graph. More precisely, if $G = (V, E)$, then a graph $G' = (V', E')$ is a **subgraph** of G precisely when $V' \subseteq V$ and $E' \subseteq E$.

Example 6

Name two subgraphs of the graph G represented in Figure 7(a).

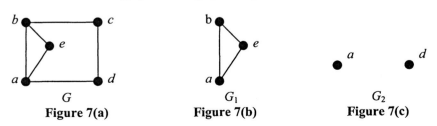

Figure 7(a) **Figure 7(b)** **Figure 7(c)**

Solution

Consider the graphs G_1 with vertices $\{a, b, e\}$ and edges $\{\{a, b\}, \{b, e\}, \{e, a\}\}$ and G_2 with vertices $\{a, d\}$ and no edges. (See Figures 7(b) and 7(c).) Both G_1 and G_2 are subgraphs of G.

A simple graph G is a **bipartite graph** if the vertex set is the union of two disjoint non-empty sets, V_1 and V_2, such that each edge is incident with one vertex in V_1 and one vertex in V_2. Notice that this means that no two vertices in V_1 are adjacent and no two vertices in V_2 are adjacent.

Example 7

Explain why the graph represented in Figure 8 is a bipartite graph.

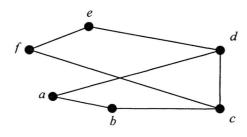

Figure 8

Solution

Consider the two sets of vertices, $V_1 = \{b, d, f\}$ and $V_2 = \{a, c, e\}$. Observe that there are no edges connecting pairs of vertices in V_1 and that there are no edges connecting pairs of vertices in V_2. Indeed, each edge in the graph is incident with a vertex in V_1 and a vertex in V_2.

In the bipartite graph in Figure 8, we observe that although vertex f is in V_1 and vertex a is in V_2 there is no edge incident with f and a. A bipartite graph in which there is an edge connecting each vertex in V_1 with each vertex in V_2 is called a **complete bipartite graph**. If V_1 has n elements and V_2 has m elements, then the resulting complete bipartite graph is denoted $K_{n,m}$. Figure 9(a) is a representation of $K_{2,3}$ and Figure 9(b) is a representation of $K_{3,5}$.

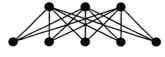

Figure 9(a)   **Figure 9(b)**

Exploratory Exercise Set 4.3

1. In this exercise we investigate graph isomorphisms between graphs G_1 and G_2.

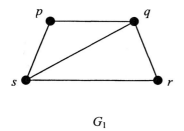   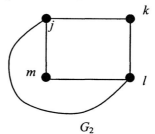

G_1 G_2

(a) Complete the following table, which describes the vertices adjacent to each vertex in graphs G_1 and G_2.

Adjacent vertices in G_1	Adjacent vertices in G_2
Vertices adjacent to p: **s and q**	Vertices adjacent to j: **k, l and m**
Vertices adjacent to q:	Vertices adjacent to k:
Vertices adjacent to r:	Vertices adjacent to l:
Vertices adjacent to s:	Vertices adjacent to m:

(b) Consider this one-to-one function from the set $\{p, q, r, s\}$ onto the set $\{j, k, l, m\}$.

$$p \rightarrow j$$
$$q \rightarrow k$$
$$r \rightarrow l$$
$$s \rightarrow m$$

Is it an isomorphism between the graphs G_1 and G_2? Explain by referring to the table from (a).

(c) Find two different one-to-one functions from the set $\{p, q, r, s\}$ onto the set $\{j, k, l, m\}$ that are graph isomorphisms.

2. In this exercise we will continue our investigations of the notion of a graph isomorphism.

(a) Suppose G_1 is a graph with the three vertices $\{a, b, c\}$ and G_2 is a graph with the three vertices $\{x, y, z\}$. There are $6 = 3 \times 2 \times 1$ one-to-one functions from the vertex set of G_1 onto the vertex set of G_2. Write them down.

(b) Graphs G_1 and G_2 are represented below. Which of the functions that you described in (a) are graph isomorphisms? Explain.

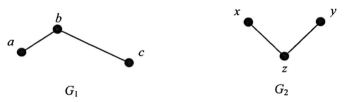

$$G_1 \qquad\qquad\qquad G_2$$

(c) Which vertex does each of the isomorphisms pair with vertex b? Can you explain why this must be true?

(d) Suppose G_1 has vertex set $\{a, b, c, d\}$ and G_2 has vertex set $\{w, x, y, z\}$. There are $24 = 4 \times 3 \times 2 \times 1$ one-to-one functions from the vertex set of G_1 onto the vertex set of G_2. Write down at least five of them.

(e) *For students who have completed the chapter on combinatorics:* Explain why we know that if graph G has n vertices and graph H has n vertices, then there are

$$n! = n \times (n-1) \times (n-2) \times \ldots \times 3 \times 2 \times 1$$

different one-to-one functions from the vertex set of G onto the vertex set of H.

3. In this exercise we will investigate Hamilton circuits for complete graphs. Choose a partner before beginning this activity.

(a) Below you will find the complete graph K_4.

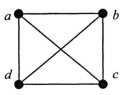

(1) Each partner should find a Hamilton circuit beginning at vertex a.

    (2) Work with your partner to find as many more distinct Hamilton circuits beginning at vertex a as possible.

    (3) Explain in words why you should be able to find six Hamilton circuits beginning at each of the vertices of K_4.

    (4) Explain why there are a total of 24 Hamilton circuits for the graph K_4.

(b) For the graph K_n there are $n! = n \times (n-1) \times (n-2) \times \ldots \times 3 \times 2 \times 1$ different Hamilton circuits.

    (1) Draw K_3 and find $3! = 3 \times 2 \times 1$ different Hamilton circuits for K_3.

    (2) Draw K_4 and find $4! = 4 \times 3 \times 2 \times 1$ different Hamilton circuits for K_4.

    (3) *For students who have completed the chapter on combinatorics*: Explain why there are $n!$ different Hamilton circuits for K_n.

(c) *For students who have completed the chapter on combinatorics*: In some applications of complete graphs, it would be useful to find all Hamilton circuits for the graph. For K_3 and K_4 this is easily done, as you have demonstrated. Let us investigate this problem further. You and your partner will want access to a calculator.

    (1) How many different Hamilton circuits can be found for the graph K_{20}?

    (2) Suppose a computer can find one such circuit for K_{20} every nanosecond (that is, it can find 10^9 Hamilton circuits every second). How many seconds would it take to find all Hamilton circuits for K_{20}?

    (3) How many years would the computer take to find all Hamilton circuits for K_{20}?

4. In this exercise we will investigate the matrix representations of isomorphic simple graphs. (Review Exploratory Exercise 5 from Section 4.2.)

    (a) Consider the function f from the vertex set of G_1 onto the vertex set of G_2.

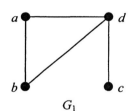
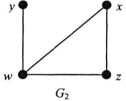

    (1) Verify that this function is a graph isomorphism.

    (2) Complete the entries in these tables to create the adjacency matrices for G_1 and G_2.

Notice that in forming the matrix for G_2, we have ordered the vertices for G_2 to correspond with the ordering of the matrices for G_1. Since we listed vertex a first in building the adjacency matrix for G_1, and function f pairs a with z, we list z first in building the adjacency matrix for G_2.

Adjacency Matrix for G_1

	a	b	c	d
a				
b				
c				
d				

Adjacency Matrix for G_2

	z	x	y	w
z				
x				
y				
w				

    (3) What do you observe about the two matrices?

(b) Now let us investigate connections between the adjacency matrices for the isomorphic simple graphs H_1 and H_2.

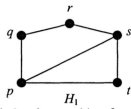

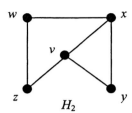

(1) Find an isomorphism from the vertex set of H_1 to the vertex set of H_2. Verify that it is one-to-one, that it is onto, and that it preserves adjacencies.

(2) Find adjacency matrices for H_1 and H_2. Choose the order for the vertices of H_2 that reflects the order imposed by your isomorphism. The vertex that your isomorphism paired with p should be listed first, and so on.

Adjacency Matrix for H_1

	p	q	r	s	t
p					
q					
r					
s					
t					

Adjacency Matrix for H_2

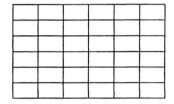

(3) What do you observe about the two matrices from (2)?

(4) Do your observations in (3) suggest the following conjecture?

> Two simple graphs are isomorphic if and only if there is an ordering of their vertices such that, with respect to that ordering, the two adjacency matrices are equal.

Though we will not pause to prove it, this conjecture is true.

5. (*Writing Exercise*) You and your study group are to meet tonight to go over this material. Your assignment is to do a better job of explaining isomorphic graphs than either the teacher or this book. Write the notes you will use in your presentation to your study group. Be sure to include examples other than ones that occur in the book. In your notes you will probably want to review functions, one-to-one functions, and onto functions before you discuss graph isomorphisms.

Exercise Set 4.3

1. Draw graphs that model the situations described in (a) through (c). After the graphs are drawn, count the number of edges in order to answer the question posed.
   (a) Five executives meet for lunch. If each executive shakes hands with each other executive, how many handshakes occur?
   (b) Three students rent a single, two-seated bicycle. If each student rides exactly one time with each other student, how many bicycle rides must occur?
   (c) Computers sit on each of four desks, and staff members from the Information Technology Department design the network so that each pair of computers is connected by a cable. How many cables must be run to complete the network?
   (d) Each of the graphs drawn in (a) through (c) is an example of what kind of graph?

2. Draw the complete graphs K_6 and K_7.

3. Sketch three different subgraphs of graph G.

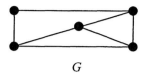

G

4. Draw the complete bipartite graphs $K_{3,3}$ and $K_{2,4}$.

5. (a) Draw a graph that models the following situation: Ben, Luis, Lois, Dell, and John negotiate to conclude a business deal. In the process, each participant makes a telephone call to each other participant.
    (b) What is the name of the graph that you drew in (a)?
    (c) Draw a graph that models the following situation: Communication links are established between computers A and C, A and D, A and E, B and C, B and D, and B and E. However, no direct links are available between A and B, C and D, C and E, or D and E.
    (d) What is the name of the graph that you drew in (c)?

6. Explain why each of the following pairs of simple graphs are not isomorphic by giving an invariant property that is not shared by the two graphs.
    (a)

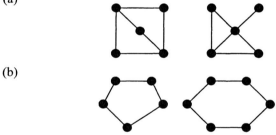

    (b)

7. (a) Will two graphs with the same number of vertices always be isomorphic? Explain.
    (b) Will two graphs with the same number of edges always be isomorphic? Explain.

8. (a) How many edges does K_2 have?
    (b) How many edges does K_3 have?
    (c) How many edges does K_4 have?
    (d) How many edges does K_5 have?
    (e) *Challenge*: Find an expression that computes the number of edges of K_n .

9. For what values of n does the complete graph K_n have an Euler circuit?

In Exercises 10 through 15, determine whether the two graphs are isomorphic. If the graphs are isomorphic, define a one-to-one and onto function between the two vertex sets that preserves the adjacency relationship in both directions. If the graphs are not isomorphic, describe an invariant property that is not shared by the two graphs.

10.

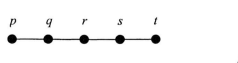

11.

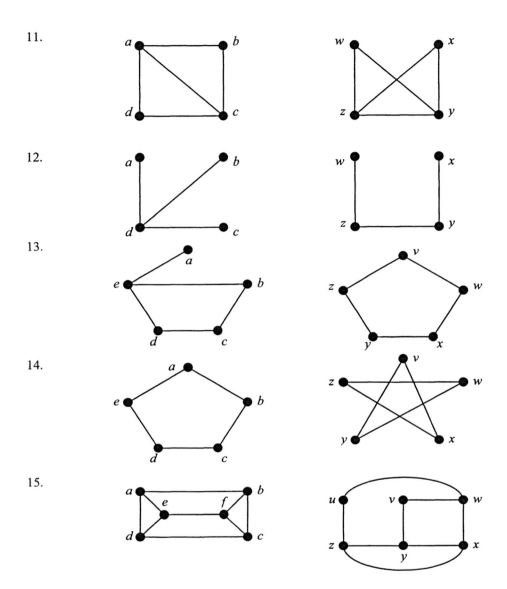

12.

13.

14.

15.

16. Sketch a complete set of non-isomorphic simple graphs with three vertices. [*Note*: Your *complete set* will be complete when you have a set of examples, no two of which are isomorphic, with the property that any other example is isomorphic to one of the examples in your set.]

17. Sketch a complete set of non-isomorphic simple graphs with 4 vertices. (See note in Exercise 16.)

18. Show that the following graphs are bipartite by finding vertex sets V_1 and V_2 so that each edge is incident with a vertex in V_1 and a vertex in V_2.

(a)   (b)

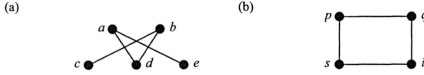

19. We will show that the graph G is not bipartite by careful consideration of vertices a, b, and c. Suppose the vertex set is partitioned into the sets V_1 and V_2 described in the definition of a bipartite graph. Give reasons for each of the following observations.

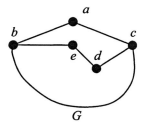

G

(a) a and b cannot both be in V_1. Why? (Similarly, they cannot both be in V_2.)
(b) a and c cannot both be in V_1. Why? (Similarly, they cannot both be in V_2.)
(c) b and c cannot both be in V_1. Why? (Similarly, they cannot both be in V_2.)
(d) Explain why this means G cannot be bipartite.

20. (a) Does $K_{2,1}$ have an Euler path?
    (b) Does $K_{2,2}$ have an Euler circuit?
    (c) Does $K_{2,3}$ have an Euler path?
    (d) Does $K_{2,4}$ have an Euler circuit?
    (e) Suppose n is odd. Explain why $K_{2,n}$ always has an Euler path.
    (f) Suppose n is even. Explain why $K_{2,n}$ always has an Euler circuit.

21. (a) Does $K_{3,1}$ have an Euler path?
    (b) Does $K_{3,2}$ have an Euler path?
    (c) Does $K_{3,3}$ have an Euler path?
    (d) Does $K_{3,4}$ have an Euler path?
    (e) Explain why $K_{3,n}$ does not have an Euler path unless $n = 2$.

22. Suppose both m and n are even. Explain why $K_{m,n}$ always has an Euler circuit.

23. Suppose m is odd. Explain why $K_{m,n}$ does not have an Euler path unless $n = 2$.

24. In describing invariant property (2) for isomorphic graphs in the discussion preceding Example 5, we indicated that the one-to-one, onto function between vertex sets gives rise to a one-to-one and onto function between edge sets of isomorphic simple graphs. Consider the isomorphism between the vertex sets of graphs G_1 and G_2 described in the following table. By completing (a) and (b) we will find the associated function between edge sets.

Function Between Vertex Sets	
Vertices of G_1	Vertices of G_2
A	Q
B	P
C	T
D	S
E	R

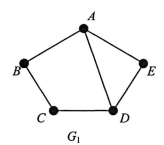

G_1

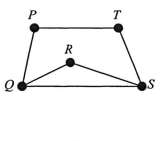

G_2

(a) Since vertex A is adjacent to vertex E in G_1, edge $\{A, E\}$ should correspond to which edge in G_2?

(b) Complete this table that describes the one-to-one function from the edge set of G_1 onto the edge set of G_2.

Function Between Edge Sets	
Edges of G_1	Edges of G_2
$\{A, E\}$	$\{Q, R\}$
$\{E, D\}$	
$\{D, A\}$	
$\{D, C\}$	
$\{C, B\}$	
$\{B, A\}$	

25. (*Proof Exercise*) Let f be an isomorphism from G_1 to G_2. In describing invariant property (3) for isomorphic graphs, we indicated that if u is a vertex of degree k in G_1, then $f(u)$ must be a vertex of degree k in G_2. Let us explore this argument in more detail.

Suppose u, a vertex in G_1, has degree k. We will show that $f(u)$, the corresponding vertex in G_2, also has degree k. Let $\{a_1, a_2, ..., a_k\}$ be the set of vertices in G_1 that are adjacent to u.

(a) $f(a_1), f(a_2), ..., f(a_k)$ are all adjacent to $f(u)$. Why?

(b) $f(a_1), f(a_2), ..., f(a_k)$ are all distinct. Why?

(c) It is now sufficient to show that $f(a_1), f(a_2), ..., f(a_k)$ are the only vertices of G_2 adjacent to $f(u)$. Why?

(d) Let w be a vertex in G_2 adjacent to $f(u)$. Then there is a vertex a of G with $f(a) = w$. Why?

(e) a is adjacent to u. Why?

(f) Thus $a \in \{a_1, a_2, ..., a_k\}$. Why? And thus $w = f(a) \in \{f(a_1), f(a_2), ..., f(a_k)\}$.

4.4 Shortest-Path Problems

The Truck Driver studies the road map in Figure 1, hoping to find the route from Alpha to Omega that will be the shortest in terms of miles traveled. The number labeling each edge of the graph is the number of miles between the cities represented by the vertices.

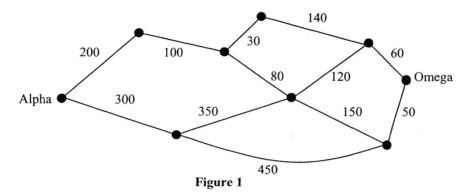

Figure 1

The Harried Traveler studies the airline map in Figure 2, hoping to find the route from Beta to Gamma that will be the shortest in terms of time required for travel. The number labeling each edge of the graph is the time required to fly between the cities represented by the vertices.

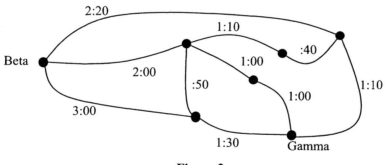

Figure 2

The Comptroller studies the graph in Figure 3, hoping to find the least expensive route to move a shipment of widgets from Kappa to Sigma. The number labeling each edge of the graph is the cost in dollars to move the shipment between the two terminals represented by the vertices.

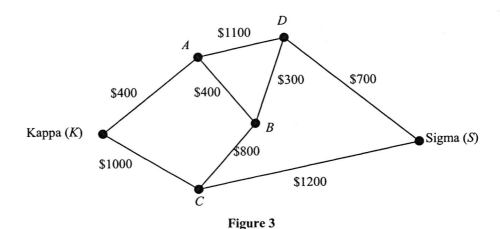

Figure 3

The graphs in Figures 1, 2, and 3 are examples of **weighted graphs**, graphs that have a number attached to each edge. The number attached to each edge of a weighted graph is called the **weight** of the edge. The **length** of a path in a weighted graph is the sum of the weights of the edges in the path. In each of the three situations described above, the problem is to find the path of least length between a specified pair of vertices.

One strategy for solving a "shortest-path" problem is obvious: Merely determine all simple paths between the two vertices, determine the length of each path, and choose the path of shortest length. In Example 1 we will use this strategy to find the shortest path from Kappa to Sigma.

Example 1

Find the shortest path for the graph from Kappa to Sigma in Figure 3 by identifying all simple paths from Kappa to Sigma and finding the length of each.

Solution

There are seven paths from Kappa to Sigma that do not repeat edges.

(K, A, D, S) with length $400 + 1100 + 700 = 2200$
(K, A, D, B, C, S) with length 3800
(K, A, B, D, S) with length 1800
(K, A, B, C, S) with length 2800
(K, C, B, A, D, S) with length 4000
(K, C, B, D, S) with length 2800
(K, C, S) with length 2200

Clearly, the path with least length is (K, A, B, D, S) with length 1800. [Note that (K, C, S) is not the "shortest path," even though it has the fewest edges.]

Unfortunately, although the solution in Example 1 is easily described conceptually, it is not so easily accomplished for graphs with a large number of vertices. When using graphs to model and solve problems, it is often easy to conceive of *some* solution to the problem but much more difficult to find a solution that is efficient enough to be implemented for graphs with many vertices and edges. Hence, much of the creative work in graph theory involves designing *efficient* algorithms that accomplish the desired tasks.

Over the years several algorithms have been developed for finding the shortest path through a weighted graph. We will examine an algorithm developed by the Dutch mathematician Edsger Dijkstra in 1959. The version we will describe will find the shortest path in simple, weighted graphs in which all the weights are positive. (Remember that simple graphs are graphs without parallel edges or loops.) Before we describe the algorithm, we will examine a geometric algorithm that provides intuition with which to understand Dijkstra's Algorithm.

A GEOMETRIC SHORTEST-PATH ALGORITHM

This algorithm will find the length of the shortest path between two vertices. It can be described with the following three steps.

Geometric Shortest-Path Algorithm:

1. Suppose the vertex at which you wish to start is a and the vertex at which you wish to end is z. Circle a.
2. Examine each vertex adjacent to a and find the weights of the corresponding edges. Choose the edge (and corresponding vertex) with the smallest weight. (If there are two or more edges incident with a with the same smallest weight, break the tie arbitrarily.) Shade the edge you choose, and circle the vertex that this edge shares with a.
3. Until z is circled:

    Examine all uncircled vertices that are adjacent to some circled vertex in the graph.

    For each uncircled vertex x being examined and each edge E incident with x and a circled vertex, find all simple paths from a to x that, in addition to x and E, contain only shaded edges and circled vertices.

    Find the length of all these paths.

    Choose the vertex x and edge E that lie in the path with the smallest length (break ties arbitrarily). Circle the vertex x and shade the edge E.

    Return to the beginning of Step 3.

   When z is circled, the shaded path from a to z is a shortest path from a to z.

Example 2

   Use the Geometric Shortest-Path Algorithm to find the shortest path from vertex a to vertex z in the weighted path in Figure 4.

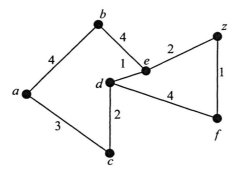

Figure 4

(continued)
Solution

Steps 1 and 2: We first circle the vertex from which we wish to start, vertex *a*. The two vertices adjacent to vertex *a* are vertices *b* and *c*. The weight of {*a*, *b*} is 4 and the weight of {*a*, *c*} is 3. Hence we shade edge {*a*, *c*} and circle vertex *c*. (See Figure 5(a).)

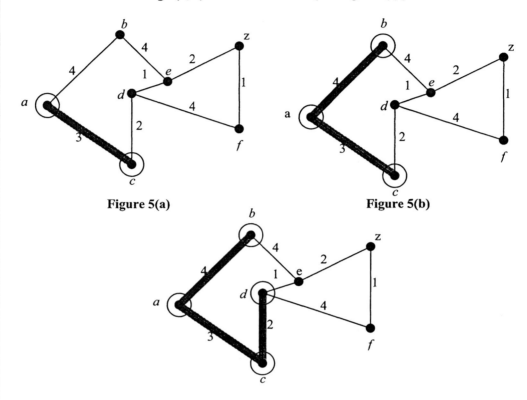

Figure 5(a)   **Figure 5(b)**

Figure 5(c)

First iteration of Step 3: The uncircled vertices adjacent to the circled vertices are *b* and *d*. The only path from *a* to *b* meeting the requirements of Step 3 is the path (*a*, *b*) of length 4. The only path from *a* to *d* meeting the requirements of Step 3 is (*a*, *c*, *d*) of length 5. Hence we shade the edge {*a*, *b*} and circle vertex *b*. (See Figure 5(b).)

Second iteration of Step 3: The vertices adjacent to the circled vertices are *d* and *e*. The only simple path from *a* to *e* that (with the exception of *e* and the edge incident to it) consists of shaded edges and circled vertices is (*a*, *b*, *e*) of length 8. Similarly, the only path from *a* to *d* meeting the requirements of Step 3 is (*a*, *c*, *d*) of length 5. Hence we shade edge {*c*, *d*} and circle vertex *d*. (See Figure 5(c).)

(continued)

Third iteration of Step 3: The vertices adjacent to the circled vertices are *e* and *f*. The simple paths from *a* to *e* that (with the exception of *e* and the edge incident to it) consist of shaded edges and circled vertices are

(a, b, e) of length 8

(a, c, d, e) of length 6

Similarly, the only path from *a* to *f* meeting requirements is (a, c, d, f) of length 9. We shade the edge $\{d, e\}$ and circle vertex *e*. (See Figure 6(a).)

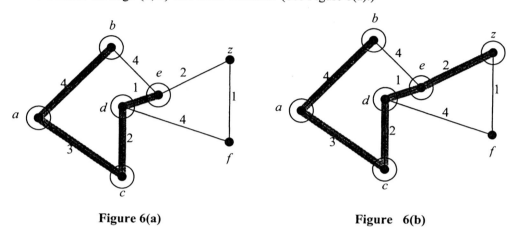

Figure 6(a)   **Figure 6(b)**

Fourth iteration of Step 3: The vertices adjacent to the circled vertices are *f* and *z*. The simple path from *a* to *f* that (with the exception of *f* and the edge incident to it) consists of shaded edges and circled vertices is (a, c, d, f) of length 9. Similarly, the path from *a* to *z* meeting requirements is (a, c, d, e, z) of length 8. The path of shortest length has length 8. We circle *z* and shade $\{e, z\}$. (See Figure 6(b).)

We see that the path of shortest length has length 8 and that a path that has this length is (a, c, d, e, z).

Although the Geometric Shortest-Path Algorithm is an improvement over the trial-and-error method described earlier, it is still cumbersome to complete. The fact that each time we prepare to find a new vertex to circle, we must compute the length of paths all the way back to the starting vertex is a severe shortcoming for very large graphs. By devising a clever way to label the vertices as they are chosen, Dijkstra created the much more efficient algorithm that bears his name.

DIJKSTRA'S SHORTEST-PATH ALGORITHM

In order to describe carefully how to perform Dijkstra's Algorithm, we need to agree on ways to keep up with what we have learned at each stage of the process. We will accomplish this by labeling each vertex in a useful way and then relabeling some of the vertices at each step. More specifically:

    (a) Initially, we will circle the vertex from which we wish to start the path. As we proceed through the graph, we will circle additional vertices.

    (b) Initially, we will label the starting vertex *a* with a 0, indicating that the shortest path from *a* to *a* is of length 0.

    (c) Initially, we will use the weight of $\{a, u\}$ as the label for all vertices *u* that are adjacent to *a*. This label represents the length of the most obvious path from *a* to *u*.

    (d) Initially, we will label each vertex that is not adjacent to *a* with ∞. This indicates that we have not yet identified a path from *a* to this vertex.

    (e) As we circle a new vertex *u,* we label all non-circled vertices *v* that are adjacent to *u* with a number that will help us find the shortest distance from *a* to *v*.

In the algorithm we will denote the weight of edge $\{u, v\}$ as weight$\{u, v\}$ and will denote the label of vertex u by $L(u)$. Dijkstra's Algorithm for finding the length of a shortest path from vertex a to vertex z can be described as follows.

Dijkstra's Shortest-Path Algorithm:

1. Circle the initial vertex, vertex a, and label it with a 0 ($L(a) = 0$).
2. For each vertex u adjacent to a, label u with weight$\{a, u\}$ ($L(u) = $ weight$\{a, u\}$). Label all other vertices with the infinity symbol.
3. Until z is circled,
    Find the non-circled vertex u with the smallest label and circle it.
    (If several such vertices have the same label, choose any one of them.)
    For all non-circled vertices v adjacent to u, relabel v with the minimum of $\{L(v), L(u) + $ weight$\{u, v\}\}$.
    Return to the beginning of Step 3.

As each vertex u is circled, its label is the length of the shortest path from a to u. In particular, after z is circled, the label of z is the length of the shortest path from a to z.

The italicized statement in the algorithm is a statement that requires proof, which we will discuss briefly later.

Example 3

Use Dijkstra's Algorithm to find the length of the shortest path from Kappa (K) to Sigma (S) in the weighted graph in Figure 3.

Solution

After Steps 1 and 2, the graph is labeled as shown in Figure 7(a). Note that the initial vertex K is labeled and circled, the vertices adjacent to K are labeled with the weights of the edge they share with K, and all other vertices are labeled with the infinity symbol.

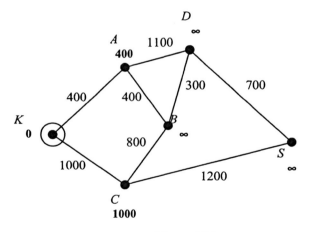

Figure 7(a)

(continued)

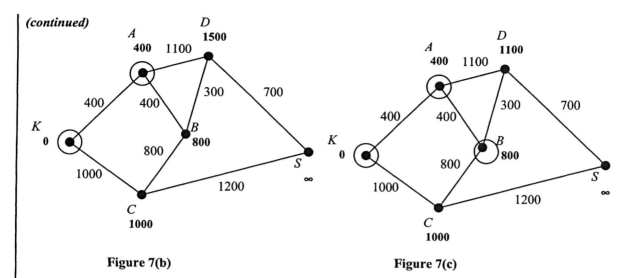

Figure 7(b)   Figure 7(c)

First iteration of Step 3: The first time we execute Step 3, we observe that vertex *A* has the smallest label and circle *A*. Since *B* and *D* are adjacent to *A*, we need to consider attaching a new label to each of them. Since both have labels of infinity, we replace the label of *B* by $L(A) + \text{weight}\{A, B\} = 400 + 400 = 800$. Similarly, $L(D)$ becomes $400 + 1100 = 1500$. See Figure 7(b).

Second iteration of Step 3: Of the vertices that have not been circled, *B* has the smallest label. We circle *B* and consider relabeling vertices *C* and *D*, which are uncircled vertices adjacent to *B*. We do not relabel *C* since $L(C) = 1000$ is less than $L(B) + \text{weight}\{B, C\}$. We do relabel *D* since 1500 is greater than $L(B) + \text{weight}\{B, D\} = 800 + 300 = 1100$. See Figure 7(c).

Carefully study Figure 8(a), (b), and (c) and follow the algorithm through three more iterations of Step 3. Make sure you understand how the vertices were labeled at each iteration. Observe that the final label of *S*, 1800, gives the length of the shortest path from *K* to *S*.

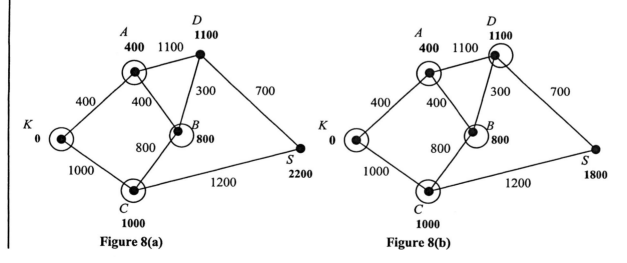

Figure 8(a)   Figure 8(b)

(continued)

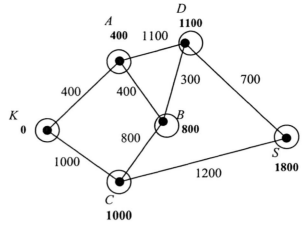

Figure 8(c)

As presently described, Dijkstra's Algorithm gives only the length of a shortest path from a to z; it does not preserve a record of a path with that minimum length. We could correct that omission by shading the appropriate edge as we circle each vertex, as in the geometric algorithm. Alternatively, we can correct that omission by adding a second label P for each vertex v that describes the predecessor of v as we build the shortest path. More precisely, revise Step 2 and Step 3 of Dijkstra's Algorithm as follows:

2. Label vertex a with 0 ($L(a) = 0$), and for each vertex u adjacent to a, label u with weight$\{a, u\}$ ($L(u) = $ weight$\{a, u\}$) and with predecessor a ($P(u) = a$).
3. Until z is circled,
   Find the non-circled vertex u with the smallest label and circle it.
   (If several such vertices have the same label, choose any one of them.)
       For all non-circled vertices v adjacent to u,
       If $L(v) > L(u) + $ weight$\{u, v\}$,
           then   (i)   let $L(v) = L(u) + $ weight$\{u, v\}$
                 (ii) let $P(v) = u$ (where $P(v)$ is the vertex that precedes v)
       Return to the beginning of Step 3.

At the end of the algorithm, we can recover the shortest path from a to z by starting with the predecessor of z and working backwards. In Figure 9 we show the results after we have taken the final labeled graph from Example 3 and labeled the vertices with their predecessors. (Compare with Figure 8(c).) Since the predecessor of S is D, the predecessor of D is B, the predecessor of B is A, and the predecessor of A is K, a shortest path is (K, A, B, D, S).

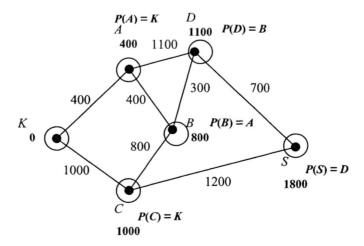

Figure 9

DIJKSTRA'S ALGORITHM WITHOUT GEOMETRIC GRAPHS

Of course, we do not have to actually relabel the vertices of geometric graphs to use Dijkstra's Algorithm; otherwise, we would not be able to implement it on a computer. We can complete the algorithm by creating a table of labels of the vertices and changing the labels as specified by the algorithm.

Example 4

Use Dijkstra's Algorithm to find a path of least length from a to z in the weighted graph in Figure 10. In this solution, do not redraw the graph at each iteration, but change the table of labels.

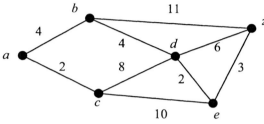

Figure 10

Solution

The initial set of labels for the vertices is found in the *Initial Labels* table, and vertex a is circled.

Initial Labels	
Vertex	$L(u)$
a	0
b	4
c	2
d	∞
e	∞
z	∞

After the first iteration of Step 3, the graph will be labeled as shown in Figure 11. See the *First Iteration* table below.

First Iteration		
Vertex	$L(u)$	$P(u)$
a	0	
b	4	a
c	2	a
d	10	c
e	12	c
z	∞	

Example 4 *(continued)*

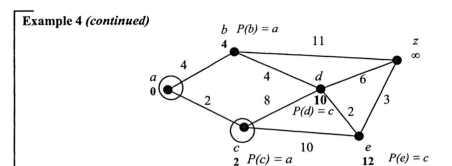

Figure 11

Now, complete the algorithm using only tables, and compare your work to the tables that follow.

Second Iteration		
Vertex	$L(u)$	$P(u)$
a	0	
b	4	a
c	2	a
d	8	b
e	12	c
z	15	b

Third Iteration		
Vertex	$L(u)$	$P(u)$
a	0	
b	4	a
c	2	a
d	8	b
e	10	d
z	14	d

Fourth Iteration		
Vertex	$L(u)$	$P(u)$
a	0	
b	4	a
c	2	a
d	8	b
e	10	d
z	13	e

Fifth iteration		
Vertex	$L(u)$	$P(u)$
a	0	
b	4	a
c	2	a
d	8	b
e	10	d
z	13	e

Example 4 *(continued)*

The length of the shortest path is 13. Since the predecessor of *z* is *e*, the predecessor of *e* is *d*, the predecessor of *d* is *b*, and the predecessor of *b* is *a*, a shortest path is (*a, b, d, e, z*).

Dijkstra's Algorithm is an example from a class of algorithms called "greedy" algorithms. The goal of the algorithm is to produce a path between a pair of vertices of minimum length. The algorithm proceeds "greedily" by choosing, at each step, the vertex that is "closest" to the starting vertex. It is not immediately clear that implementing this greedy strategy at each step will produce the minimum-length path at the end of the process. However, a fairly sophisticated induction argument shows that indeed Dikstra's Algorithm always produces the length of the shortest paths.

Exploratory Exercise Set 4.4

1. Imagine that you have before you a map of a subway system. Describe a model involving a weighted graph that could be used to solve the following problems. For each problem, specify the vertices and edges and the weights that are used to label the edges.
   (a) Find the least amount of time required to travel between two locations in the system.
   (b) Find the minimum distance that can be traveled between two locations in the system.
   (c) Find the minimum fare that must be paid to travel between two locations in the system if the fare on a trip is the sum of the fares between stops intermediate to the two locations.

2. Work with a partner to explore how to start both the Geometric Shortest-Path Algorithm discussed in this section and Dijkstra's Algorithm. Work independently to complete (a) and (b) and then discuss your results.
   (a) Suppose we wish to use the Geometric Shortest-Path Algorithm for finding the shortest path from A to Z in this weighted graph. In Step 1 we first circle A. Find the next vertex to circle and the first edge to shade, and explain your reasoning.

   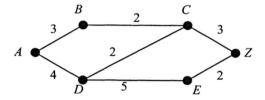

   (b) Suppose we wish to use Dijkstra's Algorithm to find the length of the shortest path in the weighted graph from (a).
      (1) Label all the vertices as required in the first two steps of Dijkstra's Algorithm. (Do not worry about predecessor labels.)
      (2) Complete the first iteration of Step 3. Find the next vertex to circle, and relabel all vertices that should be relabeled.

3. Work with a partner to explore how to complete the iterated steps of both the Geometric Shortest-Path Algorithm and Dijkstra's Algorithm. Work independently to complete (a) and (b) and then discuss your results.
   (a) In the weighted graph below you will find that the Geometric Shortest-Path Algorithm for finding the shortest path has been started and Step 3 has been completed twice.
      (1) What are the two paths whose lengths must be determined in order to choose the next vertex to circle?
      (2) Complete Step 3 again to determine the next vertex to circle and the next edge to shade.

   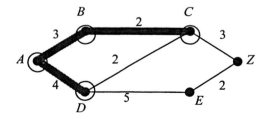

(b) In the weighted graph below you will find that Dijkstra's Algorithm has been started and Step 3 has been completed three times. Complete Step 3 again to determine the next vertex to circle and to relabel the appropriate vertices.

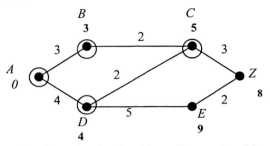

4. At this point it may not be clear why the algorithms discussed in this section are superior to the method in which we list all possible simple paths, compute their lengths, and then choose the smallest. In this problem, please work in pairs to find the shortest simple path from A to Z in three ways and then compare the three techniques.

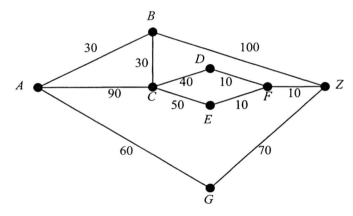

(a) *Solution 1*: We will first do an exhaustive search, finding all simple paths from A to Z.
   (1) Find all simple paths that begin with A and terminate when they reach Z.
      (*Hint*: There are 15 such paths.)
   (2) Find the length of each path from (a).
   (3) Describe a shortest path from A to Z and give its length.
(b) *Solution 2*: Use the Geometric Shortest-Path Algorithm to find a shortest path from A to Z and its length.
(c) *Solution 3*: Use Dijkstra's Algorithm to find the length of a shortest path from A to Z.
(d) Discuss the advantages of the shortest-path algorithms for a large, complex graph. (Imagine a graph describing a corporate compute network with 500 vertices and 800 edges.)

5. Work in pairs. Have each member of r pair sketch a weighted graph with at least six vertices and designate a Start Vertex and a Finish Vertex. Exchange graphs and use each of the algorithms of this section to find a shortest path from Start to Finish and its length.

6. (*Writing Exercise*) Your little brother finds you using the Geometric Shortest-Path Algorithm while doing your homework on this section. He wants you to explain what you are doing. Write a conversation that you might have with him to explain the Geometric Shortest-Path Algorithm. As you explain, work through an example of your own creation.

Exercise Set 4.4

In Exercises 1 and 2 use the Geometric Shortest-Path Algorithm to find the length of the shortest path from a to z in the weighted graph, and then find a path with this length.

1.
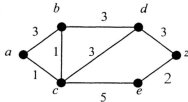

2.
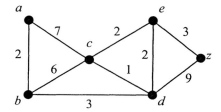

3. Use Dijkstra's Algorithm as described before Example 3 to find the length of the shortest path in Exercise 1.

4. Use Dijkstra's Algorithm as described before Example 3 to find the length of the shortest path in Exercise 2.

5. Find the length of a shortest path from Alpha to Omega in Figure 1 in this section.

6. Find the length of a path that takes the least flying time to get from Beta to Gamma in the weighted graph from Figure 2 in this section.

7. Consider the following weighted graph.

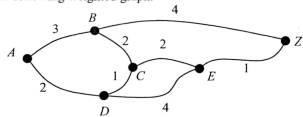

   (a) Find all simple paths from *A* to *Z*.
   (b) Find the length of each path from (a).
   (c) Describe the shortest path from *A* to *Z* and specify its length.
   (d) Use Dijkstra's Algorithm together with predecessor labels to find the shortest path and its length.

8. Use Dijkstra's Algorithm as described before Example 3 to find the length of a shortest path in this weighted graph.

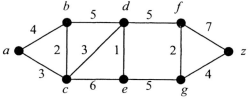

9. Below you see the first steps in the using of Dijkstra's Algorithm for finding the shortest path from a to z in a weighted graph. These first steps are illustrated by labeling a graph and recording the information in a table. Complete the algorithm for finding the length of a shortest path and a shortest path using a table for the remaining steps.

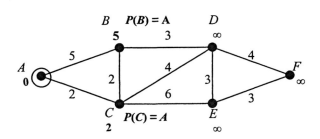

Initial Labels

Vertex	$L(u)$	$P(u)$
A	0	
B	5	A
C	2	A
D	∞	
E	∞	
F	∞	

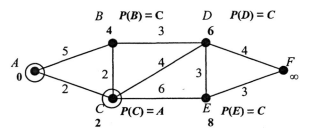

First Iteration

Vertex	$L(u)$	$P(u)$
A	0	
B	4	C
C	2	A
D	6	C
E	8	C
F	∞	

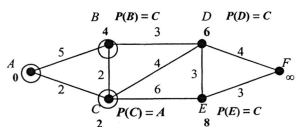

Second Iteration

Vertex	$L(u)$	$P(u)$
A	0	
B	4	C
C	2	A
D	6	C
E	8	C
F	∞	

10. Find the shortest path for the weighted graph in Exercise 1 using a table to record the changes in the labels of the vertices.

11. A company has branches in the cities of Alpha (α), Beta (β), Omega (ω), Gamma (γ), Kappa (κ), and Sigma (σ). The table below gives airline fares for direct flights between all pairs of these cities. (A hyphen indicates that there is no flight between the cities.) Create a weighted graph summarizing this information, and use one of the algorithms of this section to create a new table giving the cheapest fare between each pair of cities.

	α	β	ω	γ	κ	σ
α	-	120	250	-	-	50
β	120	-	-	240	90	-
ω	250	-	-	100	-	270
γ	-	240	100	-	130	150
κ	-	90	-	130	-	290
σ	50	-	270	150	290	-

12.

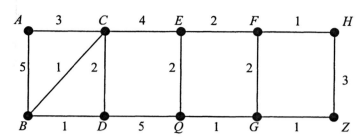

   (a) Describe in words how to modify the algorithms of this section to find the shortest path that goes from A to Z through Q.

   (b) Find the length of a shortest path from A to Z that passes through Q, and describe one such path.

13. When we have completed Dijkstra's Algorithm for finding the length of a shortest path from a to z in a weighted graph, each vertex that has been circled is labeled with the length of the shortest path from a to that vertex. Use this fact to find the shortest distance from a to each vertex in the following graphs.

   (a) The graph of Exercise 2.

   (b) The graph of Exercise 8.

4.5 Planar Graphs and Euler's Theorem

As the illustration to the left suggests, the general contractor for Emerick Construction Company has a problem. She is building three new homes that must be connected to three utilities, gas, water, and electricity. In Figure 1 we see a graph that indicates one possible way she can run her utility lines. However, she prefers to run her lines so that none of the utility lines cross. This will ensure that in subsequent excavation of one of the lines, the other lines are not disturbed. Attempt to draw the graph in Figure 1 so that none of the edges cross.

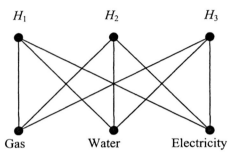

Figure 1

Did you notice that the graph in Figure 1 is the complete bipartite graph, $K_{3,3}$? Were you successful in drawing $K_{3,3}$ without edges crossing edges? In this section we will study graphs that can be represented by a drawing in the plane in which none of the edges cross.

Definition: A graph is **planar** if it can be drawn in the plane so that no pair of edges cross. (A pair of edges cross if the line segments or arcs representing them intersect at some point other than an endpoint of both edges.) Such a drawing is called a **planar representation** of the graph.

Example 1

Is the complete graph with four vertices, K_4, in Figure 2(a) planar?

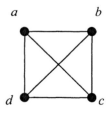

Figure 2(a)

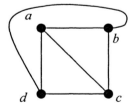

Figure 2(b)

Solution

In Figure 2(a), the edges $\{a, c\}$ and $\{b, d\}$ intersect in a point that is not an endpoint of the edges. However, K_4 can also be represented as shown in Figure 2(b). In this representation, no pair of edges intersect. Therefore, K_4 is planar.

Example 2

In Figure 3(a) we see a sketch of a cube with twelve edges and eight vertices. We can think of this sketch as being a graph. Is this graph a planar graph?

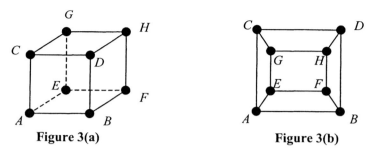

Figure 3(a)   Figure 3(b)

Solution

The graph in Figure 3(a) is a planar graph because it can be redrawn as seen in Figure 3(b) with no pair of edges crossing.

The notion of a cycle will be useful in our discussion in Sections 4.5 and 4.6 and in Chapter 5.

Definition: A **cycle** is a simple circuit in which only one vertex (the initial vertex) is repeated and it is repeated exactly once.

In Figure 4 (*a, b, c, a*) is a cycle, whereas (*a, c, b, e, d, b, a*) is a simple circuit that is not a cycle because vertex *b* is repeated.

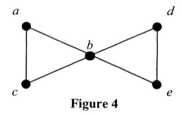

Figure 4

Observe that a planar representation of a graph separates the plane into regions (including one unbounded region). In Figure 5, the graph separates the plane into regions *A, B, C,* and *D,* with *D* being the unbounded region. If the graph is connected and has a cycle, each region can be associated with a cycle that forms its boundary. In Figure 5, the region *A* is associated with the cycle (*a, b, c, a*), and the unbounded region *D* is associated with the cycle (*a, b, d, a*). Observe also that in a planar representation of a graph, removing an edge from the cycle that bounds a region merges two regions into one. In Figure 5, removing edge {*c, d*} merges regions *B* and *C*.

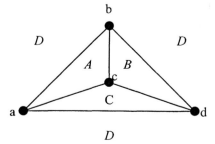

Figure 5

A very important theorem, named after the Swiss mathematician Leonhard Euler, gives a relationship among the numbers of edges, vertices, and regions of a planar graph.

Theorem 13: Euler's Theorem for Planar Graphs: In a planar representation of a connected, planar graph, the relationship among r, the number of regions, v, the number of vertices, and e, the number of edges, is $v - e + r = 2$.

Example 3
(a) Verify Euler's Theorem for the planar graph in Figure 5.
(b) Verify Euler's Theorem for the planar graph in Figure 3(b).

Solution
(a) Look again at Figure 5. There are four regions, four vertices, and six edges. Thus $r = 4$, $v = 4$, and $e = 6$ so
$$v - e + r = 2$$
$$4 - 6 + 4 = 2$$
(b) In Figure 3(b) there are six regions, eight vertices, and twelve edges.
$$8 - 12 + 6 = 2$$

We will discuss a proof of Euler's Theorem later, but first let us use it to show that the utility graph $K_{3,3}$ in Figure 6 is not planar.

Example 4
Use Euler's Theorem to show that the graph in Figure 6 is not planar.

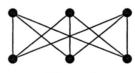

Figure 6

Solution
We will give an argument using contradiction. Suppose that there is a planar representation P for the graph in Figure 6. Since $K_{3,3}$ has six vertices and nine edges, P must have six vertices and nine edges. By Euler's Theorem, the number of regions created by the representation P can be found by solving for r in
$$v - e + r = 2$$
$$6 - 9 + r = 2$$
$$r = 5$$
Thus the planar representation P would divide the plane into five regions. The contradiction will arise by counting the edges used to bound these regions in two ways. Let N be the sum of the number of edges used on the boundary of each region of the planar graph P. [For example, for the graph in Figure 5, N would be 3 (edges bounding region A) + 3 (edges bounding region B) + 3 (edges bounding region C) + 3 (edges bounding region D) = 12.] Since there are no cycles in $K_{3,3}$ with fewer than four edges, each region of P must be bounded by at least four edges. Thus, if we let N be the sum of the numbers of edges used on the boundary of the regions, $N \geq 4 \times 5 = 20$. On the other hand, each edge is part of the boundary of at most two regions. Since there are nine edges, $N \leq 2 \times 9 = 18$. Since N cannot satisfy both of these inequalities, we have a contradiction, and the utilities graph $K_{3,3}$ must not be planar.

By a similar argument, we can show that the complete graph with five vertices, K_5, is not planar (see Exercise 12). Clearly, then, no graph that contains a copy of $K_{3,3}$ or K_5 is planar. Somewhat surprisingly, these two graphs, $K_{3,3}$ and K_5, completely characterize all non-planar graphs. In 1930 the Polish mathematician Kazimierz Kuratowski proved that a graph G is non-planar if and only if it contains K_5 or $K_{3,3}$ (or a graph closely related to these graphs) as a subgraph (see the discussion before Exercise 15).

Euler's Theorem is proved by induction on the number of edges in a planar graph. We will not prove Euler's Theorem, but we will look at two examples that reveal the heart of the proof.

Theorem 13: Euler's Theorem for Planar Graphs: In a planar representation of a connected, planar graph, the relationship among r, the number of regions, v, the number of vertices, and e, the number of edges, is

$$v - e + r = 2$$

Preparation for Theorem 13:
Examples: Careful thought about several examples will give important insight into the induction proof. In order to complete the Basis Step, we will need to think about what will happen when there is a single edge in the graph. There are two possible ways that a connected, planar graph can have a single edge. Sketch both of them, and verify Euler's formula in these cases.
In the Inductive Step, we will need to find a connection between graphs with k edges and graphs with $k + 1$ edges. We will find it useful to consider two cases. We will consider a graph with no cycles (called in Chapter 5 a **tree**) as the first case. The second case will involve a graph with a cycle.
Case 1, A graph with no cycles: Examine what happens to the formula $v - e + r = 2$ when we delete a special edge from a graph that has no cycles. The graph in Figure 7(a) has no cycles; observe that it does not divide the plane into multiple regions, so $r = 1$. The edge z is incident with vertex u. Further, u is not the endpoint of any other edge. When we delete edge z and vertex u (see Figure 7(b)), we reduce e by 1, and we reduce v by 1. Since we do not change the number of regions (r is still 1), $v - e + r$ is not changed by this deletion. Verify that $v - e + r = 2$ both before and after the deletion for the graphs in Figure 7.

Figure 7(a)   **Figure 7(b)**

Case 2, A graph with a cycle: In the graph of Figure 8(a), the edge z is one of the edges in the cycle (q, r, s, t, q). Observe that when edge z is deleted, the number of regions is also reduced by 1. (See Figure 8(b).) Since e is reduced by 1 and r is reduced by 1, the expression $v - e + r$ is unchanged by this deletion. Verify that $v - e + r = 2$ both before and after the deletion.

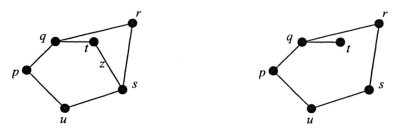

Figure 8(a)   **Figure 8(b)**

An interesting consequence of Euler's Theorem for Planar Graphs arises when we consider simple polyhedra as graphs. Remember that a **polyhedron** is a closed, connected surface in space that is formed by polygonal regions. A **simple polyhedron** has the additional property that there are no holes through the solid formed by the surface and its interior. The surfaces in Figures 9(a), (b), and (c) are simple polyhedra. The surface in Figure 9(d) is not simple because of the hole. Polyhedra can easily be regarded as graphs. For instance, in the polyhedron in Figure 9(a), the vertices are the points $\{A, B, C, D, E, F, G, H\}$, and the edges are line segments that bound the polygonal regions forming the surface. The polygonal regions forming the surface are called **faces**.

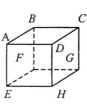

Figure 9(a)

Figure 9(b)

Figure 9(c)

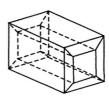

Figure 9(d)

Although the polyhedra can clearly be regarded as graphs, it is not at all clear that they are planar graphs, since their usual representation is as a graph in space. In fact, however, all simple polyhedra can be represented as graphs in the plane so that the faces of the polyhedra are in one-to-one correspondence with the regions in the planar representation. Hence we have Euler's Theorem for simple polyhedra:

Theorem 14: Euler's Theorem for Simple Polyhedra: In a simple polyhedron, the relationship among f, the number of faces, v, the number of vertices, and e, the number of edges, is

$$v - e + f = 2$$

To understand why a polyhedron like those in Figures 9(a), (b), and (c) is a planar graph, imagine that the polyhedron is made of very elastic material. Imagine cutting a hole in one of the faces of the polyhedron and stretching the hole very large. In your mental experiment, crush the polyhedron down through that hole onto a table top and trace the edges of the polyhedron onto the table top. The resulting graph is a planar representation for the polyhedron. Observe that the face of the polyhedron in which we cut the hole corresponds to the unbounded region of the planar graph.

Example 5
    (a) Verify Euler's Theorem for Simple Polyhedra for the polyhedron in Figure 10(a).
    (b) Represent this polyhedron as a planar graph.
    (c) Verify Euler's Theorem for Planar Graphs for this graph.

Figure 10(a)

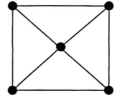

Figure 10(b)

Solution

    (a) The polyhedron in Figure 10(a) (called a quadrilateral pyramid) has five faces, eight edges, and five vertices. Hence $v - e + f = 2$ since $5 - 8 + 5 = 2$.

(continued)

(b) Imagine that the polyhedron in Figure 10(a) is made of elastic material and sits on a table. We cut a hole in the base (the bottom) of the polyhedron. Imagine stretching the hole and crushing the polyhedron down through the hole onto the table. If we then trace the edges of the polyhedron onto the table, we get a planar graph similar to the one found in Figure 10(b).

(c) The planar graph in Figure 10(b) has five regions, eight edges, and five vertices. Hence $v - e + r = 2$.

Exploratory Exercises Set 4.5

1. Work with a partner to consider these versions of the utility problem posed at the beginning of this section.
    (a) Can we connect two houses to two utilities without crossing utility lines? Explain.
    (b) Can we connect three houses to two utilities without crossing utility lines? Explain.
    (c) Can we connect four houses to two utilities without crossing utility lines? Explain.
    (d) Can we connect five houses to two utilities without crossing utility lines? Explain.
    (e) Can we connect n houses to two utilities without crossing utility lines? Explain.

2. Work with a partner to consider these versions of the utility problem posed at the beginning of this section.
    (a) Can we connect two houses to three utilities without crossing utility lines? Explain.
    (b) Can we connect three houses to three utilities without crossing utility lines? Explain.
    (c) Can we connect four houses to three utilities without crossing utility lines? Explain.
    (d) When $n \geq 3$, can we connect n houses to three utilities without crossing utility lines? Explain.

3. In each of the following problems, use Euler's Theorem to find the required information; then sketch a graph that satisfies the stated conditions. After completing the sketches, compare them with those of a partner.
    (a) A connected, planar graph has five regions and six vertices. How many edges does it have?
    (b) A connected, planar graph has seven edges and six vertices. Into how many regions does it divide the plane?
    (c) A connected, planar graph has three regions and five edges. How many vertices does it have?
    (d) A connected, planar graph has the same number of vertices as edges. Into how many regions does it divide the plane?

4. (*Writing Exercise*) In this writing exercise you will make progress toward completing the proof of Euler's Theorem for Planar Graphs.
    - Figure 11 shows a sketch of part of a graph G that has no cycles. Suppose that vertex u and edge z are removed to create graph G'. Explain carefully why the expression $v - e + r$ would be the same for both graphs.

Figure 11

- Figure 12 shows a sketch of part of a graph G that has a cycle. Suppose that edge z is removed to create graph G'. Explain carefully why the expression $v - e + r$ would be the same for both graphs.

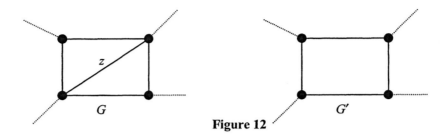

Figure 12

Exercise Set 4.5

---

1. Verify Euler's Theorem for these planar graphs by completing the table. Do not forget to count the unbounded region of the plane formed by the graph.

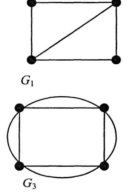

G_1

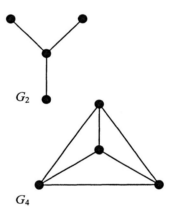

G_2

G_3

G_4

Graph	Vertices (v)	Edges (e)	Regions (r)	$v - e + r$
G_1				
G_2				
G_3				
G_4				

In Exercises 2 through 5 the graphs are planar graphs. Draw the graphs without edge crossings.

2.

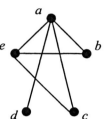

3.

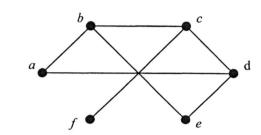

4.

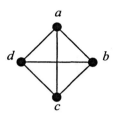

5.

6. Verify Euler's Theorem for Planar Graphs for the graph in
   (a) Exercise 2
   (b) Exercise 3
   (c) Exercise 4
   (d) Exercise 5

7. Suppose that a connected, planar graph has eight vertices, each of degree 3. Into how many regions is the plane divided by a planar representation of this graph?

8. Suppose that a connected, planar graph has six vertices, each of degree 4. Into how many regions is the plane divided by a planar representation of this graph?

9. Suppose that a connected, planar graph has 30 edges. If a planar representation of this graph divides the plane into 20 regions, how many vertices does this graph have?

10. In this graph, identify
    (a) A cycle of length 3
    (b) A cycle of length 5
    (c) A simple circuit that is not a cycle

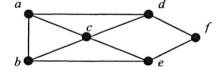

11. This connected, planar graph divides the plane into three regions, A, B, and C.
    (a) What is the cycle that forms the boundary of A?
    (b) How many edges are in the cycle that forms the boundary of A?
    (c) What is the cycle that is associated with the unbounded region C?
    (d) How many edges are in the cycle associated with C?

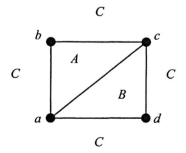

12. (*Proof Exercise*) Complete the following argument that shows that the complete graph on five vertices, K_5, is not planar:

    Suppose there is a geometric representation P of K_5 that is planar.
    (a) Find the number of edges of K_5 (and hence P).
    (b) Find the number of vertices of K_5 (and hence P).
    (c) Use Euler's Theorem to determine the number of regions r into which P divides the plane.
    (d) Let N be the sum of the number of edges used on the boundary of each region of P. Since each region is bounded by at least three edges, then by using our result from (c),
    $$N \geq ?$$
    (e) Since each edge bounds at most two regions, we can also use our result from (c) to determine that
    $$N \leq ?$$
    (f) Explain carefully the contradiction that ensures that there is no planar representation for K_5.

13. Verify Euler's Theorem for these simple polyhedra by completing the following table.

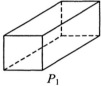 

Polyhedron	Edges (e)	Vertices (v)	Faces (f)	$v - e + f$
P_1				
P_2				
P_3				
P_4				

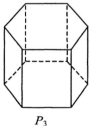

P_1

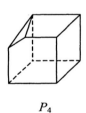

P_2

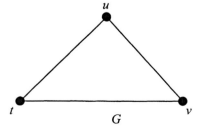

P_3

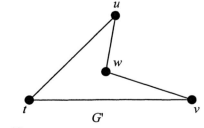

P_4

14. (*Proof Exercise*) Complete the Basis Step of an induction proof of Euler's Theorem. That is, suppose a graph has one edge. Show that $v - e + r = 2$.
    (a) There are two graphs that have exactly one edge. Draw both of them.
    (b) Show that $v - e + r = 2$ for both graphs.

*A new graph G' can be obtained from a graph G by removing an edge {u, v} and adding a new vertex w with edges {u, w} and {w, v}. (See Figure 13.) This operation is called an **elementary subdivision**.*

u

t G v

u

w

t G' v

Figure 13

With the notion of an elementary subdivision, we can state clearly the theorem of Kuratowski that characterizes planar and non-planar graphs:

Kuratowski's Theorem: A graph is non-planar if and only if it contains as a subgraph a graph isomorphic to $K_{3,3}$, K_5, or a graph obtained by a sequence of elementary subdivisions from $K_{3,3}$ or K_5.

In Exercises 15 through 16, explain how the graph is obtained by a sequence of elementary subdivisions from $K_{3,3}$.

15.

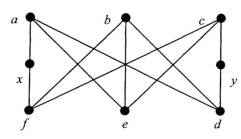

16.

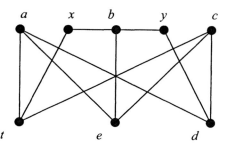

17. Show that this graph is obtained from K_5 by a sequence of elementary subdivisions.

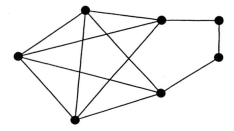

18. Show that this graph contains a subgraph isomorphic to $K_{3,3}$.

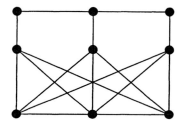

19. Explain why each of the graphs in Exercises 15 through 18 is not a planar graph.

20. Explain why K_n is not a planar graph when $n \geq 5$.

4.6 Coloring Graphs

The Office Manager has a problem. He must schedule regular weekly meetings for several problem-solving teams of professionals in his office, but his job is complicated by the fact that a number of the professionals are on more than one team. The Office Manager needs to use as few meeting times as possible to accommodate all the teams, while not scheduling any two teams that share members at the same time.

Because the Office Manager's problem is a discrete problem involving relationships, we may well hope to find a way to use graphs to solve it. In building a graph to model a particular applied situation, we must always determine what the vertices will represent and what the edges will represent. It often happens that we use edges to describe relationships between the objects we use as vertices. In this instance, we will use the teams in the office as the vertices, and we will use edges between the vertices to indicate that the teams share a member.

Example 1

In the office there are seven professionals, Bob, Lou, Sal, Ted, Sue, Jane, and Dick, who have been placed on five teams:
Team 1: Bob, Lou
Team 2: Bob, Sue, Jane
Team 3: Lou, Sal, Ted, Sue
Team 4: Jane, Dick
Team 5: Sue, Jane, Dick
Remember that the Office Manager wishes to schedule meeting times so that teams with shared members do not meet at the same time; remember also that he wishes to use as few meeting times as possible. Use a graph to model the problem that the Office Manager faces. If possible, use the model to find a solution.

Solution

In Figure 1(a) we see a graph with a vertex representing each team and an edge incident with pairs of vertices that represent teams that share a team member.

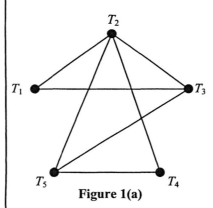

Figure 1(a)

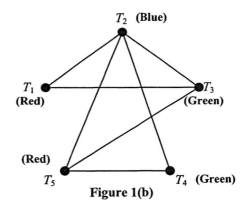

Figure 1(b)

We wish to assign a meeting time to each of the five teams so that there are no conflicts; that is, no person has two meetings at the same time. For simplicity's sake, let us create labels for the meeting times. For example,

(continued)

Red:	9:00 – 10:00
Blue:	10:00 – 11:00
Green:	11:00 – 12:00
Yellow:	1:00 – 2:00

We will have successfully solved the problem if we can economically label the vertices of the graph in Figure 1(a) with colors (meeting times) so that no two adjacent vertices have the same label. In Figure 1(b) we find that we can label the graph with three labels: Red, Blue, and Green. Further, we are not able to label the graph with only two labels. Because Bob is on both Team 1 and Team 2 (vertices T_1 and T_2 are adjacent), they must meet at different times (have different labels). Further, since Team 3 shares Lou with Team 1 and Sue with Team 2, the vertex corresponding to Team 3 must differ in color from both Team 1 and Team 2. The solution in Figure 1(b) uses the smallest possible number of meeting times.

A number of different problems can be solved by building a graph that models the problem and then labeling the vertices of the graph in such a way that no two adjacent vertices have the same label. Without loss of generality, we can think of the labels as being colors. For example, in the discussion above we named the possible time periods with colors. Hence, in this section, we will examine this problem:

Given a simple graph, what is the smallest number of colors that can be used to assign a single color to each vertex so that no two adjacent vertices have the same color?

Definitions:
- A **coloring** of a simple graph is an assignment of a color to each vertex in such a way that no two adjacent vertices have the same color.
- The **chromatic number** of a simple graph is the least number of colors necessary for a coloring of the graph.

Example 2
Find the chromatic number of the graphs G and H in Figure 2.

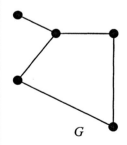

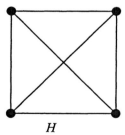

G

H

Figure 2(a)

Figure 2(b)

(continued)
Solution

In Figure 3(a) we find a coloring of graph G from Figure 2(a) that uses three colors. However, it is not necessary that vertices v_3 and v_4 have different colors, so Figure 3(b) shows a coloring of graph G that uses just two colors. Since vertices v_1 and v_2 must have different colors, the chromatic number of graph G is 2.

Graph H is the complete graph K_4. In Figure 3(c) we find a coloring of graph H that uses four colors. Each vertex of H is adjacent to each of the other three vertices of H. Hence if any two of the vertices had the same color, we would not have a coloring of the graph. Thus the chromatic number of H is 4.

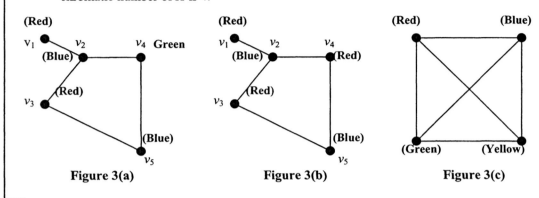

Figure 3(a)   Figure 3(b)   Figure 3(c)

The following theorems can be easily deduced about graph colorings.

Theorem 15: The chromatic number of the complete graph K_n is n.

Preparation for Theorem 15:
Review the appropriate definitions and answer these questions.
(1) How many vertices does K_n have?
(2) If two vertices are adjacent, what do we know about the colors that can be assigned to them.

Proof of Theorem 15:
We can color the graph with n colors by using a different color for each of the n vertices. However, any two vertices in a complete graph are adjacent; hence no pair of vertices may use the same color. Therefore, the chromatic number is n.■

Theorem 16: A simple connected graph has chromatic number 2 if and only if the graph is bipartite.

Preparation for Theorem 16:
Definition: A bipartite graph is a graph whose vertices can be partitioned into non-empty sets V_1 and V_2 so that each edge is incident with a vertex in V_1 and a vertex in V_2. In other words, there are no edges incident with pairs of vertices in V_1 and no edges incident with pairs of vertices in V_2.
Reflections: Remember that to prove an "if and only if" statement, we must prove two statements:
    If the chromatic number of a simple, connected graph is 2, then the graph is bipartite.
    If a simple, connected graph is bipartite, then the graph has chromatic number 2.
How shall we connect the fact that the graph can be colored with two colors with the need to partition the collection of vertices into two subsets. Think a bit on this issue; then go to Exercises 18 and 19, where we will prove Theorem 16.

Theorem 17: Suppose graph G consists of a cycle of length n. If n is even, then the chromatic number of the graph is 2. If n is odd, then the chromatic number is 3.
Preparation for Theorem 17: In Exploratory Exercises 3 and 4, we will work through some examples related to this theorem. The reasoning that underlies the proof of Theorem 17 will be more apparent.

It follows, then, that if a graph contains a cycle of odd length, the chromatic number of the graph is ≥ 3.

APPLICATIONS OF GRAPH COLORING

Much of the initial work on graph coloring was done as a consequence of investigations related to coloring maps such as those found in geographical references. In coloring countries on a map, it is customary to use different colors for countries that share a common border. This convention makes the maps much easier to read. The question arises: "What is the smallest number of colors needed to color a map?" In Figure 4(a) we see a map that can be colored with three colors, and in Figure 4(b) we see a map that can be colored with four colors but not with three.

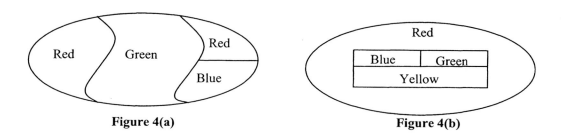

Figure 4(a)   Figure 4(b)

The statement of the map-coloring problem sounds suspiciously like the statement of the graph-coloring problem. Indeed, by regarding each country as a vertex and connecting two vertices if their countries share a common boundary, we can produce a graph called the **dual graph** of the map. [*Note*: Two countries that touch only at a single point are not regarded as sharing a common boundary.] If we are able to find a coloring of the dual graph of a map, it will provide a coloring of the map.

Example 3

Find the dual graph of the map in Figure 5(a). Color the dual graph with the minimum number of colors possible. Then display the coloring of the map that corresponds to the coloring of the graph.

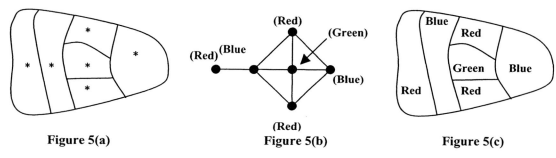

Figure 5(a)   Figure 5(b)   Figure 5(c)

Solution

In Figure 5(b) we have the dual graph of the map in Figure 5(a). Notice that a vertex has replaced each country. In Figure 5(b) we also find a coloring of the graph using three colors. Since there is a cycle of odd length in this graph, Theorem 17 ensures that this is the smallest number of colors that can be used. In Figure 5(c), we see the associated coloring of the map.

Questions about the minimum number of colors needed to color any possible map began to be asked in the middle of the nineteenth century. An early conjecture was that it is possible to color any possible map drawn on the plane using at most four colors. During the next century there were many unsuccessful attempts to prove this result, including one proof that was accepted as correct for 11 years before a flaw was found. In 1976 two American mathematicians, Kenneth Appel and Wolfgang Haken, proved the following result.

Theorem 18: (The Four-Color Theorem) The chromatic number of a simple planar graph is no greater than 4.

Since every map in the plane can be colored using the coloring of its dual planar graph, this verifies the conjecture that every map can be colored using at most four colors.

Appel and Haken's proof of the Four-Color Theorem was unusual in the way that they used computers to complete the proof. They first showed that the proof could be completed by analyzing a large but finite number of cases and then used the computer to examine those cases. To complete their work, they used over 1000 hours of computing time by what was a "high-speed" computer in 1976.

Consider the following problems that can be modeled using graphs and solved by finding colorings of the graphs.

Example 4

Communication agencies in North America assign television channels 2 through 13 to stations in such a way that stations within 150 miles of one another are not assigned the same channel. Describe a graph that models this situation, and indicate how a graph coloring might be useful in the process of assigning channels.

Solution

Construct a graph with one vertex for each station. Connect two vertices with an edge provided that their stations are located within 150 miles of one another. Find a coloring of the graph. If we then assign the same channel to all stations whose vertices have the same color, the assignment will satisfy the 150-mile rule. The chromatic number of the graph is the minimum number of channels that will accommodate all of the stations.

Example 5

Canisters of chemical products are to be shipped on rail cars to a disposal site. Government regulations prohibit shipping chemicals that might react violently with each other in the same rail car. How should the chemicals be distributed among the rail cars? What is the smallest number?

Solution

Construct a graph with one vertex for each canister. Connect two vertices with an edge provided that the canisters cannot travel in the same rail car because of possible violent reaction. Find a coloring of the graph. Canisters whose vertices have the same color can travel in the same rail car. The chromatic number of the graph is the smallest number of rail cars that can be used to transport the chemicals.

Exploratory Exercise Set 4.6

1. Work in pairs. Discuss how to use a graph to model the situations described in (a) and (b), create the graph model, and use it to solve the problem.
   (a) A pet store owner prepares to display seven species of fish $\{A, B, C, D, E, F, G\}$ in several tanks. For a variety of reasons (including the fact that some of the species dine on others), A cannot live with B, C, or E; B cannot live with A, C, or G; C cannot live with A, B, D, E, or F; and D cannot

live with *C*, *F*, or *G*. What is the smallest number of aquariums that the pet store owner can use to display his fish?

(b) At the end of the school day at T.J. Johnson Elementary School, there are a variety of jobs that teachers must complete: check the halls (*H*), sell lunch tickets (*L*), patrol the bus loading area (*B*), monitor the playground (*M*), check the restrooms (*R*), call parents of detained children (*C*), and lower the flag (*F*). On each day of the week, the principal assigns a small group of teachers to stay after school and complete these tasks. However, for reasons of timing and location, some tasks cannot be done by the same teacher. In the following table, an *X* in a specific row and column indicates that the task of the row and the task of the column cannot be completed by the same teacher. What is the minimum number of teachers required to complete the tasks?

Tasks	*H*	*L*	*B*	*M*	*R*	*C*	*F*
H		*X*	*X*	*X*		*X*	
L	*X*		*X*	*X*			
B	*X*	*X*		*X*	*X*		
M	*X*	*X*	*X*		*X*		
R			*X*	*X*		*X*	
C	*X*				*X*		*X*
F						*X*	

2. Each member of a group of four students should draw a map with at least six countries, and each member should then copy each of the other three maps in the group.
    (a) Group Member 1 should color each of the four maps using as few colors as possible.
    (b) Group Member 2 should draw the dual graph for each of the four maps.
    (c) Group Member 3 should color each of the four dual graphs with as few colors as possible.
    (d) Group Member 4 should translate the coloring of the four dual graphs to create a coloring of the original maps.
    (e) Compare and discuss the results. Determine the chromatic number of each graph/map.

3. In general it is quite hard to find the chromatic number of a graph and its associated coloring. However, in certain cases, it is quite easy. Consider graphs that consist of a cycle.
    (a) We first consider graphs that consist of a cycle with an even number of vertices.
        (1) Find the chromatic number of this cycle with four vertices.

        (2) Find the chromatic number of a cycle with six vertices.
        (3) Find the chromatic number of a cycle with eight vertices.
        (4) What can be said in general about the chromatic number of a cycle with an even number of vertices?
    (b) We now consider graphs that consist of a cycle with an odd number of vertices.
        (1) Find the chromatic number of a cycle with three vertices.

        (2) Find the chromatic number of a cycle with five vertices.
        (3) Find the chromatic number of a cycle with seven vertices.
        (4) What can be said in general about the chromatic number of a cycle with an odd number of vertices?
        (5) What does the result in (4) say about the chromatic number of a simple graph that contains a cycle with an odd number of vertices?

4. In completing Exploratory Exercise 3(a), you may have noticed a pattern that can be used to color any connected graph that does not have cycles with an odd number of edges. The following algorithm describes how to color such a graph with the colors Red and Blue. (Hence the chromatic number of any graph that does not have cycles with an odd number of edges is 2.) At each step in this algorithm we will tabulate a set C consisting of all vertices that have already been colored. Initially set C is empty.

    Step 1: Pick a vertex V and color it Red. Place V in C.

    Step 2: Identify all vertices not in C that are adjacent to vertices in C. Color each of them Blue and place them all in C.

    Step 3: Identify all vertices not in C that are adjacent to vertices in C. Color each of them Red and place them all in C.

    Step 4: Repeat Steps 2-4 until C contains all the vertices of the graph.

(a) This graph is a simple graph with no cycles of odd length. We have done the first three steps of the algorithm. Use the algorithm to complete the coloring.

Initially, $C = \{\ \}$.
Step 1:
    $C = \{a\}$

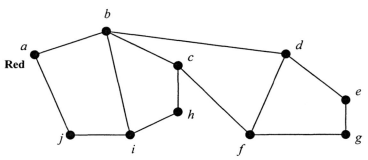

Step 2: $C = \{a, b, j\}$

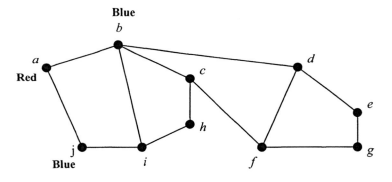

Step 3: $C = \{a, b, j, c, i, d\}$

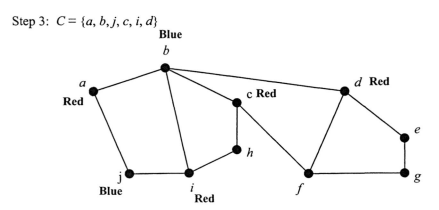

(b) Use the algorithm to color this graph with two colors.

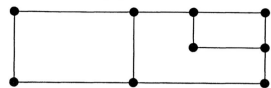

(c) Suppose vertices u and w in Figures 6(a) and (b) are to be colored Red in Step 3 of the algorithm.
   (1) How do you know that u is not adjacent to vertex v that was colored Red in Step 1. (That is, how do you know that the situation pictured in Figure 6(a) cannot occur?)
   (2) How do you know that u and w are not adjacent to each other? (That is, how do you know that the situation pictured in Figure 6(b) cannot occur?)

   Note: These two observations provide the basis for the proof that this algorithm always works.

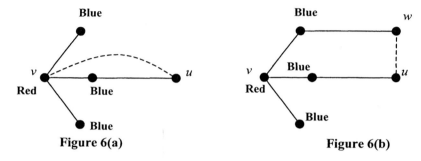

Figure 6(a)   **Figure 6(b)**

5. (*Proof Exercise*) Although the chromatic number of a graph can be quite hard to find, the following theorem provides an upper bound on the size of the chromatic number.

   Theorem: If D is the maximum of the degrees of the vertices of a simple graph, then the graph can be colored with $D + 1$ colors. (That is, the chromatic number $\leq D + 1$.)

(a) For each of these graphs:
   (1) Find D.
   (2) Find the chromatic number of the graph.
   (3) Compare the chromatic number to $D + 1$.

G   G'

(b) Provide reasons for the following steps in the proof of this theorem. We will use a proof by induction on the number of vertices of the graph.
   (1) *Basis Step:* If a simple graph has only one vertex, the degree of that one vertex is 0. Draw a picture of this case, and explain why the theorem is true in this case.
   (2) *Preparation:* There are two simple graphs that have two vertices. Draw both of them, and explain why the theorem must be true of graphs with two vertices.
   (3) *Preparation:* There are four simple graphs that have three vertices. Draw each of them, and explain why the theorem must be true of graphs with three vertices.

(4) *Inductive Step:* Now, let us assume that the theorem is true for all simple graphs with k vertices and suppose G is a simple graph with $k + 1$ vertices. Let V be a vertex from G. Delete V, together with all edges adjacent to V, to create a graph G'.

- How many vertices does G' have?
- If D is the maximum of the degrees of the vertices of G, what do we know about the maximum of the degrees of the vertices of G'?
- Explain how we know that G' can be colored with $D + 1$ colors.
- Notice that the coloring of G' provides a coloring for all vertices of G except V. What is the largest number of vertices from G that can be adjacent to V?
- Explain why we know we can use one of the $D + 1$ colors to color V.

6. (*Writing Exercise*) Review the examples and exercises of this section, and then write two problems of your own creation that can be solved using graph-coloring methods. After writing the problems, write solutions to the problems, explaining each step carefully.

Exercise Set 4.6

In Exercises 1 through 6 use the labels red (R), blue (B), green (G), and yellow (Y) to find a coloring of the graph. Use as few colors as possible.

1.

2.
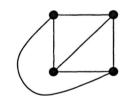

3.
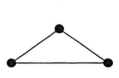

4.
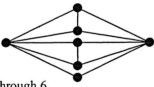

5.
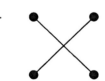

6.

7. Find the chromatic number of each graph in Exercises 1 through 6.

8. (a) Sketch a graph with a chromatic number of 1.
   (b) Describe all simple graphs with a chromatic number of 1.

In each of Exercises 9 through 12, you will find a map.
   (a) Color the map using the smallest number of colors possible.
   (b) Find the dual graph for the map.
   (c) Find the coloring of the planar graph that corresponds to the coloring of the map.

9.

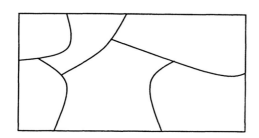

10.

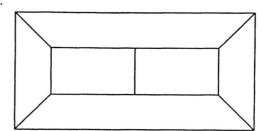

11.

12.

Create a graph model for each of Exercises 13 through 16 and solve the problem using the model.

13. A naturalist at the zoo prepares displays in which several animals occupy the same habitat. However, because some of the animals eat others or in some other way disrupt their lives, certain animals cannot occupy the same display. Currently, the naturalist wishes to display rabbits, squirrels, horned owls, blue jays, bobcats, foxes, and cardinals. The rabbits and squirrels cannot reside with the owls, bobcats, or foxes. In addition, the horned owls cannot live with the blue jays or cardinals. The blue jays cannot live with the cardinals. What is the smallest number of displays that the naturalist can use to house these animals?

14. In preparation for an accreditation visit, six committees are formed from the faculty of Washington Middle School. How many different meeting times are necessary to ensure that no person has two meetings at the same time? The committees are C_1 = {Nelson, Marino, Garcia}, C_2 = {Marino, Jones, Wheeler}, C_3 = {Nelson, Wheeler, Garcia}, C_4 = {Jones, Wheeler, Garcia}, C_5 = {Nelson, Marino}, C_6 = {Marino, Wheeler, Garcia}.

15. Canisters of highly active chemicals are to be shipped by train. Environmental regulations stipulate that two chemicals cannot be shipped in the same freight car if the two chemicals react explosively. In the table, an X in a specific row and column indicates that the chemical of the row reacts explosively with the chemical of the column. If we have one canister of each chemical, how many different freight cars must be used to ship the chemicals?

Chemicals	A	B	C	D	E	F	G
A		X	X		X		X
B	X		X	X	X	X	
C	X	X					
D		X			X		
E	X	X		X			
F		X					X
G	X					X	

16. How many different channels are needed for seven radio stations if stations are not permitted to use the same channel when they are within 150 miles of each other. The table gives the distances between pairs of stations.

Stations	1	2	3	4	5	6	7
1	-	10	44	178	65	210	120
2	10	-	58	156	79	234	98
3	44	58	-	227	112	193	181
4	178	156	227	-	201	341	31
5	65	79	112	201	-	199	148
6	210	234	193	341	199	-	297
7	120	98	181	31	148	297	-

17. (a) What is the chromatic number of the complete graph on three vertices, K_3?
    (b) What is the chromatic number of K_4?
    (c) What is the chromatic number of K_5?
    (d) What is the chromatic number of K_6?
    (e) Why are the results of (c) and (d) not a violation of the Four-Color Theorem.

18. (*Proof Exercise*) Give reasons for the proof of this result:
    If the chromatic number of a simple, connected graph is two, the graph is bipartite.
    (a) Since the chromatic number of the graph is 2, the smallest number of colors that can be used to color the graph is two. Suppose we have a coloring of the graph using the two colors Red and Blue. Let V_1 be the vertices that are colored Red and let V_2 be the vertices colored Blue. Why are both V_1 and V_2 non-empty?
    (b) Why do we know that there are no edges incident with pairs of vertices in V_1?
    (c) Why do we know that there are no edges incident with pairs of vertices in V_2?
    (d) How do we know the graph is bipartite?

19. (*Proof Exercise*) Give reasons for the proof of this result:
    If a simple, connected graph is bipartite, then the graph has chromatic number 2.
    (a) Since the graph is bipartite, its vertices can be partitioned into two special disjoint, non-empty sets, V_1 and V_2. We can assign colors to the vertices by assigning the vertices of V_1 the color Red and the vertices of V_2 the color Blue. How do we know that this is a coloring? (That is, how do we know that with this assignment, no pair of adjacent vertices are colored the same color?)
    (b) How do we know that we cannot color all the vertices of the graph with a single color?
    (c) How do we know that the chromatic number of the graph is 2?

Trees and Directed Graphs

Chapter

5

5.1 Directed Graphs

The Highway Inspector has been assigned to a new district, and she puzzles over the map of her new assignment as she sloshes down coffee at the Pancake Parlor. The map that puzzles her appears in Figure 1. She is perplexed because the road system she is to inspect consists of a series of one-way streets, all of whose directions are indicated by the arrows in Figure 1. Can she find an economical way to inspect these roads? Can she travel over each section of road without traveling over any one section more than once?

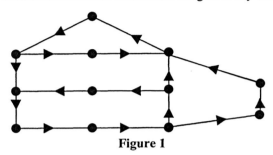

Figure 1

The graph shown in Figure 1 differs from graphs found earlier in this chapter in that each edge has been given a direction. Figure 1 is a geometric representation of a directed graph.

Definition: A **directed graph** $G = (V, E)$ consists of V, a non-empty, finite set whose elements are called vertices, and E, a set of *ordered* pairs of elements of V whose elements are called edges. A directed graph is also called a **digraph**.

If u and v are vertices of G, and (u, v) is an edge of G, then we say u is the **initial vertex** of the edge of (u, v) and that v is the **terminal vertex** of (u, v). We also say that the edge is from u to v. The edge (u, u) describes a loop whose initial and terminal vertices are the same. Observe that in our definition of a directed graph, we do not allow multiple edges from vertex u to vertex v.

Example 1

Describe the directed graph represented in Figure 2 in terms of its vertex set V and its set of edges E.

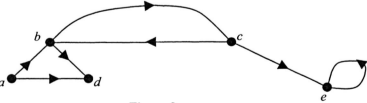

Figure 2

Solution

The vertex set of the graph in Figure 2 is $V = \{a, b, c, d, e\}$. The set of edges $E = \{(a, b), (a, d), (b, c), (b, d), (c, b), (c, e), (e, e)\}$.

In order to answer the question of the Highway Inspector, we will need to refine the idea of the degree of a vertex and the idea of a connected graph.

Definitions:

- In a directed graph, the **indegree of a vertex** v is the number of edges for which v is the terminal vertex.
- The **outdegree of a vertex** v is the number of edges for which v is the initial vertex.
- Let n be a positive integer. A **path from a to b** in a directed graph of length n is an alternating sequence of vertices and edges $a, (a, x_1), x_1, (x_1, x_2), x_2, (x_2, x_3), \ldots, (x_{n-2}, x_{n-1}), x_{n-1}, (x_{n-1}, b)$. Notice that in all but the first edge, the initial vertex of the edge is the same as the terminal vertex of the edge that precedes it. Notice also that we lose no information if we omit the edges and describe the path by listing the vertices in order: $(a, x_1, x_2, x_3, \ldots, x_{n-1}, b)$.

With little difficulty, the definition of a path from a to b in a directed graph can be modified to yield the definitions of a circuit, a simple path or circuit, an Euler path or circuit, and a Hamilton path or circuit in a directed graph.

Definitions:

- A **simple path in a directed graph** is a path in which no directed edge is repeated.
- An **Euler path in a directed graph** is a simple path that includes each directed edge of the graph.
- A directed graph is **strongly connected** provided that for each pair of vertices a and b, there is a path from a to b.
- A directed graph is **weakly connected** if, when we ignore the directions of the edges, the underlying graph is connected.

Example 2

Consider the directed graph in Figure 3.
(a) Compute the indegree and the outdegree of vertex v.
(b) Find a path from u to w.
(c) Explain why this graph is not strongly connected.
(d) Explain why this graph is weakly connected.

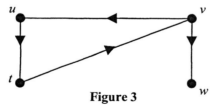

Figure 3

Solution

(a) The indegree of v is 1 and the outdegree is 2.
(b) The path $(u, t), (t, v), (v, w)$
(c) There is no path from w to v.
(d) If we ignore directions on the edges, there is a path between each pair of vertices. For instance, if we ignore direction, the edge (v, w) provides a path from w to v.

With these issues resolved, we can state the following theorems, which are very closely related to Theorems 10 and 11 of Section 4.2.

Theorem 1: If the indegree equals the outdegree for each of the vertices of a weakly connected directed graph, then there is an Euler circuit for the directed graph.

Theorem 2: Let G be a weakly connected directed graph that has vertices for which the indegree is not equal to the outdegree.

(a) If there are precisely two vertices, v and w, for which the indegree is not equal to the outdegree, and the indegree of v exceeds the outdegree of v by one while the outdegree of w exceeds the indegree of w by one, then G has an Euler path from w to v.

(b) Otherwise, there is no Euler path for the directed graph.

Looking back at Figure 1, we see that the road system studied by the Highway Inspector is weakly connected. However, we see that the Highway Inspector does not have an economical path through this road system since there are five vertices for which the indegree does not equal the outdegree.

Example 3

Determine whether the directed graphs in Figure 4 have Euler paths or circuits, and if so, find one.

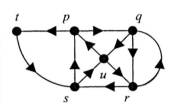 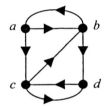 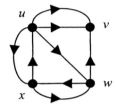

Figure 4(a)   Figure 4(b)   Figure 4(c)

Solution

Since the indegree and outdegree for each vertex of the directed graphs in Figure 4(a) are equal, there is an Euler circuit for this graph. One circuit is $((p, q), (q, r), (r, s), (s, p), (p, t), (t, s),$ $(s, u), (u, r), (r, q) (q, u), (u, p))$.

In Figure 4(b), the indegree and outdegree of vertices b and c are equal. We can find an Euler path from a to d since the outdegree of vertex a is one more than the indegree while the indegree of vertex d is one more than the outdegree. An Euler path is $((a, b), (b, d), (d, c), (c, b),$ $(b, a), (a, c), (c, d))$.

In Figure 4(c), the outdegree of vertex u is 4 while the indegree of u is 1. There is no Euler path for this directed graph.

TOURNAMENTS

During the recent hotly contested season of the Chatham County Lacrosse League, each of the five teams played the other four teams, resulting in ten games. The results are found below.

Ants defeated Bats.	Bats defeated Sloths.	Sloths lost to Robins.	Robins lost to Gnats.
Ants defeated Sloths.	Bats lost to Robins.	Sloths beat Gnats.	
Ants lost to Robins.	Bats lost to Gnats.		
Ants lost to Gnats.			

In Figure 5 we have represented this information in a directed graph. We represented the teams as vertices and directed an edge from vertex P to vertex Q if Team P defeated Team Q. In the graph the Ants are represented by vertex A, the Bats by vertex B, the Sloths by S, the Robins by R, and the Gnats by G.

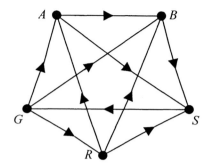

Figure 5

Because no team won all its games, there is a controversy about which team is best. Mel, sportswriter for the *Castleberry Times*, writes that the Ants were clearly the best team since the Ants beat the Bats who beat the Sloths who beat the Gnats who beat the Robins. Said another way, (A, B, S, G, R) is a directed path from A to R that includes all the vertices in the graph (a Hamilton path). Melba, writer for the *Chatham Herald*, argues that the Bats were clearly the best team since the Bats defeated the Sloths who beat the Gnats who beat the Robins who beat the Ants. [(B, S, G, R, A) is a Hamilton path through the graph, and its initial vertex is B.] Cheryl Tukesberry, who played for the Robins, points out that the Robins beat the Ants who beat the Bats who beat the Sloths who beat the Gnats. Perhaps if we turn to graph theory, we can understand this confusing situation.

When each of the n competitors in a league play every other, the competition is called a **round-robin competition**. Such a competition can be represented by a directed version of the complete graph K_n, where the vertices are the competitors and we direct the edge from P to Q if P defeats Q. A directed graph formed by giving a direction to each edge of the complete graph K_n is called a **tournament graph**. In a tournament graph, we define vertex P to be a **winner** if there is a Hamilton path for the graph that begins with P. The following question then arises.

> In which round-robin competitions can some competitor claim to be a winner on the basis of the results of the competition?

The following theorem answers the question by asserting that in every tournament graph, there is at least one winner.

Theorem 3: In a tournament graph with two or more vertices, there is always a Hamilton path.

> *Preparation for the proof of Theorem 3:*
> *Examples:* We will induct on the number of vertices in the tournament graph. If there are two vertices, we have the directed graph in Figure 6, and the path consisting of A, the edge from A to B, and B is a Hamilton path.

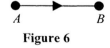

Figure 6

Reflection: Let $k \geq 2$ and assume that every tournament graph with k vertices has a Hamilton path. We will use this statement to prove that *a tournament graph with $k + 1$ vertices has a Hamilton path*. The crucial question then may be "How do you use a Hamilton path in a tournament graph with k vertices to find a Hamilton path in a tournament graph with one more vertex? Figure 7(a) represents a tournament graph G with four vertices. If we delete vertex x, together with all edges incident with x, we get the tournament graph G' in Figure 7(b) with only three vertices. The path (a, b, c) is a Hamilton path for G'. Now, can we use this Hamilton path for G' to find a Hamilton path for G? Since (b, x) and (x, c) are directed edges in G, we get the desired Hamilton path for G by replacing (b, c) by these edges in the path for G'. (a, b, x, c) is a Hamilton path for G.

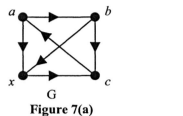

Figure 7(a)

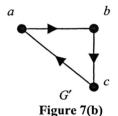

Figure 7(b)

Had the vertex x been the initial vertex for all the edges for which x is an endpoint, we would have needed to proceed differently. In this case we would place an edge involving x at the beginning of the path for G'. Similar considerations would need to be given to the case in which vertex x is the terminal vertex for all the edges for which x is an endpoint. Our argument in the Inductive Step will deal with these three cases.

Proof of Theorem 3:

Basis Step: If a tournament graph G has two vertices A and B, then the theorem is true. The required Hamilton path is the path consisting of the two vertices and the connecting edge.

Inductive Step: Now, suppose we can find a Hamilton path for all tournament graphs with k vertices for some $k \geq 2$. We need to show that this ensures that we can find a Hamilton path for an arbitrary tournament graph G with $k + 1$ vertices. Choose one of the vertices of G, which we will call x. Remove x and all edges incident with x from the graph to create a new tournament graph G' with k vertices. Let $(v_1, v_2, v_3, \ldots, v_{k-1}, v_k)$ be the Hamilton path in G'. Remember that since the original graph G is a tournament graph, there is an edge from v_i to x or from x to v_i for each i in the set $\{1, \ldots, k\}$. We examine three cases:

Case 1: If x is the initial vertex for all edges incident with it in G, then (x, v_1) is in G, so $(x, v_1, v_2, v_3, \ldots, v_{k-1}, v_k)$ is a Hamilton path in G.

Case 2: If x is the terminal vertex for all edges incident with it, then $(v_1, v_2, v_3, \ldots, v_{k-1}, v_k, x)$ is a Hamilton path in G.

Case 3: If vertex x is not described in either Case 1 or Case 2, let r be the largest integer from the set $\{1, 2, 3, \ldots k\}$, so that (v_r, x) is an edge in G. Since $r < k$, then $(v_1, v_2, \ldots, v_r, x, v_{r+1}, \ldots, v_k)$ is a Hamilton path in G. ∎

We have established that in every round-robin competition there is at least one team that can claim to be a winner, namely the initial vertex of the Hamilton path that is guaranteed to exist. However, in our discussion of the Chatham County LaCrosse League, we saw that there were several teams that could make that claim. [Indeed, we will see in Exercise 12 that each of the five teams in that league could claim to be a winner.] When do the results of a round-robin competition guarantee a unique ranking of all the participating teams? That is, when are we guaranteed that the tournament graph has a unique Hamilton path? One answer to this question is given by Theorem 4. We will discuss this theorem further in Exercises 17 through 19.

Theorem 4: A tournament graph has a unique Hamilton path if and only if the graph has no simple circuits.

Example 4

In the tournament graph in Figure 8:

(a) Find a winner and a ranking of the competitors by finding a Hamilton path for the tournament graph.

(b) Determine whether this ranking is unique.

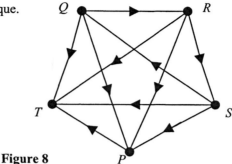

Figure 8

Solution

(a) Observe that (Q, R, S, P, T) is a Hamilton path that establishes Q as a winner and establishes a ranking Q-R-S-P-T.

(b) Because the graph has a simple circuit (Q, R, S, Q), we know from Theorem 4 that the ranking we found in (a) is not unique. Indeed, we can find another Hamilton path, (R, S, Q, P, T), that establishes that R is also a winner and provides the ranking R-S-Q-P-T. Can you find other Hamilton paths in the tournament graph of Figure 8?

Exploratory Exercise Set 5.1

1. (a) In this directed graph, find
   (1) an example of a simple path that is not an Euler path.
   (2) an example of an Euler path.

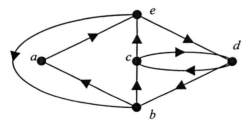

   (b) Work in pairs. Each partner should write a careful definition of each of the following terms and, when possible, find an example of each concept in this directed graph. After all definitions have been written and examples found, compare your definitions and examples with those of your partner.
   Define these terms for a directed graph:
   (1) Circuit
   (2) Simple circuit
   (3) Euler circuit
   (4) Hamilton path
   (5) Hamilton circuit

2. Review the material on relations in Section 2.2.

(a) Observe that a relation on a set S can be represented as a directed graph by regarding the elements of S as vertices and including (a, b) as a directed edge if a is related to b. Draw a directed graph that represents each of these relations.

(1) The relation $R = \{(a, b), (a, a), (a, c), (c, d)\}$ on the set $\{a, b, c, d\}$

(2) The relation "is the same sex as" on the set $\{$Fred, Louise, June, Tom, Max$\}$

(b) Observe that a directed graph with vertices V defines a relation on the set V where vertex a is related to vertex b provided that (a, b) is an edge in the directed graph. For each of the following graphs, find the associated relation. Be sure to indicate the set on which the relation is defined and the ordered pairs in the relation.

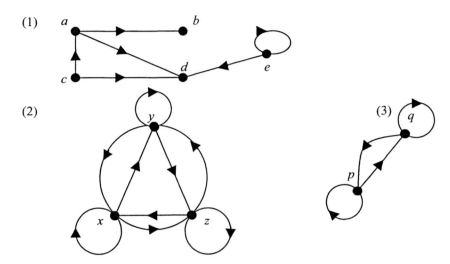

(c) Work with a partner. Review the concepts of reflexive relation, symmetric relation, and equivalence relation in Section 2.2.

(1) Make up or find an example of a reflexive relation, and draw the associated directed graph.

(2) What characteristic do the directed graphs of reflexive relations always have?

(3) Make up or find an example of a symmetric relation, and draw the associated directed graph.

(4) What characteristic do the directed graphs of symmetric relations always have?

(5) Make up or find an example of an equivalence relation, and draw the associated directed graph.

(6) What characteristic do the directed graphs of equivalence relations always have?

(d) Compare your work with your partner.

3. (*Proof Exercise*) In the proof of Theorem 4 in Section 4.1, we describe a method for finding a trip for the Highway Inspector on a map on which all towns have even degree. In this exploration you will work with a partner to translate that method into a method for finding an Euler circuit in a weakly connected, directed graph.

(a) Rewrite the proof of Theorem 4 using the language of directed graphs to describe a method for finding an Euler circuit for a weakly connected, directed graph in which the indegree of each vertex is equal to the outdegree of the vertex. When you are finished, you will have proved Theorem 1.

(b) Use Theorem 1 to determine whether this directed graph has an Euler circuit.

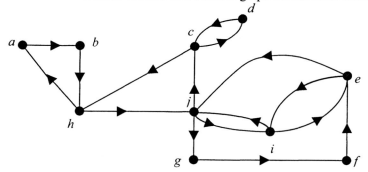

(c) Use your work from (a) to find an Euler circuit for the graph from (b).
(d) Explain briefly how you would modify your work from (a) to describe how to find an Euler path for a directed graph of the sort described in Theorem 2(a).

4. In Exploratory Exercise 5 of Section 4.2, we investigated using an adjacency matrix to represent a graph. An **adjacency matrix for a directed graph** is a square table of numbers with one row for each vertex and one column for each vertex. Each entry of the table is either 0 or 1. If (v_i, v_j) is an edge in the directed graph, then we place a 1 in the ith row and j^{th} column; otherwise, we place a 0 there. Below we find a directed graph G and its associated adjacency matrix. Notice that since (v_1, v_2) is an edge in G, a 1 appears as the entry in the first row, second column of the matrix. Likewise, since (v_2, v_1) is not an edge in G, 0 appears in the second row, first column of the matrix.

	v_1	v_2	v_3	v_4	v_5
v_1	0	1	0	0	1
v_2	0	0	1	0	1
v_3	1	0	0	0	0
v_4	0	0	0	0	1
v_5	1	1	0	0	0

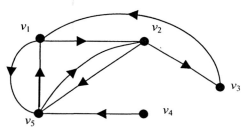

(a) Sketch a directed graph with vertices v_1, v_2, v_3, and v_4 that corresponds to the information in the adjacency matrices. Assume the rows and columns are labeled in the order v_1, v_2, v_3, and v_4.
(1)   (2)

0	1	1	1
0	0	1	1
1	0	1	0
0	0	1	0

0	1	0	0
1	0	1	0
0	1	0	1
0	0	1	0

(b) For each of these directed graphs, choose an ordering for the vertices and form the associated adjacency matrix.
(1)   (2)

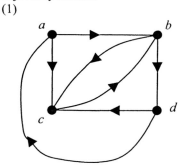

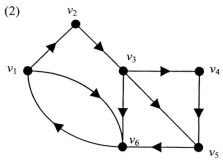

5. (*Writing Exercise*) Write a careful discussion comparing and contrasting the concepts of graph and directed graph. In your discussion, be certain to deal with as many of these issues as possible.
    - When we ignore directions on the edges of a directed graph, is the result a graph?
    - We have powerful theorems that determine whether a graph or a connected graph has Euler paths and circuits. What are the similarities and differences between the two sets of theorems.
    - In what way do the concepts of strongly connected and weakly connected provide a natural comparison between graphs and directed graphs?

Exercise Set 5.1

In Exercises 1 and 2 list the vertices and edges for the directed graphs.

1. 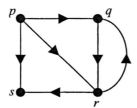   2.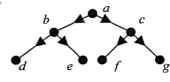

3. Sketch directed graphs with at least four vertices and six edges with the following characteristics.
    (a) Weakly connected but not strongly connected
    (b) Strongly connected
    (c) Neither weakly connected nor strongly connected

In Exercises 4 and 5 complete the table showing the indegree and outdegree of each vertex in the directed graph.

4.

Vertex	Indegree	Outdegree
a		
b		
c		
d		
e		

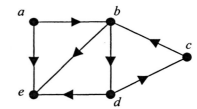

5.

Vertex	Indegree	Outdegree
p		
q		
r		
s		
t		
u		

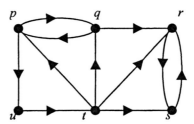

6. Determine whether the directed graphs in Exercises 4 and 5 have either an Euler path or an Euler circuit. Explain.

7. Many of the hallways at O. C. Welch Middle School are very narrow, so the principal decides to permit walking in only one direction in all but the very widest hallways. The arrows in this graph show the permissible directions in each hallway. Study the graph and answer the questions that follow.

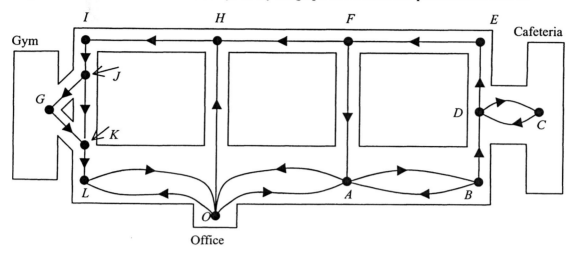

(a) Find a simple path from the office to the gym.
(b) Find a simple path from the gym to the cafeteria.
(c) Find a simple circuit that begins and ends at the gym and includes a stop by the office.
(d) Can the principal start at the office and make an inspection tour of the hallways? On her inspection tour, she would like to walk each hallway in each permissible direction exactly once. Explain your answer.

In Exercises 8 through 10 determine whether the graph has an Euler path or circuit, and if it does, find one.

8.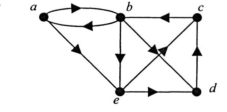

9.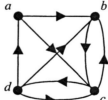

10.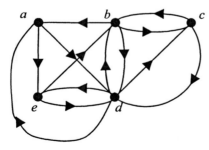

11. Ferrante eats only meat and potatoes. He prefers pork over fish and chicken, beef over pork and chicken, chicken over mutton and fish, and mutton over beef. Draw a directed graph that represents this situation.

12. For the tournament described in Figure 5, explain how the Gnats might claim to be a winner of the competition. Explain how the Sloths might claim to be a winner.

13. Consider the following results of a round-robin competition between four teams.
    Champs defeat Chumps.   Chumps defeat Lumps.   Lumps defeat Bumps.
    Champs defeat Lumps.   Chumps lose to Bumps.
    Champs lose to Bumps.
    (a) Draw a tournament graph representing this competition.
    (b) Which teams may claim to be winners?

14. Suppose every team of the PeeWee Football League plays every other team exactly one game.
    Suppose the Colts beat every other team, the Jets lose to every other team, the Bills beat only the Jets
    and the Patriots, and the Dophins beat everyone but the Colts.
    (a) Represent this competition with a tournament graph.
    (b) Find all winners.
    (c) Is there a unique winner?
    (d) Is there a unique ranking?

In Exercises 15 and 16 find all Hamilton paths for the tournament graph.

15.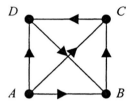

16.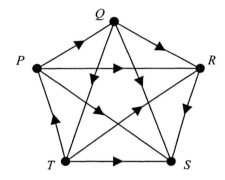

*Theorem 4 asserts that whether or not a tournament graph has a simple circuit determines whether there is
a unique Hamilton path in the tournament graph. As a consequence, whether the tournament graph has a
simple circuit determines whether the tournament has a unique ranking. In Exercises 17 through 20 we
will try to understand this connection better by looking at some illustrative examples that focus on the issue
of when the winner in a tournament graph is unique.*

17. The graph below records partial results for a tournament in which Team A beats Team B which beats
    Team C which beats Team A. Notice that (A, B, C, A) describes a circuit.

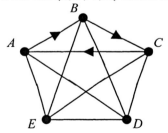

    (a) Complete the graph with edges that indicate that each of A, B, and C defeats Team D and E.
         Direct the edge between D and E in any way that you choose.
    (b) Find all Hamilton paths in the resulting graph.
    (c) Which of the five teams are winners?

18. The graph below records partial results for a tournament in which Team *A* beats
    Team *B* which beats Team *C* which beats Team *D* which beats Team *A*. Notice that (A, B, C, D, A)
    describes a circuit.

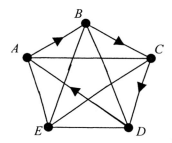

(a) Complete the tournament graph with edges that indicate that each of *A*, *B*, *C*, and *D* defeat Team
    E. Direct all other edges in any way that you choose.
(b) Find all Hamilton paths in the resulting tournament graph.
(c) Which of the five teams are winners?

19. Complete this conjecture on the basis of our observations from Exercises 17 and 18.
    - Suppose *T* is the set of teams in a tournament.
    - Suppose $S = \{v_1, v_2, \ldots, v_k\}$ is a subset of *T* and $(v_1, v_2, \ldots, v_k)$ is a circuit in the tournament
      graph.
    - Suppose each team in *S* defeats every team in $T - S$.
        Then …

20. Suppose that (A, B, C, D, E) and (B, C, A, D, E) are Hamilton paths in a tournament graph. [*Note*: This
    means that both *A* and *B* are winners.]
    (a) Record this information by directing the appropriate edges in this tournament graph.

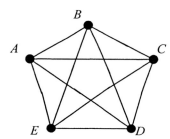

(b) Explain why the existence of these two Hamilton paths ensures that the tournament graph has a
    simple circuit.

5.2 An Introduction to Trees

Aunt Matilda began to study her family's history. Shortly thereafter she organized what she learned in the graph in Figure 1. Two generations ago, grandfather Ethan immigrated to America and had a son James and a daughter Anne. In her representation of the family, Matilda preserves parent/child relationships by representing each child below his or her parent in the graph.

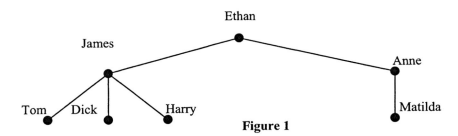

Figure 1

In 1857 Arthur Cayley wished to represent possible bonds in hydrocarbon molecules (molecules formed from hydrogen and carbon atoms). Recognizing that each carbon atom (*C*) would be bonded with four other atoms and that each hydrogen atom (*H*) would be bonded with exactly one atom, Cayley studied diagrams like the one found in Figure 2(a). This figure could also be represented as the graph in Figure 2(b), in which the vertices represent the carbon and hydrogen atoms and the edges represent bonds between atoms.

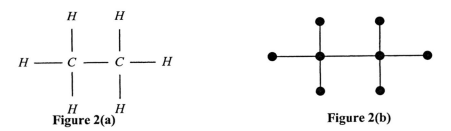

Figure 2(a)   Figure 2(b)

Both Cayley and Aunt Matilda were using a special kind of graph called a tree. Trees are used to model a wide variety of problems, including the representation of organic molecules in chemistry, possible decision paths in management science, and the organization of stored data in computer science. Surprisingly, Aunt Matilda's application provides much of the vocabulary that is used in the discussion of trees.

Definition: A **tree** is a connected graph with no cycles.

Example 1

The two graphs in Figure 3 are examples of trees. In each graph:
(a) Identify all simple paths from vertex u to vertex v.
(b) Count the number of vertices and edges.

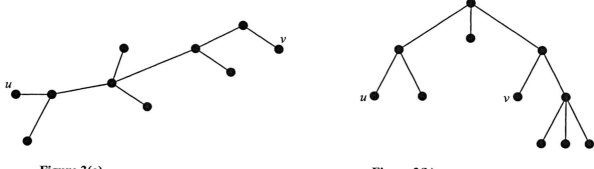

Figure 3(a)   Figure 3(b)

Solution

(a) In both trees there is exactly one simple path from vertex u to vertex v.
(b) In the tree of Figure 3(a), there are 10 vertices and 9 edges; in Figure 3(b) there are 11 vertices and 10 edges.

In each of the trees of Example 1, we saw that there was a unique simple path connecting the vertices u and v. In fact there is a unique simple path connecting each pair of vertices. We also saw that the number of edges is one less than the number of vertices. These observations made for specific examples of trees are true for all trees.

Theorem 5: In a tree there is a unique simple path between each pair of vertices.

Theorem 6: In a tree with n vertices, there are precisely $n - 1$ edges.

Preparation to prove Theorem 5:

Definitions: In proving both Theorems 5 and 6, we will give careful attention to the defining characteristics of a tree. That is, we will use heavily the facts that a tree is connected and has no cycles.

Reflection: In this proof we will use the fact that if a graph has a simple circuit (a circuit in which no edges repeat), then it will also have a cycle (a simple circuit in which no vertices repeat other than the initial and terminal vertex of the circuit). Although this fact may seem intuitively obvious (if the vertex x appears multiple times in the circuit, remove all edges and vertices between the first and last occurrence of x), we will give a careful verification of this fact in Exercise 19. Then, in Exercise 20, we will have the opportunity to review the method of proof by contradiction as we prove Theorem 5.

Preparation for Theorem 6: We will prove this theorem by induction on the number of vertices in the graph. We will prove the proposition $P(n)$: If a tree has n vertices, then it has $n - 1$ edges. A tree with one vertex will have no edges since an edge would create a cycle, and hence $P(1)$ is true. A tree with two vertices will have a single edge incident with the vertices. Hence $P(2)$ is true. Think about a connected graph with three vertices. If there is no cycle, how many edges must it have?

The hypothesis of the inductive step will be "A tree that has k vertices has $k - 1$ edges." We must show that if this hypothesis is true for positive integer k, then a tree with $k + 1$ vertices has k edges.

Proof of Theorem 6:

Basis step: A tree with one vertex has no edges. Hence $P(1)$ is true.

Inductive step: Suppose each tree with k vertices has $k-1$ edges. Let T be a tree with $k+1$ vertices. Let $x_1, x_2, x_3, \ldots, x_{r-1}, x_r$ be the vertices in a simple path in T of maximal length. Observe that x_{r-1} is the only vertex to which x_r is adjacent for the following reasons.

If x_r is adjacent to any one of $x_1, x_2, \ldots, x_{r-2}$, this creates a simple circuit and hence mandates the existence of a cycle. (See Exercise 19.) However, T has no cycles.

x_r cannot be adjacent to vertices of T not on the path since the path has maximal length.

Thus x_r has degree 1. Let $\hat{T}$ be the subgraph of T formed when vertex x_r and the edge $\{x_{r-1}, x_r\}$ are deleted from T.

Suppose u and v are vertices in $\hat{T}$. By Theorem 5, there is a unique simple path in T from u to v. Further, since x_r has degree 1, x_r is not on this path. Thus the path is a path from u to v in $\hat{T}$, so $\hat{T}$ is connected. Since T has no cycles, $\hat{T}$ has no cycles. Thus $\hat{T}$ is a tree with k vertices. By inductive hypothesis $\hat{T}$ has $k-1$ edges. Thus T has k edges. ∎

In fact, even stronger results can be proved with additional work. The properties observed in Theorems 5 and 6, together with connectivity, completely characterize trees.

Theorem 7: The following statements about a graph T with n vertices are equivalent:
1. T is a tree.
2. T is connected and there is a unique simple path between each pair of vertices.
3. T is connected and has $n-1$ edges.

ROOTED TREES

In many applications we need a particular kind of tree called a rooted tree.

> **Definition:** A **rooted tree** is a directed graph formed from a tree by designating a specific vertex as the root and directing each edge away from the root. [By Theorem 5 there is a unique path from the root vertex to each other vertex. Give edge $\{u, v\}$ the direction (u, v) if u lies on the path from the root vertex to v. Otherwise, direct the edge as (v, u).]

Example 2
Find the rooted tree that is formed from the tree in Figure 4(a) when
(a) vertex u is chosen as the root.
(b) vertex x is chosen as the root.
Solution
(a) See the rooted tree in Figure 4(b).
(b) See the rooted tree in Figure 4(c).

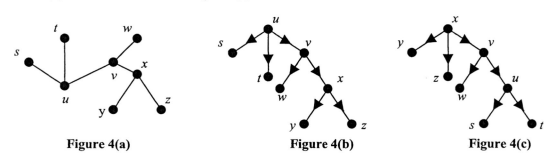

Figure 4(a)   Figure 4(b)   Figure 4(c)

Although a rooted tree is a directed graph, we often omit the direction arrows on edges when providing a geometric representation. Rather, we understand the root to be the vertex at the top of the tree and the edges to be directed away from the root. Figure 5(a) shows a rooted tree represented with its direction arrows, and Figure 5(b) shows the same rooted tree without direction arrows.

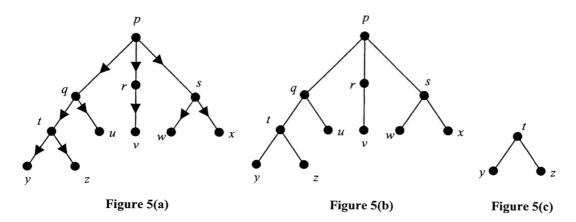

Figure 5(a)   Figure 5(b)   Figure 5(c)

Because rooted trees can be used to model family trees, we use terminology that comes from this application. If v is a vertex in a rooted tree other than the root, the **parent** of v is the unique vertex u such that there is a directed edge from u to v. When u is the parent of v, we say v is the **child** of u. If v and w have the same parent, we say v and w are **siblings**. If vertex x is a vertex in the unique path from the root to y (other than y itself), x is said to be an **ancestor** of y and y is a **descendant** of x. If z is a vertex in a rooted tree, then the **subtree with z as the root** is the subgraph of the tree consisting of z and its descendants and all edges incident with its descendants.

Example 3

In the rooted tree in Figures 5(a) and (b):
(a) Identify the root of the tree.
(b) Identify the parent of vertex u.
(c) Identify the children of vertex s.
(d) Identify the siblings of u.
(e) Identify the ancestors of vertex t.
(f) Identify the descendants of vertex q.
(g) Draw the subtree of the tree in Figures 5(a) and (b) with t as its root.

Solution

(a) The root of the tree is p.
(b) The parent of vertex u is vertex q.
(c) The children of vertex s are vertices w and x.
(d) The sibling of vertex u is vertex t.
(e) The ancestors of vertex t are vertices q and p.
(f) The descendants of vertex q are vertices $t, u, y,$ and z.
(g) The subtree of the tree in Figures 5(a) and (b) with t as its root is found in Figure 5(c).

Exploratory Exercise Set 5.2

1. In this exercise you and the members of your group will explore the basic concepts related to trees.
    (a) Each member of the group should draw a tree with at least eight vertices. On the tree, label one vertex as u and a second vertex as v.
        (1) Each member of the group should exchange trees with another member of the group, and the group should work together to complete the following table.

Name of Group Member	Number of Vertices	Number of Edges	Number of Simple Paths from u to v

  (2) What theorems of this chapter do these investigations confirm?
    (b) Each member of the group should draw a rooted tree with at least eight vertices. After the rooted tree is drawn, label the vertices, making sure that one vertex is labeled as C. On a separate piece of paper, identify the following objects:
        (1) The root of the tree
        (2) The parent of C
        (3) The children of C
        (4) The siblings of C
        (5) The ancestors of C
        (6) The descendants of C
    (c) Exchange your rooted tree from (b) with another group member, and ask that member to identify the objects from (b) on your tree.
    (d) As a group, discuss each of the rooted trees drawn in (b), attempting to determine correct answers for those trees for which any members of the group do not agree.

2. In a single elimination tournament, each competitor plays until a game is lost. We can model such tournaments using a tree. Figure 6(a) is a tree representing a single elimination tournament involving the four teams A, B, C, and D. In the first round A defeats B, and D defeats C; in the second round D defeats A to win the tournament. Figure 6(b) is a representation of a single elimination tournament involving the five teams. T wins this tournament. Observe that with four teams, three games are necessary to complete the competition and that with five teams, four games are necessary to complete the competition.

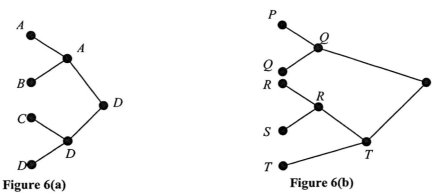

Figure 6(a)   Figure 6(b)

    (a) Draw a tree representing a single elimination tournament with 2 competitors.
    (b) Draw a tree representing a single elimination tournament with 3 competitors.
    (c) Draw a tree representing a single elimination tournament with 6 competitors.

(d) Draw a tree representing a single elimination tournament with 7 competitors.

(e) Draw a tree representing a single elimination tournament with 8 competitors

(f) Complete the following table recording the number of games required to complete a single elimination tournament.

Number of Competitors	Number of Games
2	
3	
4	3
5	4
6	
7	
8	

(g) Make a conjecture about the number of games that must be played to complete a single elimination tournament with n players.

(h) Verify your conjecture by drawing a tree representing a single elimination tournament with 9 competitors.

(i) On the basis of your conjecture, how many games must be played to determine a winner in a single elimination chess tournament in which there are 500 entries?

3. The English chemist Arthur Cayley first worked with trees in 1857 when he was studying saturated hydrocarbons. A saturated hydrocarbon is a molecule involving n atoms of carbon (C) and $2n + 2$ atoms of hydrogen (H) and is represented by chemists by the symbol C_nH_{2n+2}. In a saturated hydrocarbon, each carbon atom is bonded to four other atoms, and each hydrogen atom is bonded to exactly one carbon atom. Hence, in the graphical representation, a vertex of degree 4 represents a carbon atom and a vertex of degree 1 represents a hydrogen atom.

Figure 7 shows the graph model for the saturated hydrocarbon C_1H_4 (methane) and the graph model for C_2H_6 (ethane).

Methane   Ethane

Figure 7

(a) Notice that in both cases the graphs are trees. By definition of a molecule, the graph representation of a molecule will be connected. By answering (1) – (5) below, explain why the graphs of saturated hydrocarbons must always be trees.

(1) In the graph for C_nH_{2n+2}, how many vertices represent carbon atoms? How many vertices represent hydrogen atoms?

(2) What is the total number of vertices in the graph for C_nH_{2n+2}?

(3) Since each vertex for a carbon atom has degree 4 and each vertex for a hydrogen atom has degree 1, what is the sum of the degrees of the vertices of a graph model of a saturated hydrocarbon?

(4) By Theorem 12 of Chapter 4, how many edges are found in the graph model for C_nH_{2n+2}?

(5) Explain why the graph model for C_nH_{2n+2} must be a tree.

(b) In trying to enumerate all saturated hydrocarbons, Cayley learned that non-isomorphic trees with n vertices of degree 4 and $2n + 2$ vertices of degree 1 correspond to saturated hydrocarbons with different chemical properties. Using this knowledge, he was able to predict the existence of new saturated hydrocarbons.

   (1) Find a graph model for the only saturated hydrocarbon with three carbon atoms and eight hydrogen atoms (C_3H_8). (This saturated hydrocarbon is called propane.)

   (2) Find a graph model for each of the two different saturated hydrocarbons with four carbon atoms and ten hydrogen atoms (C_4H_{10}).

4. (*Proof Exercise*) Mathematical reasoning often becomes more clear when discussed with others. Each member of your group of four students should spend 10 minutes carefully rereading the proof of Theorem 6. As you read, make notes of questions that arise as you attempt to understand the proof (one example: What is a path of maximal length in a graph?). Then discuss your questions as a group and resolve as many of them as possible.

5. (*Writing Exercise*) Write a careful discussion comparing and contrasting the concepts of graph and tree. In your discussion, be certain to deal with as many of these issues as possible.

   - Is every tree a graph?
   - Is every graph a tree?
   - The concepts of paths, circuits, degree of a vertex, connectedness, and loops are all important in the context of general graphs. Are there special things that can be said when we attempt to understand these concepts for trees?
   - We spent quite a bit of energy determining whether a general graph has an Euler path or circuit. What can you say about when a tree has an Euler path or circuit?
   - What can you say about when a tree has a Hamilton path or circuit?

Exercise Set 5.2

In Exercises 1 through 6 determine whether the graph is a tree. If the graph is not a tree, explain why it is not a tree.

1.

2.

3.

4.

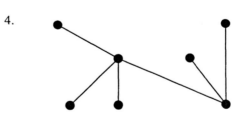

5.

6.

7. For each of the graphs in Exercises 1 through 6 that you identified as a tree, count the number of vertices and the number of edges. What theorem do these investigations confirm?

In Exercises 8 and 9 find the following in the rooted tree.
   (a) The root
   (b)   The parent of vertex X
   (c) The children of vertex Y
   (d) The siblings of vertex Z
   (e) The ancestors of vertex Q
   (f) The descendents of vertex W
   (g) The subtree of the tree with R as its root

8.   9.

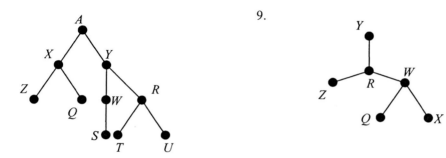

10. Rooted trees are often used to represent the structures of organizations. In this tree describing the organization of a county school board, determine the following:
   (a) The ancestors of the principal of Rosewood Elementary School.
   (b) The parent of the principal of West High School
   (c) The children of the Assistant Superintendent of Operations.
   (d) The root

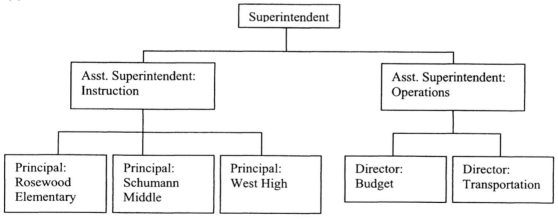

11. In the rooted tree from Exercise 8, draw the subtree that is rooted at
   (a) *Y*
   (b) *Z*

12. Redraw this tree as a rooted tree
   (a) with *R* as the root
   (b) with *T* as the root

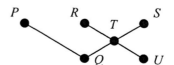

13. (a) How many vertices are there in a tree with 13 edges?
    (b) How many edges are there in a tree with 11 vertices?

14. In a tree, the number of vertices is always one more than the number of edges. Draw a graph that is not a tree in which the number of vertices is one more than the number of edges.

15. For which values of n is the complete graph K_n a tree?

16. Explain why the parent of a vertex in a rooted tree is unique.

17. The reproductive process of bees is somewhat unusual in that a male bee has only a mother, whereas a female bee is the product of mating between a mother and and a father. Draw a rooted tree giving ancestors for four generations for a male bee. Assume that there is no mating between generations.

18. Review the notion of isomorphic graphs from Section 4.3. All trees with two vertices are isomorphic to the tree in Figure 8(a), and all trees with three vertices are isomorphic to the tree in Figure 8(b).
    (a) Draw two trees with four vertices that are not isomorphic to one another.
    (b) Draw three trees with five vertices so that no pair of them are isomorphic.
    (c) Find the largest possible set of trees with six vertices so that no pair of them are isomorphic.

Figure 8(a)   **Figure 8(b)**

19. (*Proof Exercise*) Give the reasons in this proof of the following statement:
    If a graph contains a simple circuit, the graph contains a cycle.

    Proof: Consider an arbitrary simple circuit from u to u described by $C = (u, e_1, v_1, e_2, v_2, \ldots, v_i, e_{i+1}, \ldots, v_j, \ldots, e_n, v_n)$ where $v_n = u$.
    (a) If no vertices other than the first vertex and the last vertex in C are repeated, we are done. Why?
    (b) Suppose that $v_i = v_j$ for two distinct integers i and j between 1 and n. Create simple circuit $\hat{C}$ from C by deleting $e_{i+1}, \ldots, v_j$ from C. If no vertices other than the first vertex and the last vertex are repeated in $\hat{C}$, then we are done. Why?
    (c) If we continue this process (find a pair of repeated vertices and delete edges and vertices between), the process must come to an end in a finite number of steps. Why?
    (d) The simple circuit remaining at the end of the process will be a cycle. Why?

20. (*Proof Exercise*) Give reasons for the following steps in the proof of Theorem 5:
    In a tree there is a unique simple path between each pair of vertices.
    Proof: We will let x and y be a pair of vertices in a tree. Since a tree is connected, there is at least one simple path from x to y. By way of contradiction, suppose that there are two distinct simple paths from x to y.
    (a) In a general graph, we must describe paths by an alternating list of vertices and edges. Why can we describe paths in a tree merely by listing the vertices along the path?
    (b) Let $C_1 = (w_1, w_2, \ldots, w_k)$ with $w_1 = x$ and $w_k = y$ be a list of vertices describing one simple path from x to y, and let $C_2 = (u_1, u_2, \ldots, u_r)$ with $u_1 = x$ and $u_r = y$ be a list of vertices describing a second, different simple path from x to y. Why is $w_k = u_r$? Why is $w_1 = u_1$?
    (c) Let t be the smallest integer so that $w_t \neq u_t$. How do we know such an integer exists? Why is $t > 1$?
    (d) Let s be the next integer beyond t with w_s equal to a vertex in C_2. How do we know there is such an integer s?

(e) Consider the circuit that starts at w_{t-1}, follows C_1 to w_s, then follows C_2 back to w_{t-1}. How do we know that this is a simple circuit?

(f) Review Exercise 19 and complete the proof by finding a contradiction.

5.3 Spanning Trees

The Traffic Engineer glances at the map in Figure 1 and then looks out the window at 8 inches of newly fallen snow. Although she cannot get all the roads in the county plowed by nightfall, she must see to it that enough roads are plowed so that a driver can drive between any two of the critical locations marked on the map by a dot.

Figure 1

Because the Traffic Engineer wishes to reach all vertices of the graph but also wishes to plow a small number of road segments, she determines to find a tree within the graph. Indeed, she may well find that a spanning tree suits her purposes.

Definition: A **spanning tree** of a graph G is a subgraph of G that includes all the vertices of G and is also a tree.

As we shall see in the next theorem, every connected graph has a spanning tree. Because a tree has no cycles, we can find a spanning tree for a connected graph by discarding edges that are part of cycles until we are left with a subgraph that has no cycles.

Example 1
      Finding a spanning tree for the graph in Figure 2.

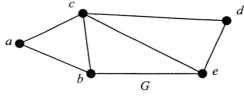

Figure 2

Solution
      The graph G in Figure 2 is a connected graph but is clearly not a tree since it has cycles such as (a, b, c, a) and (e, c, d, e). If we discard segment $\{a, b\}$, the graph will still be connected, will still contain all of the vertices of G, but will have one fewer cycle. Similarly, if we discard $\{c, d\}$, we will have one fewer cycle. Finally, if we remove $\{b, e\}$, the resulting subgraph still has all the vertices of G but has no cycles remaining. The result is a spanning tree. See Figure 3.

(continued)

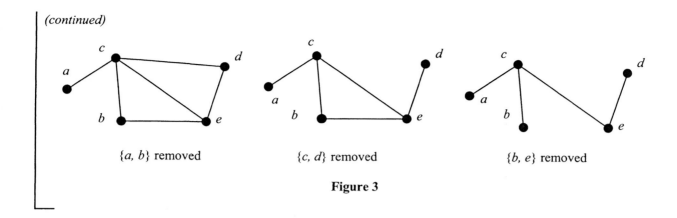

{a, b} removed   {c, d} removed   {b, e} removed

Figure 3

Notice that as we removed edges to destroy cycles, we made choices. For instance, we could have initially removed edge {a, c} instead of {a, b}. Consequently, a single graph will typically have several different spanning trees. For instance, if we removed the sequence of edges {a, c}, {c, e}, {c, d}, we would get the spanning graph shown in Figure 4.

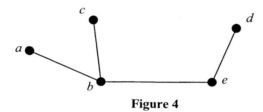

Figure 4

These investigations bring us to the following theorem.

Theorem 8: A graph is connected if and only if it has a spanning tree.

> *Preparation for Theorem 8*: We have two statements to prove:
> (1) If a graph has a spanning tree, then it is connected.
> (2) If a graph is connected, then it has a spanning tree.
> The first statement should follow easily from the definitions that assert that trees are connected and that spanning trees contain all the vertices of their graphs. In our proof of Statement (2), we will use the fact that if we remove an edge from a cycle in a connected graph, the resulting subgraph continues to be connected. This fact may seem geometrically clear to you, but we invite you to complete a careful argument in Exercise 15. This insight, together with the reasoning used in the Example 1, should enable us to establish Statement (2).

Proof of Theorem 8:
Proof of (1): Suppose that graph G has a spanning tree T. If u and v are any two vertices of G, they are also vertices of T. Since T is a tree, T is connected and there is a simple path from u to v in T. Since T is a subgraph of G, the path from u to v in T is also a path from u to v in G. Thus G is connected.
Proof of (2): Suppose that G is connected. If G has no cycles, then G is a tree and hence is its own spanning tree. If G has cycles, create a subgraph G' by removing an edge from a cycle of G. Observe that G' is connected (see Exercise 15) and contains all of the vertices of G. If G' has cycles, remove an edge again. Continue this process, and, in a finite number of steps, the subgraph created will have no cycles, will be connected, will contain all the vertices of G, and hence will be a spanning tree for G. ∎

BREADTH-FIRST SPANNING TREE SEARCH ALGORITHM

For graphs with a small number of vertices, it is quite easy to find several spanning trees merely by inspection. For slightly larger graphs, we might be guided by the proof of Theorem 8 and find a spanning tree by finding cycles and destroying them by removing edges. However, for larger graphs, it is important to have a more systematic algorithm. One such algorithm is called the Breadth-First Spanning Tree Search Algorithm.

We will give a geometric description for this algorithm that finds a spanning tree for a connected graph G. In the algorithm we arbitrarily choose a vertex with which to start (in our description of the algorithm, we will start with vertex a). We will build the spanning tree in steps, at each step adding those vertices that are adjacent to the vertices added in the previous step. As we proceed, we will shade the edges and vertices we have added. At the end of the process, the edges and vertices shaded will form the spanning tree.

Breadth-First Spanning Tree Search Algorithm:
1. Initially,
   Let STEP = 0.
   Label vertex a with STEP.
2. Label all unlabeled vertices adjacent to vertices labeled STEP with STEP + 1.
   For each vertex v labeled STEP + 1, choose exactly one edge e incident with v and some vertex labeled STEP. Shade the edge and its vertices.
   Let STEP = STEP + 1.
3. Until all vertices are shaded, return to (2). Otherwise, the vertices and edges that are shaded form a spanning tree for the graph.

Example 2
   Use breadth-first search to find a spanning tree for the connected graph in Figure 5.

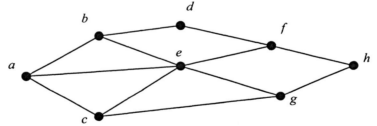

Figure 5

Solution
   On the left we will follow the iterations of (2), keeping up with the values of STEP. On the right we will draw the graph, shading the vertices and edges as they are added to the spanning tree that is being built.

Initially, vertex a is labeled 0, as seen in Figure 6(a). STEP = 0.

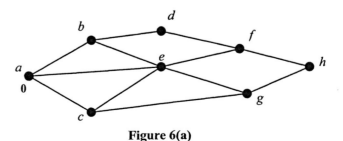

Figure 6(a)

First Iteration of (2): Since STEP is 0, vertices *b*, *e*, and *c* are labeled STEP + 1 = 1.

For vertex *b*, we shade edge {*a, b*} and its vertices. Similarly, we shade {*a, e*} and {*a, c*} and their vertices.

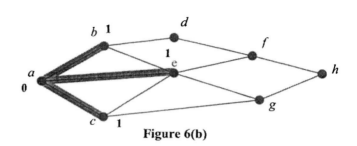

Figure 6(b)

Second Iteration of (2): Since STEP is 1, vertices *d*, *f*, and *g* are labeled 2. We shade edges {*b, d*} and {*e, f*}, as well as vertices *d* and *f*. In regards to vertex *g*, we must choose between edges {*e, g*} and {*c, g*}, both of which are incident with *g* and vertices labeled 1. We randomly choose {*e, g*} and shade this edge and its vertices. See Figure 6(c). STEP becomes 2.

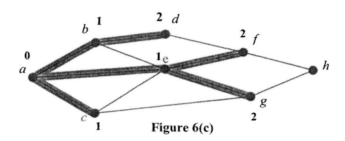

Figure 6(c)

Third Iteration of (2): Since STEP is 2, vertex *h* is labeled 3. We must choose between edges {*f, h*} and {*g, h*}. We randomly choose {*f, h*} and shade this edge and its vertices. The subgraph formed by the shaded vertices and edges in Figure 6(d) is a spanning tree for the graph in Figure 5.

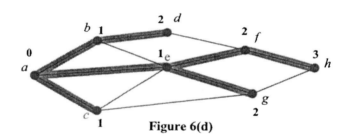

Figure 6(d)

The general principles of breadth-first search are used in many problem situations in which one must visit each vertex of a graph in an attempt to accomplish some goal. Hence, breadth-first search algorithms can be used for purposes other than looking for spanning trees of connected graphs. For instance, notice that in the algorithm just described, the label attached to each vertex is the length (in number of edges) of the shortest path between *a* and the vertex. (Indeed, the breadth-first spanning tree search is a special case of the shortest path algorithm studied in Section 4.4 where the weight of each edge is 1.)

MINIMUM SPANNING TREES

If we return to the office of the Highway Engineer, we find that she has refined her deliberations. She now stares at a map of the road system that not only indicates which roads connect which key locations but also indicates the number of hours it will take to plow each section of the road. (See Figure 7.) In other words, she now is looking at a weighted graph. Further, she has determined that she desires not just a spanning tree but a spanning tree with the special property that the sum of the "plow times" of the road segments in the tree is a minimum.

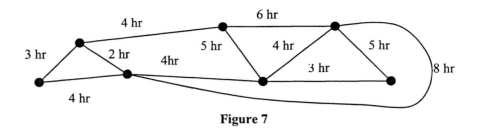

Figure 7

In a weighted, connected graph, the **weight of a subgraph** is the sum of the weights of the edges in the subgraph. A **minimum spanning tree of graph** *G* is a spanning tree of *G* with the smallest possible weight.

One obvious way to find a minimum spanning tree is to find all spanning trees in a graph and then find the tree or trees with the least weight. However, this strategy would require that we find all spanning trees of a specific graph, an onerous task when the graph is very large. In 1957 Robert Prim designed an algorithm that builds a minimum spanning tree *T* edge by edge. The algorithm can be described as follows, where *T* is the minimum spanning tree that we are forming step-by-step for a connected graph *G* with *n* vertices.

Prim's Algorithm:
    (1) Initially let *T* be the tree consisting of an edge of minimum weight together with the vertices with which the edge is incident
    (2) Let *L* be the set of all edges that share exactly one vertex with a vertex in *T*. Choose from *L* an edge of minimum weight and include the edge and its vertices in *T*.
    (3) If *T* contains *n* – 1 edges, *T* is a minimum spanning graph for *G* and we are done. Otherwise return to Step 2.

Observe that in Step 2 we consider adding only edges that are incident with exactly one vertex in T. This ensures that as we add edges to *T*, we do not create a cycle in *T*, and hence *T* continues to be a tree as we add edges and their vertices.

Example 3
    Use Prim's Algorithm to find a minimum weighted spanning tree for the graph in Figure 8.

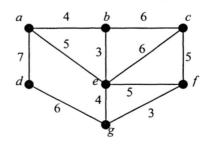

Figure 8

Solution
    The graph in Figure 8 has seven vertices, so each spanning tree will have six edges. Because the minimum weight among all edges is 3, we can start with either edge {*b, e*} or edge {*f, g*}. We will start with {*b, e*} and trace the growth of *T*. At the beginning of each step, we will shade the edges of *T* and use dashes to indicate the edges of *L*.

Initially, we see that T consists of $\{b, e\}$ and its vertices. (See Figure 9(a).)

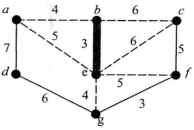

Figure 9(a)

Iteration 1 of (2):

In Figure 9(a) $L = \{\{b, a\}, \{b, c\}, \{e, a\}, \{e, c\}, \{e, g\}, \{e, f\}\}$. We may choose either $\{e, g\}$or $\{b, a\}$ of minimum weight 4. We randomly choose $\{b, a\}$. As shown in Figure 9(b), T now consists of $\{\{b, e\}, \{b, a\}\}$ and their vertices.

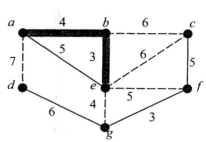

Figure 9(b)

Iteration 2 of (2):

In Figure 9(b), $L = \{\{a, d\}, \{b, c\}, \{e, c\}, \{e, f\}, \{e, g\}\}$. Observe that $\{e, a\}$ is no longer in L since both of its vertices are in T. The inclusion of $\{e, a\}$ in T would create a cycle. From L we choose $\{e, g\}$ of minimum weight 4. As shown in Figure 9(c), T now consists of $\{\{b, e\}, \{a, b\}, \{e, g\}\}$ and their vertices.

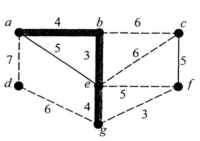

Figure 9(c)

Iteration 3 of (2):

In Figure 9(c), $L = \{\{a, d\}, \{b, c\}, \{e, c\}, \{e, f\}, \{g, d\}, \{g, f\}\}$. From L we choose $\{g, f\}$ of minimum weight 3. In Figure 9(d), T consists of $\{\{b, e\}, \{b, a\}, \{e, g\}, \{g, f\}\}$ and their vertices.

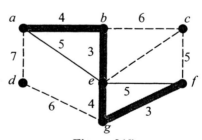

Figure 9(d)

Iteration 4 of (2):

In Figure 9(d), $L = \{\{a, d\}, \{b, c\}, \{e, c\}, \{g, d\}, \{f, c\}\}$. From L we choose $\{f, c\}$ of minimum weight 5. In Figure 9(e), T consists of $\{\{b, e\}, \{b, a\}, \{e, g\}, \{g, f\}, \{f, c\}\}$ and their vertices.

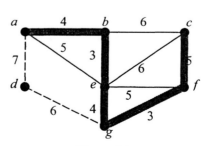

Figure 9(e)

Iteration 5 of (2):
   In Figure 9(e), $L = \{\{a, d\}, \{g, d\}\}$.
   From L we choose $\{g, d\}$ of minimum
   weight 6. In Figure 9(f), T now consists
   of $\{\{b, e\}, \{b, a\}, \{e, g\}, \{g, f\}, \{f, c\},$
   $\{g, d\}\}$ and their vertices. At this point T
   contains 6 edges, so we are done.

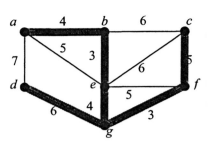

Figure 9(f)

T is a minimum spanning tree for G with weight $3 + 4 + 4 + 3 + 5 + 6 = 25$.

Prim's Algorithm is a "greedy" algorithm (as was Dykstra's Algorithm). At each step it selects the "best" edge (the edge with the smallest weight). Such a greedy strategy does not always produce the best possible overall solution, but it does in the case of Prim's Algorithm. We will omit the induction proof that Prim's Algorithm always produces the spanning tree of minimum weight.

Exploratory Exercise Set 5.3

1. Work in pairs. For this graph, each individual in the pair should find a spanning tree using the method of (a) and a spanning tree using the method of (b). Then individuals should compare their work.

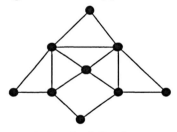

   (a) Find a spanning tree by completing the following steps until there are no more cycles.
       (1) Find a cycle.
       (2) Remove an edge from the cycle.
   (b) Find a spanning tree using the breadth-first search algorithm.

2. Work in a group to explore the notion of spanning trees.
    (a) Consider the problem of finding spanning trees for this graph.

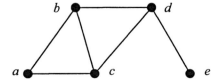

       (1) For 5 minutes each member of the group should work independently and find as many spanning trees for this graph as possible.

(2) The group should examine together the examples produced by group members and ascertain whether each of them is a spanning tree.

(3) Now work together until you have found all possible spanning trees for this graph. (*Hint*: There are eight of them.)

(b) Suppose weights are added to the edges of the graph from (a) to produce a weighted graph.

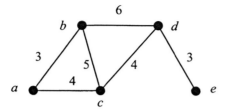

(1) Find the sum of the weights of the edges of each of the eight spanning trees from part (3) of (a).

(2) Use your work from (1) to find a minimum spanning tree.

(3) Use Prim's Algorithm to find a minimum spanning tree.

3. Kruskal's Algorithm provides an alternative approach to finding a minimum spanning tree for a connected, weighted graph. As we built the spanning tree in Prim's Algorithm, we were careful at each step to choose an edge connecting a new vertex to a vertex already chosen. This ensured that the subgraph created by the process was connected. Somewhat surprisingly, in Kruskal's Algorithm we proceed as follows:

(1) Make a list of the edges ordered by weight, and choose an edge of minimum weight. Let subgraph *T* consist of the edge and its vertices.

(2) From those edges not yet chosen, choose an edge of minimum weight that does not form a cycle when added to *T*, and add the edge and its vertices to *T*.

(3) If *T* does not yet have *n* − 1 edges, return to Step (2). Otherwise, *T* is a minimum spanning tree for the graph.

The surprise is that even though the edges are chosen without regard to connectivity, at the conclusion of the algorithm the resulting subgraph is connected; indeed, it is a tree. (This fact, of course, requires a proof that we omit at this point.)

(a) Work with a partner to use Kruskal's Algorithm to find a minimum spanning tree for the weighted graph in Figure 10.

(1) Step 1: We first need to list the edges in order by weight and choose an edge of minimum weight. We will choose edge {*a*, *b*} of weight 1. Complete this list, and shade {*a*, *b*} and its vertices on a copy of the graph in Figure 10.

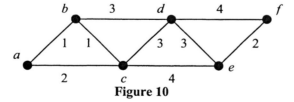

Figure 10

Edge	{*a*, *b*}	{*b*, *c*}	{*a*, *c*}	{*e*, *f*}	{*b*, *d*}				
Weight	1	1	2	2	3				

(2) Step 2: From the edges not yet chosen, choose an edge of minimum weight. Observe that two edges never form a cycle, and then shade this edge on your copy of the graph.

(3) Step 2 again: Explain why edge {*a*, *c*} cannot be the next edge chosen. From the edges not yet chosen, choose an edge of minimum weight, and shade the edge and its vertices.

(4) Continue to do Step 2 until you have chosen and shaded five edges.

(5) Verify that the edges and vertices you have shaded form a spanning tree for the graph in Figure 10. Can you find another spanning tree with a lesser weight?

(b) You and your partner should each draw a connected, weighted graph with at least eight edges.

(1) Using your partner's graph, find a minimum spanning tree using Prim's Algorithm.

(2) Using your partner's graph, find a minimum spanning tree using Kruskal's Algorithm.

[*Note*: At each step we sometimes choose from several different edges of the same weight. Just as in Prim's Algorithm, the minimum spanning tree produced by Kruskal's Algorithm is not unique.]

4. (*Writing Exercise*) Spanning trees are used to make the process of distributing information from one computer on a computer network to other computers more efficient. Do library research to differentiate between the terms *unicasting* and *multicasting,* and then write a paragraph differentiating between the terms and giving a non-technical description of how spanning trees are used to facilitate Internet multicasting.

Exercise Set 5.3

In Exercises 1 and 2 find three spanning trees for each graph.

1.

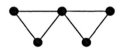

2.

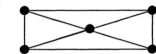

In Exercises 3 and 4 use breadth-first search to find a spanning tree for the graph.

3.

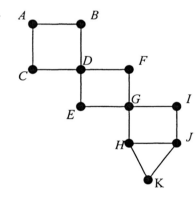

4.

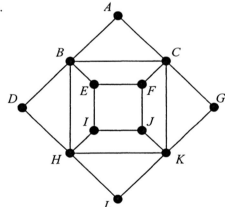

5.   The vertices in the graph that follows represent the regional offices of a certain international corporation. An edge between two vertices represents a direct computer link between two offices. The corporation wishes to upgrade some of these computer links to be completely secure in such a way that any two offices can communicate using secure links. However, because such upgrades are very expensive, the corporation wishes to upgrade with minimum cost. Suggest a way in which this can be done.

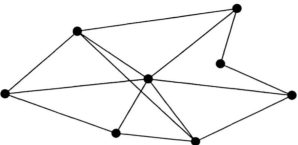

6.   Find a spanning tree for the road system in Figure 1.

7.   Draw the spanning tree that results from performing the Breadth-First Search Spanning Tree Algorithm on:
     (a) K_3 (b) K_4 (c) K_5
     (d) Describe the spanning tree that results from performing the Breadth-First Search Spanning Tree Algorithm on K_n.

8.   Draw a spanning tree that results from performing the Breadth-First Search Spanning Tree Algorithm on:
     (a) $K_{2,2}$ (b) $K_{2,3}$ (c) $K_{3,3}$

In Exercises 9 through 11 use Prim's Algorithm to find a minimum spanning tree for the weighted graph.

9.   10.

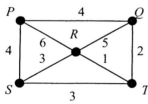

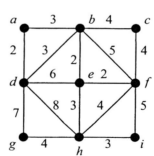

11.

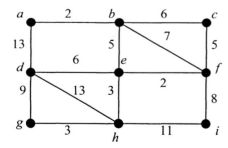

12. Assist the Highway Inspector by finding a spanning tree in Figure 7 that can be plowed in the minimum possible time. What is that time?

13. National Paper Corporation has a large plantation on which it grows many thousands of acres of trees for pulp. It has a network of primitive dirt roads within the plantation connecting facilities in which it stores equipment and other supplies. The manager of the plantation wishes to upgrade some of the roads to enable workers to reach each facility in bad weather. However, the manager also wants to pave as few kilometers of road as possible. The weighted graph shows the road map, together with the lengths of the segments in kilometers. Describe a solution to the problem.

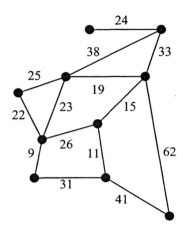

14. A **maximum spanning tree** in a weighted graph is a spanning tree with the largest possible weight.
    (a) Modify Prim's Algrorithm so that it will find a maximum spanning tree.
    (b) Use the modified Prim's Algorithm to find a maximum spanning tree for the graph in Exercise 9.

15. (*Proof Exercise*) Complete the steps in the proof of this theorem.
    Suppose G is a connected graph with a cycle. Let G' be the subgraph of G created by removing an edge of that cycle. Show that G' is connected.

    Proof: Suppose that $(v_1, v_2, \ldots, v_{r-1}, v_r, v_{r+1}, v_{r+2}, \ldots, v_k)$ is a cycle in G and that we remove edge $\{v_r, v_{r+1}\}$ to create G'. Let u and v be any pair of vertices in G'.

    (a) Explain why there is a path C from u to v in G.
    (b) If edge $\{v_r, v_{r+1}\}$ is not an edge in C, then there is a path from u to v in G'. Explain why.
    (c) If edge $\{v_r, v_{r+1}\}$ is an edge in C and v_r precedes v_{r+1} in C, describe how to modify C to create a path from u to v in G'. (Observe that a similar modification will work if v_{r+1} precedes v_r in C.)
    (d) Explain how we know G' is connected.

5.4 Binary Trees and Tree Traversals

In many cases, computer programmers are able to solve a problem only because the information for the problem has been organized using data structures based on trees. In order to get a sense of the role that trees play in these computer solutions, we will look at two problems that use trees to organize information in preparation for computer solution. Though there are a wide variety of problems that can be addressed using data structures based on trees, we will look at the problems of computing arithmetic expressions and creating alphabetized lists of words. But first we must learn the language of binary trees.

BINARY TREES

In the event that each parent in a rooted tree has zero, one, or two children, the rooted tree is called a **binary tree**. We will represent binary trees by placing each child below and to the left or right of the parent. We will follow the convention of identifying a child to the left of the parent as a **left child** and a child to the right of the parent as a **right child**. In so doing, we will have placed a natural order on the children of each parent and hence will have an **ordered binary tree**. In an ordered binary tree, we can define the **left subtree of a vertex** as the subtree consisting of the left child of the vertex, the descendents of that left child, and all edges incident with these vertices. The **right subtree of a vertex** can be defined similarly.

Example 1

In the ordered binary tree in Figure 1:
(a) Identify the left child of *b*.
(b) Identify the right child of *f*.
(c) Identify the right subtree of *a*.
(d) Identify the left subtree of *e*.
(e) Identify the right subtree of *b*.

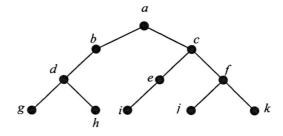

Figure 1

Solution
(a) The left child of *b* is *d*.
(b) The right child of *f* is *k*.
(c) The right subtree of *a* consists of vertices *c, e, f, i, j, k* with connecting edges.
(d) The left subtree of *e* consists of the vertex *i*.
(e) The right subtree of *b* is the empty tree consisting of no vertices and no edges.

REPRESENTING ARITHMETIC EXPRESSIONS WITH BINARY TREES

In the conversation that follows, we will use * to represent multiplication and ↑ to represent exponentiation. Arithmetic operations such as + or * or ↑ are called **binary operations** because they describe an operation that is to be performed on two numbers or expressions. These numbers or expressions on which the operations are performed are called **operands**. We can express an arithmetic operation such as + or * or ↑ in a binary tree by representing the operation as a vertex and the two operands as left and right children of the operation. Thus $x + y$ would be represented as in Figure 2(a) and $x \uparrow 2$ as in Figure 2(b).

Figure 2(a)   **Figure 2(b)**

When an operand is an expression, the operand is represented as a subtree. Thus $2 + x*y$ would be represented as in Figure 3(a), and $3(x + y)\uparrow 2$ would be represented by the tree in Figure 3(b).

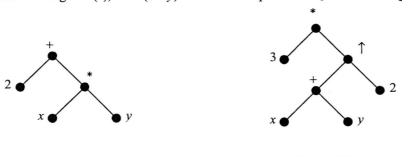

Figure 3(a)   **Figure 3(b)**

We build a binary tree that represents an expression from the bottom up. We build a subtree for the first operation performed and then continue to build subtrees for successive operations until all have been represented. The last operation performed is the root of the tree.

We will need to remember the rules for order of precedence of arithmetic operations.

Order of Precedence of Operations: In an arithmetic or algebraic expression, perform operations in the following order:
1. Operations enclosed in parentheses
2. Exponentiations from right to left
3. Multiplications and divisions from left to right
4. Additions and subtractions from left to right

Example 2
In the algebraic expression below, we will represent exponentiation with an up arrow $\uparrow$. What is the order in which the operations are performed?
$(x - y)/(x + y)^2$ or $(x - y)/(x + y)\uparrow 2$

Solution
The operations are performed in the following order:
The first operation is the subtraction $x - y$.
The second operation is the addition $x + y$.
The third operation is the exponentiation $(x + y) \uparrow 2$.
The fourth operation is the division $(x - y)/(x + y) \uparrow 2$.

Example 3

Build a binary tree that represents $(x - y)/(x + y) \uparrow 2$, the expression discussed in Example 2.

Solution

The first operation performed in this expression is $x - y$, and a tree for this operation is found in Figure 4(a). A tree for $x + y$ is found in Figure 4(b). The tree in Figure 4(c) incorporates the tree for $x + y$ and represents $(x + y) \uparrow 2$. The last operation performed is division, and this operation is represented as the root of the tree in Figure 4(d). Figure 4(d) represents the expression $(x - y)/(x + y) \uparrow 2$ as a tree.

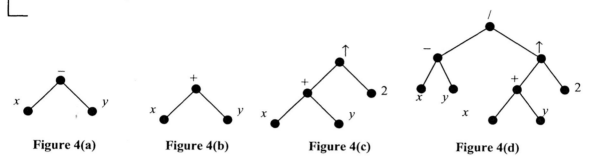

Figure 4(a)   **Figure 4(b)**   **Figure 4(c)**   **Figure 4(d)**

TREE TRAVERSALS

Once a problem is represented using a tree, we must have an algorithm to manipulate the tree to accomplish our purposes. Depending on our application, there are many such algorithms available. We will examine two algorithms designed to list each vertex on a tree exactly one time. Such algorithms are called tree traversal algorithms. We will examine the inorder traversal and the postorder traversal of a binary tree.

The **inorder traversal** of the binary tree in Figure 5 visits the vertices in this order: b then a then c. More generally, when the inorder traversal reaches a subtree with parent vertex a, it lists the vertices in the left subtree of a (if there is a left subtree), then a, and then the vertices in the right subtree of a (if there is a right subtree).

Figure 5   **Figure 6**

Example 4

List the vertices of the binary tree in Figure 6 in the order in which they are visited in the course of an inorder traversal.

Solution:

1. Starting at the root, we wish to list the vertices in the left subtree of the root (the subtree with parent q), then the root p, and then the vertices in the right subtree of the root (the subtree with parent r).

2. In the subtree with parent q, we wish to list the vertices in the left subtree, then q, and then the vertices in the right subtree of q. Hence the first three vertices listed are **s, q, t**.

(continued)
3. We then list the root, *p*.
4. As we list the elements in the subtree with parent *r*, we list the left subtree (the subtree with parent *u*), then *r*, and then the right subtree (which does not exist).
5. In the subtree with parent *u*, we wish to list the vertices in the left subtree, then *u*, and then the vertices in the right subtree of *u*. Hence the next three vertices listed are *v, u, w*.
6. Finally, we list *r*.

The inorder traversal of the tree in Figure 6 lists the vertices in this order: *s, q, t, p, v, u, w, r*.

The **postorder traversal** of the binary tree in Figure 5 visits the vertices in this order: *b* then *c* then *a*. More generally, when the postorder traversal reaches a subtree with parent vertex *a*, it lists the vertices in the left subtree of *a* (if there is a left subtree), then the vertices in the right subtree of *a* (if there is a right subtree), and then *a*.

Example 5

List the vertices of the binary tree in Figure 6 in the order in which they are visited in the course of a postorder traversal.

Solution:
1. Starting at the root, we wish to list the vertices in the left subtree of the root (the subtree with parent *q*), then the vertices in the right subtree of the root (the subtree with parent *r*), and then the root *p*.
2. Hence the first three vertices listed are *s, t, q*.
3. As we list the elements in the subtree with parent *r*, we list the left subtree (the subtree with parent *u*), then the right subtree (which does not exist), and then *r*.
4. In the subtree with parent *u*, we wish to list the vertices in the left subtree of *u*, then the vertices in the right subtree of *u*, and then *u*. Hence the next three vertices listed are *v, w, u*.
5. Then we list *r*.
6. Finally, we list the root **p**.

The postorder traversal of the tree in Figure 6 lists the vertices in this order: *s, t, q, v, w, u, r, p*.

TRAVERSING TREES THAT REPRESENT EXPRESSIONS

Once an expression has been represented in a binary tree, the expression can be evaluated by doing a traversal of the tree. The expression obtained by doing an inorder traversal of the tree is called the **infix form** of the expression. The expression obtained by doing a postorder traversal of the tree is called the **postfix form** of the expression.

Example 6

Figure 7(a) represents the expression $(x - y)/(x + y) \uparrow 2$ as a binary tree, and Figure 7(b) represents the expression $x - y/(x + y \uparrow 2)$ as a binary tree. Find the infix form of the expressions by completing inorder traversals of the trees.

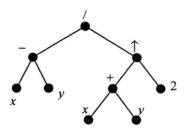

Figure 7(a)

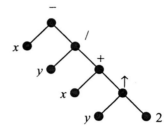

Figure 7(b)

Solution

The inorder traversals of the vertices of the trees in Figure 7(a) and 7(b) both yield the expression $x - y / x + y \uparrow 2$. The expression $x - y / x + y \uparrow 2$ is called the infix form of both expressions in Example 6.

Observe that we do not necessarily recover the original expression from the inorder traversal of the representation of the expression in a tree. Indeed, the infix forms of several different algebraic expressions may be the same. This ambiguity makes the infix form of the expression less useful than the postfix form of the expression in some areas of computer science. In order to return an expression equivalent to the original expression, as we do an inorder traversal of the tree, we would need to include parentheses in the expression each time the traversal encounters an operation. Had we done this as we completed an inorder traversal of the tree in Figure 7(a), we would have obtained the expression $((x - y)/((x + y)\uparrow 2))$. This is equivalent to the expression represented in Figure 7(a).

Example 7

Figure 7(a) is a representation of the expression $(x - y)/(x + y)\uparrow 2$ using a binary tree. Find the postfix form of the expression by completing a postorder traversal of the tree.

Solution

The postorder traversal of the vertices of the tree in Figure 7(a) yields

$$x\,y - x\,y + 2 \uparrow /$$

This is called the postfix form of the expression.

The postfix form of an expression is sometimes called **reverse Polish notation** after the Ukrainian mathematician Jan Lukasiewicz (who taught in Poland and then in Ireland during a career that spanned the first half of the twentieth century). An expression in reverse Polish notation (in postfix form) can be evaluated without ambiguity even in the absence of parentheses. In postfix form, a binary operation such as + or / follows its two operands. To evaluate an expression in postfix form, proceed from left to right until you encounter an operation. Perform the operation on the two operands immediately to the left of the operation, and place the result back into the expression in the place of the two operands and operator.

Example 8

Evaluate the following expressions written in postfix form (reverse Polish notation):

(a) 2 3 + 7 *   (b) 7 3 2 * – 2 ↑ 8 4 / +

Solution

(a) In evaluating the expression from (a), the first operator we encounter in moving from left to right is +. We apply this operator to the two numbers that precede the +, namely 2 and 3, and we place the resulting 5 back into the expression. The rest of the evaluation is shown below:

2 3 + 7 *
5 7 *
5 7 *
3 5

(b) 7 **3 2 *** – 2 ↑ 8 4 / +
7 6 – 2 ↑ 8 4 / +
7 6 – 2 ↑ 8 4 / +
1 2 ↑ 8 4 / +
1 2 ↑ 8 4 / +
1 8 4 / +
1 **8 4 /** +
1 2 +
1 2 +
3

Because reverse Polish notation (postfix represention) can be evaluated without the need for parentheses, certain classes of calculators require users to enter expressions using this notation. In computer science there are also some areas in which postfix representation and prefix representation (see Exercise 17) are more useful than infix representation.

ALPHABETIZING LISTS WITH BINARY TREES

Suppose we wish to alphabetize a list of words using a tree. The following algorithm will organize the words on a binary tree so that we can easily produce an alphabetized list from the tree. We can use a slight modification of the algorithm to search for a word in a list after the list has been organized using a tree, and hence the algorithm is sometimes called the Binary Search Tree Algorithm.

Binary Search Tree Algorithm

1. Choose a word from the list and use it to label the root of the tree. Create a left child and a right child for the root.

2. Choose a new word WORD from the list. Start at the root of the tree, and construct a directed path through the tree until you find a vertex that does not have a label. Determine directions at each vertex along this path by comparing WORD with the label of the vertex and using the following rules:

   If WORD precedes the label of the vertex in alphabetical order, go to the left child.
   If WORD succeeds the label of the vertex in alphabetical order, go to the right child.

3. Label the vertex with WORD and create a left child and a right child for the vertex.

4. If all words from the list are placed on the tree, we are done.

   Otherwise, return to Step 2.

Observe that as we use the words in the list to label the vertices of the tree, we are labeling in such a way that when a word is placed at a specific vertex, it follows all the words in the left subtree of that vertex and precedes all words in the right subtree of that vertex.

Example 9

Use this algorithm to store the following words on a binary tree: *gum, arrow, ray, set, fan, add, zoo.*

Solution

Step 1: Label the root of the tree with the word *gum* and create a left child and a right child.

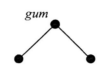

Step 2: Since in alphabetical order
   arrow precedes *gum,*
label the left child of *gum* with *arrow* and create a left child and a right child.

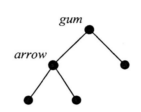

Since there are still unused words on the list, we do Step 2 again. Since in alphabetical order
   ray follows *gum,*
label the right child of *gum* with *ray* and create a left child and a right child.

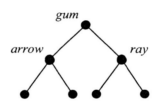

Since there are still unused words on the list, we do Step 2 again. Since in alphabetical order
   set follows *gum*
   set follows *ray*
label the right child of *ray* with *set* and create a left child and a right child.

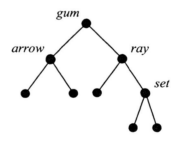

Since there are still unused words on the list, we do Step 2 again. Since in alphabetical order
   fan precedes *gum*
   fan follows *arrow*
label the right child of *arrow* with *fan* and create a left child and a right child.

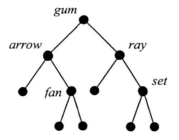

(continued)

Since there are still unused words on the list, we
do Step 2 again. Since in alphabetical order
>   *add* precedes *gum*
>   *add* precedes *arrow*
label the left child of *arrow* with *add* and create a
left child and a right child.

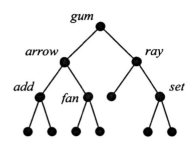

Now, use the algorithm to place *zoo* on the tree and verify that the final tree looks like Figure 8
(where we have removed vertices that did not receive labels).

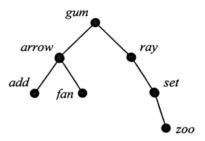

Figure 8

Example 10
> List the vertices of the binary tree in Figure 8 in the order in which they are visited in the course
> of an inorder traversal.

Solution:
> 1. Starting at the root, we list the vertices in the left subtree of the root (the subtree with parent
>   *arrrow*), the root *gum*, and then the vertices in the right subtree of the root (the subtree with
>   parent *ray*).
> 2. In the subtree with parent *arrow*, we wish to list the vertices in the left subtree of arrow,
>   then *arrow*, and then the vertices in the right subtree of *arrow*. Hence the first three vertices
>   listed are *add, arrow, fan.*
> 3. We then list the root *gum.*
> 4. As we list the elements in the subtree with parent *ray*, we list the left subtree (which does
>   not exist), then *ray*, and then the right subtree with parent *set.*
> 5. In the subtree with parent *set*, we wish to list the vertices in the left subtree (which does not
>   exist), then *set*, then the vertices in the right subtree of *set*. Hence the next three vertices
>   listed are *ray, set, zoo.*
>
> The inorder traverse of the tree in Figure 8 lists the vertices in this order: *add, arrow, fan, gum,*
> *ray, set, zoo.*

Observe that when we do an inorder traversal of the tree we created in Example 9, the list is returned to us in alphabetical order. Indeed, this is always the case. Hence, one computer solution to the problem of sorting a list into alphabetical order involves representing the list as a tree using the binary search tree algorithm and then completing an inorder traversal of the resulting binary tree.

Exploratory Exercise Set 5.4

1. In this section we reviewed the order of precedence for operations written in conventional arithmetic notation.
   (a) In the table that follows, we have described the process of computing the expression
   $4 * (3 + 4) / 2 \uparrow 2$ using the rules for order of precedence. We identified that the operation $3 + 4$ should be performed first since it is in parentheses and then listed the other three operations in order.

	Operation	Resulting Expression
First	$(3 + 4)$	$4 * 7 / 2 \uparrow 2$
Second	$2 \uparrow 2$	$4 * 7 / 4$
Third	$4 * 7$	$28 / 4$
Fourth	$28 / 4$	7

   Carefully evaluate each of the following arithmetic expressions using the rules for order of precedence of operations. Construct a table like this one to record your work.
   (1) $3 + (4 / 2) \uparrow 2 + 5$
   (2) $4 * (2 + 3) \uparrow 2 + 3 - 4$
   (3) $16 / (1 + 1) \uparrow 2$
   (4) $(6 + 4 * (4 + 2)) / (4 + 3 * 2)$
   (b) The binary tree below represents the expression discussed in (a). Observe that the root of the tree is the division, the last operation performed.

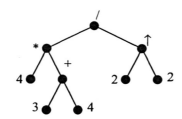

   Draw a tree representing each expression in (1) through (4) of (a).

2. Both you and your partner should draw a binary tree with at least seven vertices. Label the vertices, making sure that one vertex is labeled D and one vertex is labeled F.
   (a) Exchange binary trees with your partner. Using your partner's tree:
       (1) Identify the left child of D if there is one.
       (2) Identify the right child of F if there is one
       (3) Identify the left subtree of D if there is one.
       (4) Identify the right subtree of F if there is one.
       (5) Discuss your answers with your partner who drew the binary tree.
   (b) Again using your partner's binary tree:
       (1) Complete an inorder traversal of the binary tree. List the vertices in the order in which they were visited.
       (2) Complete a postorder traversal of the binary tree. List the vertices in the order in which they were visited.
       (3) Discuss your answers with your partner who drew the binary tree.

3. In this section we learned that to compute an arithmetic expression written in postfix form, we proceed from left to right until we encounter an operation and then apply the operation to the two operands that precede the operation. In the table that follows, we have described the process of computing the postfix expression 4 2 + 7 * 2 3 + + using the rules for order of precedence. We identified that the operation 4 + 2 should be performed first since 4 and 2 are the two operands that precede our first operation, +.

	Operation	Resulting Expression
First	4 + 2	6 7 * 2 3 + +
Second	6 * 7	42 2 3 + +
Third	2 + 3	42 5 +
Fourth	42 + 5	47

Carefully evaluate each of the following arithmetic expressions written in postfix form. Construct a table like this one to record your work.
   (a) 2 4 + 6 3 − 5 * 3 + *
   (b) 4 7 3 2 2 / + * −
   (c) 2 4 + 1 3 −* 4 2 / 3 − +
   (d) 6 3 2 − 4 8 2 / + + *

4. Consider the expression (2 + 3 * 5) − (4 − 6 / 2) and a binary tree that represents this expression.

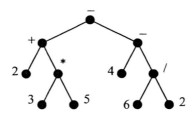

   (a) Complete an inorder traversal of the binary tree.
      (1) List the vertices in the order in which they were visited. Compare the resulting infix form with the original expression. Are the expressions equivalent if one uses the usual rules for order of precedence?
      (2) Again, complete an inorder traversal of the binary tree. Each time you encounter a binary operation, enclose the operation and its two operands in parentheses. [For example, after traversing the left subtree of the root, you will have (2 + (3 * 5)).] Compare the resulting expression with the original expression. Do they provide the same result if you compute them using the usual rules for order of precedence?
   (b) Complete a postorder traversal of the tree.
      (1) List the vertices in the order in which they were visited.
      (2) The resulting expression is in postfix form. Evaluate it as a postfix expression. Do you get the same result that you would get with the original expression using the usual rules for order of precedence?
   (c) In (b) we see that we can write an arithmetic expression in postfix form by completing a postorder traversal of the binary tree for the expression and recording the order in which we reached the vertices. Write the expressions from Exploratory Exercise 1(a) in postfix form by completing a postorder traversal of the binary trees created in Exploratory Exercise 1(b).

5. (*Writing Exercise*) A seventh-grade class has been studying rules of precedence for arithmetic operations. You are invited to present a short enrichment lecture on evaluating expressions given in postfix form. Write your lecture and be sure to include examples not given in this text.

Exercise Set 5.4

In the binary trees of Exercises 1 and 2, identify:
- (a) *The left subtree of B*
- (b) *The right child of E*
- (c) *The left child of F*
- (d) *The right subtree of D*

1.

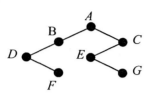

2.

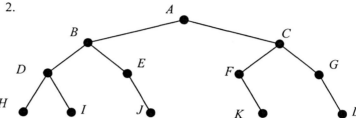

3. Construct the binary tree for the following arithmetic expressions. Remember that the last operation performed is the root of the binary tree.
   - (a) $(3 - 5) * (12 / 3)$
   - (b) $(2 + 3) * 4$
   - (c) $2 + 4 * 3 - 8 / 4$
   - (d) $(c - d / e) + x * y$
   - (e) $((6 - 3) * 3) + 7)/((5 - 1) + 8)$
   - (f) $((a - b) / c)(d + e / f)$

For each of the trees in Exercises 4 through 7:
- (a) *Complete an inorder traversal, and list the vertices in the order visited.*
- (b) *Complete a postorder traversal, and list the vertices in the order visited.*

4.

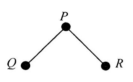

5.

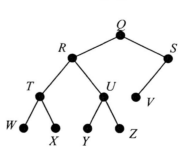

6.

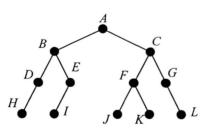

7.

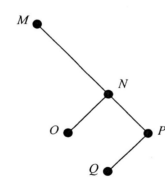

8. Evaluate the following expressions written in postfix form.
   - (a) $3\ 6 - 7 * 3\ +$
   - (b) $3\ 4\ 5\ 2\ 2 / + * -$
   - (c) $2\ 3 + 3\ 7 - - 3 * 4 +$
   - (d) $4\ 4 + 1\ 2 - * 6\ 3 / 3 - +$

9. Write each of the arithmetic expressions in Exercise 3 in postfix form by completing a postorder traversal of the associated binary tree and recording the order in which the vertices are visited.

10. Lists of words have been organized on these trees using the Binary Search Tree Algorithm. Use inorder traversal to place the lists of words in alphabetical order.

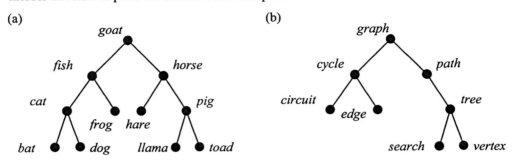

11. (a) Use the Binary Search Tree Algorithm to place the list of names {Manuel, Bob, Tran, Rita, Sida, Ric, Pat, Ida} as labels on the vertices of a binary tree.
    (b) Use an inorder traversal to produce an alphabetized version of this list.

12. Construct binary trees for the following arithmetic expressions written in postfix notation (reverse Polish notation).
    (a) $a\,c*d-d+$
    (b) $d\,e-b+a\,c-f+*$

13. Construct a binary tree for which the inorder listing of vertices is B, D, A, F, E, C and the postorder listing of vertices is D, B, F, E, C, A.

14. Construct a binary tree with at least three vertices in which the postorder listing of vertices is the same as the inorder listing of vertices.

*Another form of traversal of a binary tree is called a preorder traversal. The **preorder traversal** of the binary tree in Figure 9 visits the vertices in this order: a then b then c. More generally, when the preorder traversal reaches a subtree with parent vertex a, it lists a, then the vertices in the left subtree of a (if there is a left subtree), and then the vertices in the right subtree of a (if there is a right subtree). The vertices in the binary tree in Figure 10 would be visited in this order: **p, q, s, t, r, u, v, w**.*

Figure 9

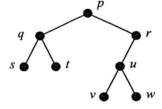

Figure 10

15. Complete a preorder traversal of these binary trees, and list the vertices in the order visited.

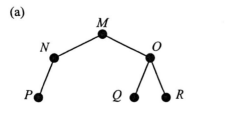

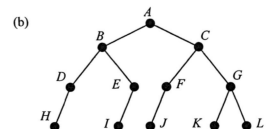

16. Complete a preorder traversal of each of the binary trees in Exercises 4 through 7, and list the vertices in the order visited.

17. When we traverse a binary tree that represents an expression using a preorder traversal, we produce an arithmetic expression that is written in **prefix form** (also called **Polish notation**). We evaluate an expression in prefix form by scanning the expression from the left until we come to an operation followed by two numbers. We then perform the operation on the two numbers, replace the numbers and operation by the result, and return to scan from the left again. Complete the evaluation of the following expressions written in prefix form.

(a) $+ + 4 * 3\ 4 + 5 / 6\ 3$
    $+ + 4\ 12 + 5 / 6\ 3$
    $+ 16 + 5 / 6\ 3$

(b) $+ * 4\ 6 - 8 * 1\ 4$
    $+ 24 - 8 * 1\ 4$

18. Use a preorder traversal of the binary trees from Exercise 3(a) through (c) and write the resulting prefix expression. Then evaluate the prefix expression and compare the results to the evaluations of the expressions in Exercise 3(a) through (c) using the rules of precedence of operations.

Combinatorics

Chapter

6

6.1 There Is More to Counting Than 1, 2, 3, …

The word *combinatorics* can look and sound rather intimidating. Combinatorics is the branch of mathematics that studies counting. Counting? Is that all it is? But why do we need a whole branch of mathematics to study something as simple as counting? In combinatorics, we often begin with fairly simple problems, such as the one in Example 1 below, that actually can be solved by simply making a list and counting how many items there are in the list. In examining simple examples, we hope to identify, often by organizing items in a particular way, patterns that enable us to count larger lists that would be impractical, or even impossible, to count by hand. The truth is that although the subject matter of this chapter involves more than what we usually think of as counting (pointing and saying 1, 2, 3, …), it is not quite as daunting as the fancy name *combinatorics* might suggest.

THE MULTIPLICATION PRINCIPLE

Example 1

Ardie stands in Happy Jack's Computer Store, perplexed at his options as he tries to put together a computer system. He needs to buy a central processing unit, or CPU, and a monitor. In his price range he can choose from two CPUs: a Granny and a HAL. He can choose from three monitors: an Audioramic, a Britevue, and a Colorsonic. Ardie first tries to list all the possible systems in no particular order, but he's not sure he has listed them all. He gets the following list of six possible systems:

> Granny-Colorsonic
> HAL-Audioramic
> HAL-Colorsonic
> Granny-Britevue
> HAL-Britevue
> Granny-Audioramic

He decides it is a pain writing out the names each time, so he comes up with the following abbreviations: *G* and *H* for the CPUs and *A*, *B*, and *C* for the monitors. This gives him a list that's easier to write down (*GC, HA, HC, GB, HB, GA*), but he's still not sure he has considered every possibility. Then Ardie has a brilliant idea! He draws the diagram in Figure 1, where the CPU choices (*G* or *H*) are shown on the left, and the monitor choices (*A*, *B*, or *C*) are shown to the right of each CPU choice. (If Ardie has happened to read Chapter 5, he might recognize that he has just drawn a special kind of graph called a **rooted tree**, where the endpoint on the left is called the **root**, and the endpoints on the right are called the **leaves**.)

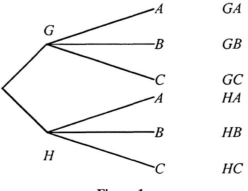

Figure 1

(continued)

By counting the leaves in his tree, Ardie is now confident that there are 6 possible systems he can purchase. Looking at the tree more closely, Ardie observes that there are 2 choices for the CPU; then, for each of the CPU choices, there are 3 choices for a monitor. Aha! There are $2 \cdot 3 = 6$ ways of building his computer system.

Ardie has discovered one of the fundamental principles of counting, the Multiplication Principle.

The Multiplication Principle:

(1) If a process can be broken down into two steps, performed in order, with n_1 ways of completing the first step and n_2 ways of completing the second step after the first step is completed, then there are $n_1 \cdot n_2$ ways of completing the process.

(2) More generally, if a process can be understood as a sequence of k steps performed in order, with n_i the possible number of ways of completing the ith step after the first $i - 1$ steps have been completed, then the number of ways of completing the process is the product $n_1 \cdot n_2 \cdot \cdots \cdot n_k$.

Example 2

The chorale at a Savannah university is planning a concert tour with performances in Chicago, then New York, and finally, back home in Savannah. The musicians can travel from Savannah to Chicago by bus or plane; from Chicago to New York by bus, plane, or train; and from New York to Savannah by plane, ship, or train. In how many ways can the transportation be arranged for this trip?

Solution

The process of arranging transportation can be broken into three steps: choosing transportation from Savannah to Chicago, choosing transportation from Chicago to New York, and choosing transportation from New York to Savannah. There are $n_1 = 2$ ways to complete the first step, $n_2 = 3$ ways to complete the second step, and $n_3 = 3$ ways to complete the third step. Hence, there are $2 \cdot 3 \cdot 3 = 18$ ways of arranging the transportation. Figure 2 shows a rooted tree that verifies this result.

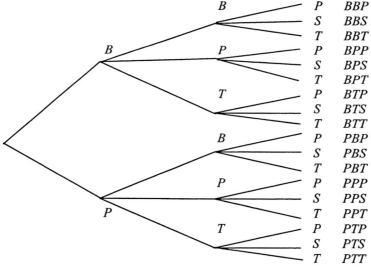

Figure 2

Even though we can draw a diagram to see all of the possible ways of arranging transportation in Example 2, the real power of the Multiplication Principle is that we can count all options without drawing a picture or even listing abbreviations for each possibility. This is especially useful when the number of possible choices is so large that it would not be practical to list them all. In the examples that follow, we do not recommend attempting to draw a rooted tree to illustrate all possible choices.

Example 3

Count the number of possible outcomes when:

(a) A coin is tossed five times.

(b) A standard die is rolled three times.

Solution

(a) The process of flipping a coin five times can be viewed in terms of five distinct steps. If we classify the outcome of each flip as either a head or a tail, there are 2 ways of completing each step. By the Multiplication Principle, there are $2 \cdot 2 \cdot 2 \cdot 2 \cdot 2 = 2^5 = 32$ possible outcomes.

(b) A standard die is a cube with 1, 2, 3, 4, 5, and 6 dots, respectively, on its six faces. Hence there are six possible results each time we complete the step of rolling the die. Thus there are $6 \cdot 6 \cdot 6 = 6^3 = 216$ possible outcomes when we roll a die three times.

Example 4

How many six-digit positive integers are there?

Solution

The first digit (the hundred thousands digit) cannot be a 0, so there are 9 choices for it. There are 10 choices for each of the other five digits, so there are a total of $9 \cdot 10 \cdot 10 \cdot 10 \cdot 10 \cdot 10 = 9 \cdot 10^5 = 900,000$ six-digit positive integers. Alternatively, we could count all positive integers with six or fewer digits (999,999), and then subtract the number of positive integers with fewer than six digits (99,999), giving a total of $999,999 - 99,999 = 900,000$.

Example 5

In the state of Panic, automobile license plates contain an arrangement of three capital letters followed by three digits. If all letters and digits may be used repeatedly, how many different arrangements are available?

Solution

We can think of the process of forming a license plate arrangement in terms of a sequence of six steps. We need to choose a letter for each of the three letter positions on the plate; then we must choose a digit for each of the three digit places. There are 26 letters that may occupy each of the three letter places, and there are 10 digits that may occupy the digit places. By the Multiplication Principle, the number of arrangements is

$$26 \cdot 26 \cdot 26 \cdot 10 \cdot 10 \cdot 10 = 17,576,000.$$

PERMUTATIONS

Example 6

Elizabeth is the curator of a museum. She has three paintings by a particular artist that she wants to display side by side on one wall of a gallery. Elizabeth wonders how many different ways there are to arrange the three paintings.

Solution

We can break down the process of arranging the paintings into three steps: First choose a painting to hang in the position on the left; then choose a painting to hang in the middle; finally, choose a painting to hang on the right. There are $n_1 = 3$ ways to choose a painting to hang on the left, since there are three paintings to choose from. After we have completed the first step, there are only two paintings left to hang, so there are $n_2 = 2$ ways to choose a painting to hang in the middle. Finally, after we have completed the first two steps, there is only one painting remaining to be hung, so there is $n_3 = 1$ way to choose a painting to hang on the right. Using the Multiplication Principle, there are $3 \cdot 2 \cdot 1 = 6$ different arrangements possible. If the paintings are labeled A, B, and C, then the rooted tree in Figure 3 illustrates the six possible arrangements.

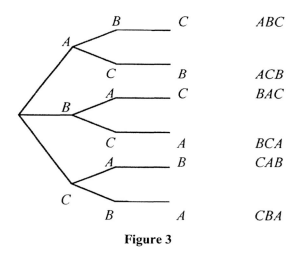

Figure 3

Ordered arrangements of a set of distinct elements, like the arrangements of Elizabeth's paintings, are called **permutations** of a set. As we saw in Example 6, the Multiplication Principle is particularly useful in counting the number of permutations of a set. Notice that in Example 6, repetitions of a painting are not allowed. For example, AAB is not a permissible arrangement, since we do not have two copies of painting A to hang. Contrast this with the license plates in Example 4, where we are allowed to repeat both letters and digits.

Example 7

In how many ways can six students line up to go outside for recess?

Solution

Clearly we need to count the ordered arrangements, or permutations, of a set of six people. As a first step, we choose a person to be first in line; there are six ways to do this. Then we can choose a second person, and only five persons are available after the first person is chosen. Similarly, the third place must be filled by one of four persons, and so on. There are $6 \cdot 5 \cdot 4 \cdot 3 \cdot 2 \cdot 1 = 720$ ways of arranging these six students.

In the previous two examples, we observed that a set of 3 distinct elements has $3 \cdot 2 \cdot 1$ permutations, whereas a set of 6 elements has $6 \cdot 5 \cdot 4 \cdot 3 \cdot 2 \cdot 1$ permutations. Notice that we would get the same numbers if we were arranging flowers or animals rather than paintings or students, as long as the objects we were arranging were all distinct. In general, we can make the following generalization.

Number of Permutations of n Objects: The number of permutations of n distinct objects is given by the product $n(n-1)(n-2) \cdot \cdots \cdot 3 \cdot 2 \cdot 1$.

The product $n(n-1)(n-2) \cdot \cdots \cdot 3 \cdot 2 \cdot 1$, for $n \geq 1$, is represented by the special symbol $n!$, which is read n **factorial**. Hence we may say that the number of permutations of a set of n objects is $n!$. The number zero factorial, represented by $0!$, is defined to be 1.

Example 8

Suppose that Elizabeth has secured two more paintings for her exhibit, for a total of five paintings available, but she still has only enough room to display three of them side by side. How many arrangements of three paintings can she make, if there are five paintings to choose from?

Solution

As before, we will break down the process into three steps: Choose the painting on the left; then choose the painting in the middle; finally, choose the painting on the right. There are now five ways to choose the painting on the left, after which there are four ways to choose the painting in the middle, and three ways to choose the painting on the right, for a total of $5 \cdot 4 \cdot 3 = 60$ possible arrangements.

An ordered arrangement of a subset of 3 elements chosen from a set of 5 elements is called a **3-permutation** of the set. The number of 3-permutations from a set of 5 elements is denoted $P(5, 3)$. In the previous example, we determined that

$$P(5, 3) = 5 \cdot 4 \cdot 3 .$$

If we wished to express $P(5, 3)$ using factorials, we could observe that $5 \cdot 4 \cdot 3$ starts out like $5!$ but is missing the last two factors, $2 \cdot 1$.

$$P(5, 3) = 5 \cdot 4 \cdot 3 = \frac{5 \cdot 4 \cdot 3 \cdot 2 \cdot 1}{2 \cdot 1} = \frac{5!}{2!} .$$

Thus, $P(5, 3) = \dfrac{5!}{2!} = \dfrac{5!}{(5-3)!}$.

An **r-permutation of a set of n elements** is an ordered arrangement of a subset of r elements chosen from the set of n elements. The number of r-permutations of a set of n elements is denoted by $P(n, r)$. Other common notations for this number include $_nP_r$ and $P_{n,r}$. By using the Multiplication Principle and reasoning as in the computation above, we discover the following fact:

The **number of r-permutations from a set of n elements** is given by

$$P(n, r) = n \cdot (n-1) \cdot (n-2) \cdot \cdots \cdot (n-r+1) = \frac{n!}{(n-r)!} .$$

Notice that this new formula confirms the fact we learned earlier in Examples 6 and 7, that the number of n-permutations from a set of n objects is $P(n, n) = \dfrac{n!}{(n-n)!} = \dfrac{n!}{0!} = n!$, since we defined $0!$ to be equal to 1.

Example 9

Count the number of 2-permutations of the set $\{a, b, c, d, e, f\}$ using the formula, and then list all of the permutations.

Solution

$P(6, 2) = \dfrac{6!}{(6-2)!} = \dfrac{6 \cdot 5 \cdot 4 \cdot 3 \cdot 2 \cdot 1}{4 \cdot 3 \cdot 2 \cdot 1} = 6 \cdot 5 = 30$. The thirty 2-permutations of this set are listed below.

ab	*ac*	*ad*	*ae*	*af*
ba	*bc*	*bd*	*be*	*bf*
ca	*cb*	*cd*	*ce*	*cf*
da	*db*	*dc*	*de*	*df*
ea	*eb*	*ec*	*ed*	*ef*
fa	*fb*	*fc*	*fd*	*fe*

THE ADDITION PRINCIPLE

Example 10

One member of the faculty of the College of Education is to be appointed to a search committee for the president of the university. That faculty member is to be selected from the nine faculty members in the Department of Early Childhood Education, *or* from the six members of the Department of Middle Grades and Secondary Education, *or* from the five members of the Department of Special Education. In how many ways can this appointment to the search committee be made?

Solution

The process of making this appointment can be completed by performing one of the following three steps: selecting a member of the Department of Early Childhood Education, selecting a member of the Department of Middle Grades and Secondary Education, or selecting a member of the Department of Special Education. However, the process is not completed by performing all three steps sequentially, so using the Multiplication Principle is not appropriate. Rather, the process is completed by performing exactly *one* of the steps, to the exclusion of the others. Hence, if no two of the departments share members, the process can be completed in $9 + 6 + 5 = 20$ ways. In solving this problem, we have happened upon the **Addition Principle**.

The Addition Principle:

(1) Suppose that X_1 and X_2 are disjoint sets, where X_1 has n_1 elements and X_2 has n_2 elements. If a process is completed by choosing one element from set X_1 or by choosing one element from set X_2, then there are $n_1 + n_2$ ways of completing the process.

(2) More generally, suppose that $\{X_1, X_2, \ldots, X_k\}$ is a collection of pairwise disjoint sets (that is, no two sets have elements in common), where set X_i has n_i elements for each integer i, $1 \le i \le k$. If a process is completed by choosing one element from exactly one of the sets in this collection, then the number of ways to complete the process is the sum $n_1 + n_2 + \cdots + n_k$.

Example 11

Jermonte needs one course to finish his core curriculum. He can choose that course from a list of three courses in the history department, from a list of four courses in political science, or from a list of six courses in philosophy. If no two of the lists share courses, in how many ways may Jermonte finish his curriculum?

Solution

Because Jermonte makes one choice from among three lists of courses that are pairwise disjoint, we use the Addition Principle to count the number of options. Jermonte may finish the core curriculum in $3 + 4 + 6 = 13$ ways.

Many problems can be solved using the Addition Principle and the Multiplication Principle in concert.

Example 12

Passwords on a school's computer system are six characters long. Either they begin with 2 letters that are followed by 4 digits or they begin with 3 letters that are followed by 3 digits. How many different passwords are possible on this system?

Solution

Let L be the total number of possible passwords. Let L_2 be the number of possible passwords that begin with exactly 2 letters, and let L_3 be the number of possible passwords that begin with exactly 3 letters. By the Addition Principle, $L = L_2 + L_3$. Using the Multiplication Principle, we observe that

$L_2 = 26 \cdot 26 \cdot 10 \cdot 10 \cdot 10 \cdot 10 = 6,760,000$ and

$L_3 = 26 \cdot 26 \cdot 26 \cdot 10 \cdot 10 \cdot 10 = 17,576,000,$

so the number of possible passwords is $6,760,000 + 17,576,000 = 24,336,000$.

PERMUTATIONS WITH STICKY LETTERS

We close this section with a nice trick to help us count permutations where two or more letters must appear in a specified order. Consider the following examples where we imagine gluing two or more of the letters together.

Example 13

How many arrangements are there of the letters in the word EQUATOR if the letter Q must be followed immediately by the letter U?

Solution

If there were no restriction on the order of the letters, there would be $7! = 5040$ permutations. Since the Q must be followed by the U, we can think of gluing the Q and the U together to make a single "letter" QU. Then we simply permute the six "letters" E, QU, A, T, O, and R, for a total of $6! = 720$ possible arrangements.

Example 14

Ann, Bob, Cal, Deb, Eve, Flo, and Guy go to see a movie. In how many ways can they sit in a single row if Ann and Bob insist on sitting next to each other?

Solution

> This is only slightly different from the previous example since we are not specifying whether Ann is on Bob's left or right. There are 6! arrangements with Ann on Bob's left and 6! different arrangements with Ann on Bob's right. Since Ann cannot simultaneously be on Bob's left and on his right, the Addition Principle tells us that the total number of arrangements is 6! + 6! = 1440.

Example 15

> How many arrangements are there of the letters in the word PITCHES in which the letter *H* is preceded immediately by *C, P, S,* or *T*?

Solution

> If we glue the *C* and the *H* together to make *CH,* there are 6! permutations. Similarly, there are 6! different permutations with each of the pairs *PH, SH,* and *TH,* for a total of $4 \cdot 6! = 2880$ possible arrangements.

Exploratory Exercise Set 6.1

1. In Ms. Johnson's class, students may choose one of three math activities (geoboards, pentominos, or tangrams), one of two spelling activities (crossword or word search), and one of two social studies activities (clay map or wood puzzle).
    (a) How many different choices of activities are possible if the students must pick one activity from each subject area?
    (b) Assign abbreviations to all of the activities, and draw a tree diagram that illustrates the different choices.
    (c) How many different choices are possible if the students must pick a total of only one activity?
    (d) Explain the principles you used in parts (a) and (c) and why you used them.

2. Split up into pairs to consider the following problem. Computer passwords in Ms. Hall's class must consist of four characters, where each character must be a digit, that is, an element of the set {0, 1, 2, 3, 4, 5, 6, 7, 8, 9}.
    (a) Have one person compute the number of possible passwords if digits may not be repeated.
    (b) Have the other person compute the number of possible passwords if digits may be repeated.
    (c) Compare results from (a) and (b) and then, together, try to determine how many passwords are possible if the only restriction is that no two *consecutive* digits may be the same.

3. A permutation of *n* distinct objects is sometimes called a rearrangement of those objects. This may be a bit misleading in that not every object must have its position changed.
    (a) List all 3! = 6 rearrangements of the letters in the word ACE (including the original arrangement).
    (b) How many rearrangements of the letters in the word ACE are there in which no letter is in its original position? Such a rearrangement is called a *derangement.*
    (c) Is ZWYX a derangement of WXYZ? Explain.
    (d) List all of the derangements of the letters in the word BEST. How many are there?

The following exercises introduce ideas that we will explore in greater detail later in the chapter.

4. Split up into teams of two or three and consider the word TEAMS.
    (a) How many three-letter words can your team form from the letters in TEAMS? (The words must be in the English language; no abbreviations or proper names are allowed.)
    (b) How many 3-permutations can be formed using letters from the word TEAMS?

(c) Consider the set {T, E, A, M, S}. Have one student list all possible 3-element subsets of this set. How many distinct 3-element subsets did you get? (Remember that {T, E, A} and {A, T, E} are equal as sets, since they contain the same elements.)

(d) List each of the 3-element subsets found in part (c) in a column, with one subset on each row. Have the other team member(s) list all possible permutations (arrangements) of the three letters in the 3-element subset in a row to the right of the subset.

(e) How many permutations are there in each row? How many rows are there? What is the total number of permutations of three letters from {T, E, A, M, S}?

5. (a) A coin is flipped five times. How many different sequences of heads and tails are possible with two heads and three tails? List them.

(b) How many different sequences with two heads and three tails are possible if no two consecutive heads are allowed? List them.

(c) In how many ways can five students be arranged in a line for a picture?

(d) In how many ways can two girls and three boys be arranged in a line if no two girls are allowed to stand next to each other? (Try naming the girls Ann and Betty and the boys Cole, Don, and Ed.)

(e) Explain how your answers to (c) and (d) are related to your answers to (a) and (b).

6. (*Writing Exercise*) Your friend is having trouble understanding when to use the Multiplication Principle and when to use the Addition Principle. Write a paragraph explaining this to him. Include a specific example of a counting problem using each principle and a third example where you use both principles.

Exercise Set 6.1

---

1. A student plans a trip from Atlanta to Boston to London. From Atlanta to Boston, he can travel by bus, train, or airplane. However, from Boston to London, he can travel only by ship or airplane.
    (a) In how many ways can the trip be made?
    (b) Verify your answer by drawing a tree diagram and counting the routes.

2. A sociology quiz contains a true-false question and two multiple-choice questions with possible answers (a), (b), (c), and (d).
    (a) In how many possible ways can the test be answered?
    (b) Verify your answer by drawing a tree diagram and counting the routes.

3. There are 6 roads between A and B and 4 roads between B and C.
    (a) In how many ways can Joy drive from A to C by way of B?
    (b) In how many ways can Joy drive a round trip from A to C through B and back?

4. Kate wants to buy an automobile. She has a choice of 2 body styles (2-door or 4-door), 2 types of transmissions (automatic or stick-shift), and 4 colors (green, red, black, or blue). In how many ways can she select an automobile?

5. (a) How many 3-digit positive integers are there?
    (b) How many 4-digit positive integers are there?
    (c) How many 5-digit positive integers are there?
    (d) How many n-digit positive integers are there, where n is a positive integer?

6. (a) How many 8-digit positive integers are there?
    (b) How many 8-digit positive integers are there in which all of the digits are even?
    (c) How many 8-digit positive integers are there with at least one odd digit?

7. A restaurant offers the following menu:

Main Course	Vegetables	Beverage
Beef	Potatoes	Milk
Ham	Green beans	Coffee
Fried chicken	Green peas	Tea
Shrimp	Asparagus	

If you must choose 1 main course, 1 vegetable, and 1 beverage, in how many ways can you order a meal?

8. Compute the following factorial numbers without using a calculator:
   (a) 4!   (b) 7!   (c) 0!   (d)1!

9. (a) How many permutations are there of the set $\{P, Q, R\}$?
   (b) Write down all of the permutations of $\{P, Q, R\}$.

10. (a) How many 2-permutations are there from the set $\{W, X, Y, Z\}$?
    (b) List all of the permutations in (a).

11. (a) In how many ways can two speakers be arranged on a program?
    (b) In how many ways can three speakers be arranged on a program?
    (c) In how many ways can four speakers be arranged on a program?

12. Evaluate each of the following permutation numbers.
    (a) $P(5, 3)$ (b) $P(6, 5)$ (c) $P(8, 1)$
    (d) $P(9, 2)$ (e) $P(7, 2)$ (f) $P(8, 7)$

13. A die is tossed, and a chip is drawn from a box containing three chips numbered 1, 2, and 3. How many possible outcomes can be obtained from this experiment? Verify your answer with a tree diagram.

14. (a) How many 3-digit numbers are there? (Remember, a 3-digit number cannot begin with 0.)
    (b) How many 3-digit numbers are there that end in 3, 6, or 9?

15. Write a simple expression for each of the following:
    (a) $P(r, 1)$ (b) $P(k, 2)$ (c) $P(r, r - 1)$
    (d) $P(k, k - 2)$ (e) $P(k, 3)$ (f) $P(k, k - 3)$

16. The license plates for a certain state display 3 letters followed by 3 digits (examples: MFT-986, APT-098). How many different license plates can be manufactured if no repetitions of letters or digits are allowed?

17. Employee ID numbers at a large factory consist of 4-digit numbers such as 0133, 4499, and 0000, where it is possible for the first digit to be zero.
    (a) How many possible ID numbers are there?
    (b) How many possible ID numbers are there in which all 4 digits are different?

18. Consider the number 35,964.
    (a) How many 3-digit numbers can be formed using digits from 35,964 if digits may be repeated?
    (b) How many 3-digit numbers can be formed using digits from 35,964 if *no* digits may be repeated?
    (c) What is the sum of all of the 3-digit numbers in part (b)?

19. How many arrangements are there of the letters in the word SQUARE if:
    (a) the Q must be followed by the U?
    (b) the R must be followed by a vowel?

20. In Ms. Ng's class of 10 students, every time Calvin and Bob sit next to each other they get into trouble. How many ways are there to arrange the 10 students in a line so that:
    (a) Calvin and Bob sit next to each other?
    (b) Calvin and Bob do not sit next to each other?

6.2 Combinations

When we last saw Elizabeth, the art museum curator in Example 6 of Section 6.1, she was computing how many arrangements of three paintings she could make from a collection of five paintings. Recall that this amounts to counting the number of 3-permutations from a set of five elements, which is

$$P(5, 3) = \frac{5!}{(5-3)!} = 5 \cdot 4 \cdot 3 = 60.$$

Example 1

Now Elizabeth has a different problem. Another museum has contacted her and wants to borrow any three of the five paintings in her collection for a new exhibit. In how many ways can Elizabeth choose three paintings to send to the other museum?

Before answering this question, let us pause to consider what makes it different from the permutation problem in Example 6 of Section 6.1. In that permutation problem, the order in which the three paintings are arranged on the wall is important to Elizabeth. Hanging painting A on the left, B in the middle, and C on the right may have a different visual impact on the viewers of the exhibit than hanging, say, C on the left, B in the middle, and A on the right. In the problem in Example 1, however, the order in which the three paintings are chosen to send to the other museum is not important. No matter whether Elizabeth sends away paintings A, C, and D or paintings C, D, and A, she is still missing the same three paintings. In the language of sets, the subset $\{A, C, D\}$ is equal to the subset $\{C, D, A\}$ since they contain the same elements. We can translate the question in Example 1 into the language of sets by asking, "How many subsets with three elements can be formed from a set with five elements?" We make the following definition to distinguish between permutations, where order is important, and subsets, also called **combinations**, where order is not important.

Definition: A subset with r elements is called an r-**combination**. The number of r-combinations from a set with n elements is denoted in either of the following ways:

$$C(n, r) \quad \text{or} \quad {}_nC_r \quad \text{or} \quad \binom{n}{r}$$

The third notation is read aloud as "n choose r," since it gives the number of ways r objects can be chosen from a (larger) set of n objects.

Using our new terminology, the question in Example 1 asks for $C(5, 3)$, the number of 3-combinations from a set of 5 elements, or $\binom{5}{3}$, "5 choose 3," the number of ways to choose three objects from a set of five. Let us now finally answer that question.

Solution to Example 1

We have already seen that, when order is important, the number of 3-permutations from a set with five elements is $P(5, 3) = 60$. If we label the five paintings as A, B, C, D, and E, then we can list the 60 permutations (see Table 1).

ABC	ACB	BAC	BCA	CAB	CBA
ABD	ADB	BAD	BDA	DAB	DBA
ABE	AEB	BAE	BEA	EAB	EBA
ACD	ADC	CAD	CDA	DAC	DCA
ACE	AEC	CAE	CEA	EAC	ECA
ADE	AED	DAE	DEA	EAD	EDA
BCD	BDC	CBD	CDB	DBC	DCB
BCE	BEC	CBE	CEB	EBC	ECB
BDE	BED	DBE	DEB	EBD	EDB
CDE	CED	DCE	DEC	ECD	EDC

Table 1

Notice that the first row of our list contains the $3! = 6$ permutations of paintings A, B, and C and that the second row contains the 6 permutations of paintings A, B, and D. In general, each row contains the 6 permutations of some three of the five paintings.

If order is not important, then each arrangement is considered to be the same as any of the other arrangements that contain the same three paintings. In other words, any two arrangements in the same row in Table 1 correspond to the same combination, or subset, of $\{A, B, C, D, E\}$. In order to count the number of ways Elizabeth can choose three paintings to ship away, we simply need to count the number of rows in Table 1. The answer is 10.

A FORMULA FOR $C(N, R)$

But why 10? We would like to see where the answer is coming from so that we may generalize this question to other examples, even examples where the numbers are so large that it is not convenient to list all of the possibilities. Notice that there are a total of 60 arrangements in Table 1, with $3! = 6$ arrangements in each row containing the same three paintings. The number of rows, therefore, must be $60/6 = 10$.

More generally, suppose we have a collection of n objects and we want to choose r of them, without regard to order. We have seen in the previous section that, when order is important, the number of r-permutations from a set with n elements is

$$P(n, r) = \frac{n!}{(n-r)!}.$$

If order is not important, then for any choice of r objects, there are $r!$ arrangements that all contain those same r objects. We can imagine constructing a rectangular table as in Table 1, with $r!$ arrangements in each row. Thus the number of ways we can choose r objects from a set of n objects is

$$C(n,r) = \frac{P(n,r)}{r!} = \frac{n!}{(n-r)!} \cdot \frac{1}{r!} = \frac{n!}{r!(n-r)!}$$

We summarize the formula for counting combinations that we have just discovered:

The $C(n, r)$ Formula: The **number of r-combinations from a set of n elements** is given by

$$C(n,r) = \frac{n!}{r!(n-r)!}.$$

Example 2

Compute the following combination numbers:
(a) $C(6, 2)$ (b) $C(5, 2)$
(c) $C(7, 3)$ (d) $C(7, 4)$
(e) $C(5, 1)$ (f) $C(6, 6)$

Solution

(a) $C(6, 2) = \dfrac{6!}{2!(6-2)!} = \dfrac{6!}{2!\,4!} = \dfrac{6\cdot5\cdot4\cdot3\cdot2\cdot1}{2\cdot1\cdot4\cdot3\cdot2\cdot1} = \dfrac{6\cdot5}{2\cdot1} = 15$

(b) $C(5, 2) = \dfrac{5!}{2!(5-2)!} = \dfrac{5!}{2!\,3!} = \dfrac{5\cdot4\cdot3\cdot2\cdot1}{2\cdot1\cdot3\cdot2\cdot1} = \dfrac{5\cdot4}{2\cdot1} = 10$

(c) $C(7, 3) = \dfrac{7!}{3!(7-3)!} = \dfrac{7!}{3!\,4!} = \dfrac{7\cdot6\cdot5\cdot4\cdot3\cdot2\cdot1}{3\cdot2\cdot1\cdot4\cdot3\cdot2\cdot1} = \dfrac{7\cdot6\cdot5}{3\cdot2\cdot1} = 35$

(d) $C(7, 4) = \dfrac{7!}{4!(7-4)!} = \dfrac{7!}{4!\,3!} = \dfrac{7\cdot6\cdot5\cdot4\cdot3\cdot2\cdot1}{4\cdot3\cdot2\cdot1\cdot3\cdot2\cdot1} = \dfrac{7\cdot6\cdot5}{3\cdot2\cdot1} = 35$

(e) $C(5, 1) = \dfrac{5!}{1!(5-1)!} = \dfrac{5!}{1!\,4!} = \dfrac{5\cdot4\cdot3\cdot2\cdot1}{1\cdot4\cdot3\cdot2\cdot1} = 5$

(f) $C(6, 6) = \dfrac{6!}{6!(6-6)!} = \dfrac{6!}{6!\,0!} = \dfrac{6!}{6!\,1} = 1$ (Recall that we define 0! to be 1.)

Example 3

Let S be a set with 6 elements. How many subsets of S are there that contain
(a) no elements?
(b) exactly 1 element?
(c) exactly 2 elements?
(d) exactly 3 elements?
(e) exactly 4 elements?
(f) exactly 5 elements?
(g) at least 4 elements?

Solution

(a) There is only one subset with no elements: the empty set. The formula verifies

$$C(6, 0) = \frac{6!}{0!(6-0)!} = \frac{6!}{6!} = 1.$$

(b) This is $C(6, 1) = \dfrac{6!}{1!(6-1)!} = \dfrac{6!}{5!} = 6$.

(c) This is $C(6, 2)$, which is equal to 15, as we saw in Example 2(a).

(continued)

(d) This is $C(6, 3) = \dfrac{6!}{3!(6-3)!} = \dfrac{6!}{3!\,3!} = \dfrac{6\cdot 5\cdot 4\cdot 3\cdot 2\cdot 1}{3\cdot 2\cdot 1\cdot 3\cdot 2\cdot 1} = \dfrac{6\cdot 5\cdot 4}{3\cdot 2\cdot 1} = 20$.

(e) This is $C(6, 4) = \dfrac{6!}{4!(6-4)!} = \dfrac{6!}{4!\,2!} = \dfrac{6\cdot 5\cdot 4\cdot 3\cdot 2\cdot 1}{4\cdot 3\cdot 2\cdot 1\cdot 2\cdot 1} = \dfrac{6\cdot 5}{2\cdot 1} = 15$.

(f) This is $C(6, 5) = \dfrac{6!}{5!(6-5)!} = \dfrac{6!}{5!\,1!} = \dfrac{6\cdot 5\cdot 4\cdot 3\cdot 2\cdot 1}{5\cdot 4\cdot 3\cdot 2\cdot 1\cdot 1} = 6$.

(g) A subset of S with at least 4 elements has either 4 or 5 or 6 elements, so by the Addition Principle, there are $C(6, 4) + C(6, 5) + C(6, 6) = 15 + 6 + 1 = 22$ subsets of S with at least 4 elements.

THE LEFTOVER PRINCIPLE

You may have noticed that some of the combination numbers you calculated in the previous examples are identical. If not, take another look at parts (c) and (d) from Example 2 and parts (c) and (e) from Example 3. Why is it that $C(7, 3) = C(7, 4)$ and $C(6, 2) = C(6, 4)$?

To answer this question, let us revisit Example 1. If Elizabeth chooses three paintings out of five to send away, then two paintings remain. Choosing paintings A, C, and D to send away has the same result as choosing paintings B and E to keep. Thus the number of ways of choosing three paintings from a set of five to send away, $C(5, 3)$, must be equal to the number of ways of choosing two paintings to keep, $C(5, 2)$. More generally, if you choose r elements from a larger set of n elements, then you have also chosen, perhaps without even realizing it, a set of $n - r$ elements: the ones that got left behind. We can state this principle mathematically as follows, and we can easily prove it using the combination formula we have developed.

Theorem 1 (The Leftover Principle): If n and r are integers and $0 \le r \le n$, then $C(n, n - r) = C(n, r)$.

Preparation for Theorem 1:
Important Formula: The key formula we will use is the $C(n, r)$ formula,

$$C(n, r) = \frac{n!}{r!(n-r)!}.$$

Example: If $n = 10$ and $r = 3$, then $n - r = 10 - 3 = 7$.

$$C(n, n-r) = C(10, 7) = \frac{10!}{7!\,3!} = \frac{10\cdot 9\cdot 8\cdot 7!}{7!\cdot 3\cdot 2\cdot 1} = \frac{10\cdot 9\cdot 8}{3\cdot 2\cdot 1} = \frac{720}{6} = 120 \text{, and}$$

$$C(n, r) = C(10, 3) = \frac{10!}{3!\,7!} = \frac{10\cdot 9\cdot 8\cdot 7!}{3\cdot 2\cdot 1\cdot 7!} = \frac{10\cdot 9\cdot 8}{3\cdot 2\cdot 1} = \frac{720}{6} = 120 .$$

Reflection: Notice that the two fractions that we obtain from the $C(n, r)$ formula, $\dfrac{10!}{7!\,3!}$ and $\dfrac{10!}{3!\,7!}$, are nearly identical. They differ only in the order of the two factorial terms in the denominators. This is true in the general case as well.

Proof of Theorem 1: From the $C(n, r)$ formula,

$$C(n, n-r) = \frac{n!}{(n-r)!(n-(n-r))!} = \frac{n!}{(n-r)!\,r!} = \frac{n!}{r!(n-r)!} = C(n, r) . \blacksquare$$

Example 4

On a history test, you are instructed to choose any 10 of 12 questions to answer. In how many ways can you choose 10 questions to answer?

Solution

There are $C(12, 10)$ ways to choose 10 questions. We could calculate this directly, or we could use the Leftover Principle to see that

$$C(12, 10) = C(12, 2) = \frac{12!}{2!10!} = \frac{12 \cdot 11 \cdot 10!}{2 \cdot 1 \cdot 10!} = \frac{12 \cdot 11}{2 \cdot 1} = 66.$$

PERMUTATIONS OR COMBINATIONS?

Even though the formulas for permutations and combinations can look frightening at first glance, many students find that they quickly catch on to, and even enjoy, using those formulas, with all those canceling factorial numbers. Instead, many students find that the most challenging aspect of working with permutations and combinations can be deciding which of the two should be used for a particular problem. Recall that permutations are arrangements, where the order of the items being arranged is important, and combinations are subsets, where the order of the elements listed is not important. In the following examples, it may be helpful to keep in mind the question "Is order important here?"

Example 5

There are 25 students in Ms. Dunbar-Brown's class.
(a) In how many ways can she select four students to be in the school Diversity Day assembly?
(b) In how many ways can she assign four students to the jobs of the day: line leader, attendance counter, calendar setter, and temperature reporter?
(c) In how many ways can she display four self-portraits on the front wall?
(d) In how many ways can she choose four students to work with base-10 blocks?

Solution

In (a) and (d) the order in which the students are selected is not important, so the answer is the combination number

$$C(25, 4) = \frac{25!}{4!21!} = \frac{25 \cdot 24 \cdot 23 \cdot 22 \cdot 21!}{4 \cdot 3 \cdot 2 \cdot 1 \cdot 21!} = 25 \cdot 23 \cdot 22 = 12,650.$$

In (b), the order is important because there are different jobs for the students. The order is important in (c) as well, since the self-portraits will be displayed on the wall. Thus the answer for both (b) and (c) is the permutation number

$$P(25, 4) = \frac{25!}{(25-4)!} = \frac{25!}{21!} = \frac{25 \cdot 24 \cdot 23 \cdot 22 \cdot 21!}{21!} = 25 \cdot 24 \cdot 23 \cdot 22 = 303,600.$$

In the following example, it may not be immediately clear whether permutations or combinations are appropriate; it may be that *neither* is appropriate. Nonetheless, there is an important connection with combination numbers. This connection is developed more fully in Exploratory Exercise 4.

Example 6

A coin is tossed six times, and the resulting sequence of heads and tails is recorded. Record *H* for heads and *T* for tails.
(a) How many sequences are possible?
(b) How many sequences with exactly two *H*'s are possible?
(c) How many sequences with exactly four *H*'s are possible?
(d) How many sequences with at least four *H*'s are possible?

(continued)
Solution

(a) There are two choices for each toss, namely H or T, so the Multiplication Principle from Section 6.1 tells us there are $2 \cdot 2 \cdot 2 \cdot 2 \cdot 2 \cdot 2 = 2^6 = 64$ possible sequences. Notice that we are allowed (in fact, required) to repeat letters in our sequence. Neither permutations nor combinations are applicable here.

(b) We could list all such sequences, such as *HHTTTT* or *TTHTHT*, and count them by hand (see Exploratory Exercise 4), but it will be helpful to make a more general observation. Notice that to get a sequence with exactly 2 *H*'s, we need only choose two of the six tosses that will land heads up, so each sequence with two *H*'s corresponds to a two-element subset of $\{1, 2, 3, 4, 5, 6\}$. For example, *HHTTTT* corresponds to $\{1, 2\}$ since the first and second tosses landed heads up, and *TTHTHT* corresponds to $\{3, 5\}$ since the third and fifth tosses landed heads up. Thus the number of sequences with exactly two *H*'s is the same as the number of two-element subsets of $\{1, 2, 3, 4, 5, 6\}$, which is just the combination number $C(6, 2) = 15$.

(c) By the same reasoning as in part (b), this is equal to $C(6, 4)$, which by the Leftover Principle is equal to $C(6, 2) = 15$.

(d) We want the number of sequences that have four or five or six *H*'s. Since these sequences are disjoint, we can use the Addition Principle to get $C(6, 4) + C(6, 5) + C(6, 6) = 15 + 6 + 1 = 22$ such sequences.

In the last example, you might be wondering why we didn't use permutations, since the order of the *H*'s and *T*'s was certainly important in counting the sequences. The reason we could not use permutations to count the sequences is that repetitions of the letters H and T were allowed in Example 6, whereas $P(n, r)$ counts the number of arrangements of r *distinct* objects from a set of n objects, with no repetitions allowed. The reason we could use combinations is that we translated the original problem about sequences of *H*'s and *T*'s into an equivalent problem about counting two-element subsets of $\{1, 2, 3, 4, 5, 6\}$. We will investigate permutations and combinations for which repetition is allowed in greater detail in Section 6.4.

COMBINATIONS ALL AROUND US

Combinations provide a wealth of opportunities for counting things that we encounter every day. Many examples come from games of chance, where combinations help determine the likelihood, or probability, that an event will occur. Consider the following examples.

Example 7
A state lottery has a game in which six balls are chosen from a set of balls numbered 1 through 46. In how many ways can six balls be chosen?
Solution
Since the order in which the balls are chosen is not important, the number of ways is

$$C(46, 6) = \frac{46!}{6!(46-6)!} = \frac{46 \cdot 45 \cdot 44 \cdot 43 \cdot 42 \cdot 41 \cdot 40!}{6 \cdot 5 \cdot 4 \cdot 3 \cdot 2 \cdot 1 \cdot 40!} = \frac{46 \cdot 45 \cdot 44 \cdot 43 \cdot 42 \cdot 41}{6 \cdot 5 \cdot 4 \cdot 3 \cdot 2 \cdot 1} = 9{,}366{,}819.$$

Example 8
One version of the card game poker involves dealing each player 5 cards from a standard deck of 52 different cards.
(a) How many different 5-card poker hands are possible?
(b) How many 5-card hands have exactly 3 spades? (There are 13 spades in a deck.)

(continued)

Solution

(a) Since the order of the cards is not important, the number of hands is

$$C(52, 5) = \frac{52!}{5!(52-5)!} = \frac{52 \cdot 51 \cdot 50 \cdot 49 \cdot 48 \cdot 47!}{5 \cdot 4 \cdot 3 \cdot 2 \cdot 1 \cdot 47!} = \frac{52 \cdot 51 \cdot 50 \cdot 49 \cdot 48}{5 \cdot 4 \cdot 3 \cdot 2 \cdot 1} = 2,598,960.$$

(b) Break this process down into two steps. First, from the 13 spades in the deck, choose 3 to be in the hand. There are $C(13, 3)$ ways to do this. Next, choose 2 more cards to complete the hand from the 39 non-spades in the deck. There are $C(39, 2)$ ways to do this. By the Multiplication Principle, the number of hands with exactly 3 spades is

$$C(13, 3) \cdot C(39, 2) = \frac{13!}{3!10!} \cdot \frac{39!}{2!37!} = \frac{13 \cdot 12 \cdot 11}{3 \cdot 2 \cdot 1} \cdot \frac{39 \cdot 38}{2 \cdot 1} = 286 \cdot 741 = 211,926$$

Example 9

Ms. Johnson's third-grade class has 15 girls and 12 boys. In how many ways can she choose a group of 4 girls and 3 boys to organize the class's recycling efforts?

Solution

Split the task of forming the group into two steps: First choose the girls and then choose the boys. There are $C(15, 4)$ ways to choose the 4 girls and $C(12, 3)$ ways to choose the 3 boys, so by the Multiplication Principle, the total number of ways to choose the group is

$$\begin{aligned} C(15, 4) \cdot C(12, 3) &= \frac{15!}{4!11!} \cdot \frac{12!}{3!9!} \\ &= \frac{15 \cdot 14 \cdot 13 \cdot 12}{4 \cdot 3 \cdot 2 \cdot 1} \cdot \frac{12 \cdot 11 \cdot 10}{3 \cdot 2 \cdot 1} \\ &= 1365 \cdot 220 \\ &= 300,300. \end{aligned}$$

Exploratory Exercise Set 6.2

1. Divide up into pairs.
   (a) Consider the set $\{A, B, C, D\}$.
      (1) Have one student list all the possible permutations of three items from the set $\{A, B, C, D\}$. Check that the number of permutations listed agrees with the formula for $P(4, 3)$.
      (2) Have the other student list all the possible combinations of three items from the set $\{A, B, C, D\}$. Check that the number of combinations listed agrees with the formula for $C(4, 3)$.
      (3) Together, write down, below each combination of $\{A, B, C, D\}$ that you listed in part (2), all of the permutations from part (1) that have the same letters as that combination. How many permutations did you write down below each combination?
      (4) On the basis of your work in part (3), complete the following equation:

$$P(4, 3) = \underline{\hspace{1cm}} \cdot C(4, 3).$$

   (b) Repeat (1) through (3) for permutations and combinations of *two* items from the set $\{A, B, C, D\}$. Switch roles so that the student who did part (1) in (a) does part (2) in (b), and vice versa. On the basis of your work in part (3), complete the following equation:

$$P(4, 2) = \underline{\hspace{1cm}} \cdot C(4, 2).$$

2. Divide into pairs. Determine whether each of the following problems involves counting combinations or permutations, and then compute your answers. Discuss your answers with your partner, and resolve any discrepancies.
    (a) There are 14 books by your favorite author in your school library. In how many ways can you pick 4 of them to check out?
    (b) There are 12 new books that have just arrived at your library. In how many ways can you choose 5 of them and display them on the "Favorite New Books" shelf?
    (c) In how many ways can a president, vice president, secretary, and treasurer be chosen from a class of 100 people?
    (d) In how many ways can a committee of 4 faculty members be chosen from a department with 15 faculty members?

3. Divide into pairs. Consider the set $\{A, B, C, D, E, F\}$.
    (a) Have one student list all of the combinations of 2 items from $\{A, B, C, D, E, F\}$.
    (b) Have the other student list all of the combinations of 4 items from the set.
    (c) According to the Leftover Principle, how should the number of sets that the first student listed compare to the number of sets that the second student listed?
    (d) Match up each 2-combination with its complementary 4-combination. (Recall from Chapter 2 that the **complement** of a set A is the set of all elements in the universal set that are not in A.)

4. (For this exercise, it may be helpful to refer to Example 6 in this section.) Divide up into pairs. Have one student do part (a) and the other student do part (b). Then work together on parts (c), (d), and (e).
    (a) If a coin is tossed six times and the resulting sequence of heads and/or tails is recorded (with H for heads and T for tails), list all possible sequences containing exactly two H's.
    (b) List all possible two-element subsets of the set $\{1, 2, 3, 4, 5, 6\}$.
    (c) Match each of the sequences listed in (a) with each of the subsets listed in (b).
    (d) (*Proof Exercise*) Carefully explain why the number of sequences listed in part (a) is equal to the combination number $C(6, 2)$.
    (e) (*Proof Exercise*) Let n and r be positive integers, with $r \leq n$. Suppose a coin is tossed n times, and the resulting sequence of heads and tails is recorded. Carefully explain why the number of possible sequences with exactly r H's is equal to the combination number $C(n, r)$.

5. (*Writing Exercise*) Your classmates have chosen you to make a video to explain the difference between combinations and permutations. Write a paragraph describing what you will say. Include a specific example of a word problem not found in this text that requires counting combinations, and another that requires counting permutations. Explain why combinations or permutations are appropriate for each problem, and explain your answer to each problem.

Exercise Set 6.2

1. (a) How many 3-permutations are there from a set with 8 elements?
    (b) How many 3-combinations are there from a set with 8 elements?
    (c) How many 4-permutations are there from a set with 8 elements?
    (d) How many 4-combinations are there from a set with 8 elements?

2. Compute the following combination numbers:
    (a) $C(4, 2)$ (b) $C(7, 5)$
    (c) $C(5, 5)$ (d) $C(7, 0)$
    (e) $C(7, 1)$ (f) $C(7, 7)$

3. If A is a set with n elements, where n is a positive integer, how many subsets of A have:
   (a) no elements?
   (b) exactly 1 element?
   (c) exactly $n - 1$ elements?
   (d) n elements?

4. Use your answers from Exercise 3 to determine the following combination numbers, where n is a positive integer.
   (a) $C(n, 0)$
   (b) $C(n, 1)$
   (c) $C(n, n - 1)$
   (d) $C(n, n)$

5. A coin is tossed 10 times and the resulting sequence of heads and/or tails is recorded. How many sequences have:
   (a) exactly 4 heads?
   (b) exactly 7 heads?
   (c) at least 8 heads?
   (d) at least 2 heads?

6. Recall that a standard deck of cards has 52 cards. How many 5-card poker hands have:
   (a) 4 queens?
   (b) 5 clubs?
   (c) 3 queens and 2 jacks?
   (d) no hearts?

7. In a state lottery game, 6 balls are chosen from a set of 48 balls, numbered 1 through 48. In how many ways can 6 balls be chosen?

8. In a multistate lottery game, 5 balls are chosen from a set of 50 white balls, and 1 additional ball is chosen from a set of 36 gold-colored balls. In how many ways can a combination of 5 white balls and 1 gold ball be chosen?

9. There are 11 alto saxophone players and 9 tenor saxophone players trying out for a jazz band. In how many ways can the band be chosen if there must be:
   (a) a total of 8 saxophone players in the band?
   (b) a total of 8 saxophone players and either 3 or 4 tenor saxophone players in the band?

10. A package of jelly beans has 15 green jellybeans and 11 red jellybeans. In how many ways can you choose a handful of 5 jellybeans in which there are:
    (a) 2 green and 3 red jellybeans?
    (b) at least 3 red jellybeans?

11. Suppose there are 2 defective computer chips and 18 nondefective chips on an assembly line. If you pick a sample of 4 chips from the 20 on the assembly line, in how many ways can you choose a sample that has:
    (a) no defective chips?
    (b) 1 defective chip and 3 nondefective chips?
    (c) 2 defective chips and 2 nondefective chips?
    (d) at least 2 nondefective chips?

6.3 Pascal's Triangle and Binomial Coefficients

Look at the triangular array of numbers in Figure 1. This simple pattern of numbers has fascinated mathematicians young and old for over 800 years. Although it is named for the French mathematician and philosopher Blaise Pascal (1623–1662), the earliest extant presentation of the triangle dates back to the eleventh-century Chinese mathematician Yang Hui. We note that although Figure 1 shows only the first several rows, Pascal's Triangle never ends! Once we understand the pattern, we can extend the triangle to include as many rows as patience or computing time will allow.

$$
\begin{array}{c}
1 \\
1 \quad 1 \\
1 \quad 2 \quad 1 \\
1 \quad 3 \quad 3 \quad 1 \\
1 \quad 4 \quad 6 \quad 4 \quad 1 \\
1 \quad 5 \quad 10 \quad 10 \quad 5 \quad 1 \\
1 \quad 6 \quad 15 \quad 20 \quad 15 \quad 6 \quad 1 \\
1 \quad 7 \quad 21 \quad 35 \quad 35 \quad 21 \quad 7 \quad 1 \\
1 \quad 8 \quad 28 \quad 56 \quad 70 \quad 56 \quad 28 \quad 8 \quad 1
\end{array}
$$

Figure 1: Pascal's Triangle

What patterns do you see in Pascal's Triangle? Let's start with a few obvious ones. Notice that each row has one more number than the row above it. In addition, each row begins and ends with a 1 and is symmetric from left to right. You might also notice a pattern in the second number in each row (from left to right or from right to left). It will be convenient to refer to individual rows of Pascal's Triangle and to individual numbers within each row. You might think that the top row would be called Row 1, but the convention with Pascal's Triangle is to begin counting with 0. In other words, the top row is referred to as Row 0, the second row as Row 1, the third row as Row 2, and so forth, so that the second number of Row n is n itself, provided that n is greater than 0.

Example 1
    (a) Can you describe in words the pattern that generates the numbers in the triangle?
    (b) Translate this description into more precise mathematical terms by filling in the blanks below:
        The sum of the kth number and the $(k + 1)$st number in Row n is equal to the _____ number in Row _____ .
    (c) Armed with this simple pattern, and the fact that each row begins and ends with a 1, generate the next two rows of Pascal's Triangle.

Solution
    (a) The sum of two neighboring numbers in one row is found below and between those two numbers in the next row.
    (b) The sum is the $(k + 1)$st number in Row $n + 1$.
    (c) Based on the pattern described in (a), the next two rows are

$$
\begin{array}{c}
1 \quad 9 \quad 36 \quad 84 \quad 126 \quad 126 \quad 84 \quad 36 \quad 9 \quad 1 \\
1 \quad 10 \quad 45 \quad 120 \quad 210 \quad 252 \quad 210 \quad 120 \quad 45 \quad 10 \quad 1
\end{array}
$$

You may wonder what this quaint little triangle, amusing though it may be, has to do with combinatorics. Perhaps the following example will suggest a connection.

PASCAL'S TRIANGLE AND COMBINATIONS

Example 2

Consider the four-element set $A = \{a, b, c, d\}$. List all of the subsets of A, and group them according to how many elements are in each subset.

Solution

There are 16 subsets in all (did you remember the empty set?), listed and grouped as requested, in Table 1. Recall from Section 6.2 that a subset is also called a combination. In addition, recall that a subset with two elements is called a 2-combination, a subset with three elements is called a 3-combination, and, in general, a subset with r elements is called an r-combination.

Table 1: Subsets of $\{a, b, c, d\}$		
Type of Subset	List of Subsets	Number of r-combinations $C(4, r)$
0-combinations	$\{\,\}$	1
1-combinations	$\{a\}, \{b\}, \{c\}, \{d\}$	4
2-combinations	$\{a, b\}, \{a, c\}, \{a, d\}, \{b, c\}, \{b, d\}, \{c, d\}$	6
3-combinations	$\{a, b, c\}, \{a, b, d\}, \{a, c, d\}, \{b, c, d\}$	4
4-combinations	$\{a, b, c, d\}$	1

Take another look at Pascal's Triangle in Figure 1. Does the last column in Table 1 look familiar? It is the same sequence of numbers that appears in Row 4 of Pascal's Triangle: 1, 4, 6, 4, 1. In other words, hidden inside Row 4 of Pascal's Triangle are the numbers that describe how many subsets (r-combinations) there are from a set with 4 elements: $C(4, 0) = 1$, $C(4, 1) = 4$, $C(4, 2) = 6$, $C(4, 3) = 4$, and $C(4, 4) = 1$. Notice that if we go from left to right, the second number in Row 4 is not $C(4, 2)$ but $C(4, 1)$. This is because the first number from left to right is $C(4, 0)$. Once again, we begin with 0 when working with Pascal's Triangle.

After seeing this remarkable connection between Row 4 of Pascal's Triangle and $C(4, r)$ combination numbers, you may wonder whether this connection exists for any other rows.

Example 3

List all of the subsets of the set $\{a, b\}$, and construct a table as in Table 1 giving the number of r-combinations of this set. In which row of Pascal's Triangle can these numbers be found?

Solution

Table 2 shows this table; the combination numbers 1, 2, and 1 can be found in Row 2 of Pascal's Triangle.

Table 2: Subsets of {*a*, *b*}		
Type of Subset	List of Subsets	Number of *r*-combinations $C(2, r)$
0-combinations	{ }	1
1-combinations	{*a*}, {*b*}	2
2-combinations	{*a*, *b*}	1

Example 4

    (a)   On the basis of our observations with Rows 2 and 4 of Pascal's Triangle, which row of Pascal's Triangle do you think will tell us about the number of subsets of the set $A = \{a, b, c, d, e, f, g\}$?

    (b)   Use that row to determine the number of 3-element subsets of *A*. (Don't try to list them.)

    (c)   Use the same row to determine the number of subsets of *A* that have *at least* 3 elements.

Solution

    (a)   Since the set *A* has 7 elements, we want to look at Row 7 of Pascal's Triangle:

$$1 \quad 7 \quad 21 \quad 35 \quad 35 \quad 21 \quad 7 \quad 1$$

           (Remember, Row 7 is the row that begins with 1   7, not the seventh row down.)

    (b)   The number of 3-element subsets of *A* is $C(7, 3) = 35$. Remember that this is not the third number of Row 7 but the *fourth* number from left to right, because the row begins with $C(7, 0)$.

    (c)   A subset with at least 3 elements has 3 or more elements. The number of these is

$$C(7, 3) + C(7, 4) + C(7, 5) + C(7, 6) + C(7, 7) = 35 + 35 + 21 + 7 + 1 = 99.$$

Notice that in part (c) of Example 4, it was particularly convenient to use Pascal's Triangle to add up several different *r*-combination numbers. On the basis of the evidence of the previous examples, you might be ready to formulate a conjecture connecting the numbers found in Pascal's Triangle with the *r*-combination numbers. (If not, try working Exploratory Exercise 1 from this section.) Fill in the blanks to complete the following conjecture.

Conjecture 1: The combination number $C(n, r)$ can be found in Row _____ of Pascal's Triangle. From left to right, it is the _____ number in that row.

(See Theorem 2 on page 288 to check that you have filled in the blanks correctly.)

    Preparation for Conjecture 1:

    Example: In addition to Examples 1, 2, and 3, let us compute $C(8, 3)$ and see where it fits in Pascal's Triangle. $C(8, 3) = \dfrac{8!}{3!\,5!} = \dfrac{8 \cdot 7 \cdot 6 \cdot 5!}{3 \cdot 2 \cdot 1 \cdot 5!} = 56$. We find 56 in Row 8 of Pascal's Triangle. It appears twice in that row, as the fourth number and the sixth number. Recall that $C(8, 3) = C(8, 5)$ by the Leftover Principle.

    Definitions: Recall the pattern from Example 1 that we use to define the numbers in Pascal's Triangle.

    (1)   Each row begins and ends with a 1.

    (2)   The sum of the *k*th number and the (*k* + 1)st number in Row *n* is equal to the (*k* + 1)st number in Row (*n* + 1).

    Reflection: How might we go about proving that the conjecture is indeed true for every row of Pascal's Triangle? After all, it has infinitely many rows. Recall that the principle of mathematical induction, discussed in Section 3.3, provides just such a way of proving infinitely many theorems at once.

Proof of Conjecture 1: To prove the conjecture using induction, we must prove two things:
   (1)   The conjecture holds for Row 1 of Pascal's Triangle.
   (2)   If the conjecture holds for Row n, then it holds for Row $n + 1$, where n is a positive integer. ■

To prove Step (1), we note that any set with one element has two subsets: one 0-element subset, the empty set, and one 1-element subset, the set itself. Indeed, Row 1 of Pascal's Triangle consists of the numbers 1 and 1.

To prove Step (2), we assume that the conjecture holds for Row n of Pascal's Triangle and try to show that it also holds for Row $n + 1$. Recall that if we know Row n of Pascal's Triangle, we can generate Row $n + 1$ by starting and ending with a 1 and by using the pattern described above: The sum of two neighboring numbers in Row n is the number between them in Row $n + 1$. To show that Row $n + 1$ of Pascal's Triangle contains the r-combination numbers $C(n + 1, r)$, we need to show that the same pattern that we use to generate Row $n + 1$ also generates the r-combination numbers $C(n + 1, r)$. In other words, we need to prove that the sum of two neighboring r-combination numbers, $C(n, r)$ and $C(n, r + 1)$, is equal to $C(n + 1, r + 1)$. For clarity, we will restate the statement to be proved as a separate conjecture.

Conjecture 2: If n and r are positive integers, with $r \leq n$, then

$$C(n, r) + C(n, r+1) = C(n+1, r+1).$$

We could use induction again to prove this statement, or we could use the combinatorial formula from Section 6.2 to prove it algebraically, but instead, we will prove it by counting something in two different ways. When we count it one way, we will get the sum on the left side of the equation, but when we count it another way, we will get the quantity on the right side of the equation. Since we were counting the same thing both times, the left and right sides must be equal. This type of argument is called a counting proof or a combinatorial proof.

   Preparation for Conjecture 2:
   Example: If $n = 6$ and $r = 3$, then $C(n, r) + C(n, r + 1) = C(6, 3) + C(6, 4) = 20 + 15 = 35$, and $C(n + 1, r + 1) = C(7, 4) = 35$. It works! (See Examples 2 and 3 from Section 6.2 for the computations.)
   Definitions: Recall that $C(n, r)$ counts the number of r-element subsets that can be formed from a set with n elements.
   Reflection: From the definition above, $C(n, r + 1)$ counts the number of $(r + 1)$-element subsets that can be formed from a set with n elements, and $C(n + 1, r + 1)$ counts the number of $(r + 1)$-element subsets that can be formed from a set with $n + 1$ elements. In the proof, we will focus on one particular element, x, of a set with $n + 1$ elements and will divide the $(r + 1)$-element subsets of that set into two groups: those that contain x and those that do not.

Combinatorial Proof of Conjecture 2: Let S be a set with n elements, and let x be some element that is *not* in S. Let $T = S \cup \{x\}$, so that the set T has $n + 1$ elements. We will count the number of $(r + 1)$-combinations of T, where r is a positive integer less than or equal to n, by dividing them into two groups: those that contain x and those that do not. Any $(r + 1)$-combination of T that contains x must be the union of $\{x\}$ and an r-combination of S. Since there are $C(n, r)$ r-combinations of S, there must also be $C(n, r)$ $(r + 1)$-combinations of T that contain x. Any $(r + 1)$-combination of T that does not contain x is also an $(r + 1)$-combination of S; there are $C(n, r + 1)$ of these. Thus the total number of $(r + 1)$-combinations of T is $C(n, r) + C(n, r + 1)$. On the other hand, the total number of $(r + 1)$-combinations of T is $C(n + 1, r + 1)$ by definition, so

$$C(n, r) + C(n, r+1) = C(n+1, r+1).$$

This proves Conjecture 2, which in turn completes the proof of Conjecture 1, which, by virtue of the proof we have just provided, we may now elevate to the rank of a theorem. ■

Theorem 2: The combination number $C(n, r)$ can be found in Row n of Pascal's Triangle. From left to right, it is the $(r + 1)$st number in that row.

Now that we have proved this surprising connection between the numbers in Pascal's Triangle and the combination numbers, what use can we make of it? In one sense, all of the types of questions we asked about combinations in the previous section can be answered with Pascal's Triangle rather than the factorial formulas for $C(n, r)$. In addition, Pascal's Triangle really comes in handy when we need to add up several combination numbers with the same value of n.

Example 5

Ms. Jones's math class is going to have an ice cream party. There are 8 toppings to choose from. How many different combinations of toppings are possible if each student must choose:
(a) exactly 2 toppings?
(b) exactly 4 toppings?
(c) no more than 2 toppings?
(d) at least 3 toppings?

Solution

(a) This is $C(8, 2)$, which we see from Pascal's Triangle is equal to 28.
(b) This is $C(8, 4) = 70$.
(c) A student can choose 0, 1, or 2 toppings, so the total number of choices is
$$C(8, 0) + C(8, 1) + C(8, 2) = 1 + 8 + 28 = 37.$$
(d) A student can choose 3, 4, 5, 6, 7, or 8 toppings, for a total of
$$C(8, 3) + C(8, 4) + C(8, 5) + C(8, 6) + C(8, 7) + C(8, 8) =$$
$$56 + 70 + 56 + 28 + 8 + 1 = 219 \text{ choices.}$$

(See Example 8 in this section for a quicker way to do part (d).)

BINOMIAL COEFFICIENTS

The numbers in Pascal's Triangle pop up in many different ways. Besides the somewhat surprising connection with combination numbers we just explored, those same numbers play a central role when we multiply out, or expand, expressions like $(x + y)^3$. The expression in parentheses has two terms and so is known as a *binomial* expression. Let us investigate what happens when we expand powers of a binomial expression. For the first example, recall that any quantity (other than 0 itself) raised to the power 0 is defined to be equal to 1.

$$(x+y)^0 = 1$$
$$(x+y)^1 = x+y = 1 \cdot x + 1 \cdot y$$
$$(x+y)^2 = x^2 + 2xy + y^2 = 1 \cdot x^2 + 2 \cdot xy + 1 \cdot y^2$$
$$(x+y)^3 = x^3 + 3x^2y + 3xy^2 + y^3 = 1 \cdot x^3 + 3 \cdot x^2y + 3 \cdot xy^2 + 1 \cdot y^3$$

What pattern do you notice in the coefficients?

For $(x+y)^0$, the coefficient is

$\qquad\qquad\qquad\qquad\qquad\qquad$ 1

For $(x+y)^1$, the coefficients are

$\qquad\qquad\qquad\qquad\qquad\qquad$ 1 1

For $(x+y)^2$, the coefficients are

$\qquad\qquad\qquad\qquad\qquad\qquad$ 1 2 1

For $(x+y)^3$, the coefficients are

$\qquad\qquad\qquad\qquad\qquad\qquad$ 1 3 3 1

Pascal's Triangle strikes again! Because of this connection with the coefficients in binomial expansions, the combination numbers $C(n, r)$, that, as you remember, are the same numbers found in Pascal's Triangle, are also known as the **binomial coefficients**.

What in the world is the connection between combination numbers, which tell us the number of subsets of a certain size, and the coefficients we get when we expand a binomial expression? Perhaps we can gain some insight by looking more closely at exactly how we obtain those coefficients.

When we expand $(x+y)^2$ using the distributive property, we obtain

$$(x+y)^2 = (x+y)(x+y) = xx+xy+yx+yy = x^2+2xy+y^2.$$

If we want a term with one x and one y, for example, we can choose the x from the $(x + y)$ factor on the left, which gives xy, or from the $(x + y)$ factor on the right, which gives yx. Once we choose where the x term comes from, we have only one choice for where the y term comes from. Thus the total number of ways to get a term with one x and one y is $C(2, 1) \cdot 1 = 2 \cdot 1 = 2$, so 2 is the coefficient of the xy term in the simplified form.

Let's try it with the third power.

$$\begin{aligned}
(x+y)^3 &= (x+y)(x+y)(x+y) \\
&= (xx+xy+yx+yy)(x+y) \\
&= xxx+xxy+xyx+xyy+yxx+yxy+yyx+yyy \\
&= x^3+3x^2y+3xy^2+y^3
\end{aligned}$$

To get an x^2y term, we must choose an x from two of the three factors of $(x + y)$ and a y from the remaining factor. If we choose x from the first two factors we get xxy, whereas choosing x from the first and third factors yields xyx, and choosing x from the last two factors gives yxx. As before, once we choose where the x's come from, we have only one choice for where the y comes from. Thus there are $C(3, 2) \cdot 1 = 3 \cdot 1 = 3$ terms with two factors of x and one factor of y, and 3 is the coefficient of x^2y.

From these examples, we can generalize this pattern for any non-negative integer power of a binomial expression, to obtain the following theorem. Before stating the theorem, we should point out that in each term of the binomial expansion, the sum of the powers of x and y is equal to the power of the binomial. For example, in the term x^2y, the power of x is 2, the power of y is 1, and $2 + 1 = 3$, which is the power of $(x + y)$ in this example. In general, if r is the power of x in a term of the expansion of $(x+y)^n$, then the power of y in that term must be $n-r$.

Theorem 3: The coefficient of $x^r y^{n-r}$ in the expansion of $(x+y)^n$ is $C(n, r)$.

Preparation for Theorem 3:

Example: Let $n = 5$ and $r = 2$. The expansion of $(x+y)^5$ is

$x^5 + 5x^4y + 10x^3y^2 + 10x^2y^3 + 5xy^4 + y^5$. The coefficient of $x^2y^{5-2} = x^2y^3$ in this expansion is 10.

On the other hand, $C(5, 2) = \dfrac{5!}{2!3!} = \dfrac{5 \cdot 4 \cdot 3!}{2 \cdot 1 \cdot 3!} = \dfrac{20}{2} = 10$ also.

Definitions: Recall that the combination number $C(n, r)$ counts the number of r-element subsets chosen from a set with n elements. In the coin-tossing example from Section 6.2 (Example 6 and Exploratory Exercise 4), we saw that the combination number $C(n, r)$ also counts the number of sequences of coin tosses with exactly r heads in n tosses.

Proof of Theorem 3: To obtain an $x^r y^{n-r}$ term in the expansion of $(x+y)^n = (x+y)(x+y)\cdots(x+y)$, we must choose r factors of $(x+y)$ from which we select the x. Since there are n total factors of $(x+y)$, there are $C(n, r)$ ways to do this. Then we must select the y from each of the remaining $n - r$ terms of $(x+y)$; there is only $C(n-r, n-r) = 1$ way to do that. Thus there are $C(n, r) \cdot 1 = C(n, r)$ terms of the form $x^r y^{n-r}$ in the binomial expansion of $(x+y)^n$, and the coefficient of $x^r y^{n-r}$ is $C(n, r)$. ∎

Example 6

    (a) Expand $(x+y)^5$ completely.

    (b) Find the coefficient of x^3y^5 in the expansion of $(x+y)^8$.

    (c) Expand $(3a-b)^4$ completely.

    (d) Find the coefficient of s^3t^4 in the expansion of $(s-2t)^7$.

Solution

    (a) From Theorem 3,

$$(x+y)^5 = C(5, 5)x^5 + C(5, 4)x^4y + C(5, 3)x^3y^2 + C(5, 2)x^2y^3 + C(5, 1)xy^4 + C(5, 0)y^5$$
$$= x^5 + 5x^4y + 10x^3y^2 + 10x^2y^3 + 5xy^4 + y^5.$$

Alternatively, we could use Row 5 of Pascal's Triangle to get the coefficients.

    (b) The coefficient of x^3y^5 is $C(8, 3) = 56$.

    (c) In order to use Theorem 3, we think of $3a$ as x and of $-b$ as y. The expansion is

$$C(4, 4)(3a)^4 + C(4, 3)(3a)^3(-b) + C(4, 2)(3a)^2(-b)^2 + C(4, 1)(3a)(-b)^3 + C(4, 0)(-b)^4$$
$$= 1 \cdot 81a^4 - 4(27a^3)b + 6(9a^2)b^2 - 4(3a)b^3 + b^4$$
$$= 81a^4 - 108a^3b + 54a^2b^2 - 12ab^3 + b^4.$$

    (d) If we think of s as x and of $-2t$ as y, then the s^3t^4 term is

$$C(7, 3)s^3(-2t)^4 = 35s^3(16t^4) = 560s^3t^4,$$ so the coefficient is 560.

This connection between binomial coefficients and Pascal's Triangle gives us a quick way to see several interesting properties. Example 7 gives us a formula for the sum of the numbers in any row of Pascal's Triangle. See Exploratory Exercise 6 to see what happens when we alternately add and subtract the numbers in a row of Pascal's Triangle.

Example 7

    Find the sum of the numbers in Row n of Pascal's Triangle.

Solution

We simply substitute 1 for both x and y in Theorem 3, which then gives

$$(1+1)^n = C(n, n) \cdot 1^n + C(n, n-1) \cdot 1^{n-1} \cdot 1 + \cdots + C(n, 1) \cdot 1 \cdot 1^{n-1} + C(n, 0) \cdot 1^n$$
$$2^n = C(n, n) + C(n, n-1) + \cdots + C(n, 1) + C(n, 0),$$

so the sum of the numbers in Row n of Pascal's Triangle is 2^n.

Recall that $C(n, 0) + C(n, 1) + C(n, 2) + \cdots + C(n, n)$ is also the total number of subsets of a set with n elements, so this gives us another (arguably easier) proof of Theorem 18 from Section 3.3, where we used a proof by induction.

> **Example 8**
>
> Remember the ice cream party Ms. Jones's math class was having in Example 5? In part (d) of that problem, we computed that if each student had to choose at least 3 of the 8 toppings, then there were $C(8, 3) + C(8, 4) + C(8, 5) + C(8, 6) + C(8, 7) + C(8, 8) = 219$ possible combinations of toppings. Show how we can use Example 7 to determine the answer more quickly.
>
> **Solution**
>
> The total number of combinations (with no restriction on the number of toppings) is $2^8 = 256$, so the number of combinations with at least 3 toppings must be 256 minus the number of combinations with less than 3 toppings, which is
>
> $$256 - (C(8, 0) + C(8, 1) + C(8, 2)) = 256 - (1 + 8 + 28) = 256 - 37 = 219.$$

TRIANGULAR NUMBERS IN PASCAL'S TRIANGLE

We conclude this section by examining Figure 2, which shows some triangular arrangements of dots and some connections with Pascal's Triangle.

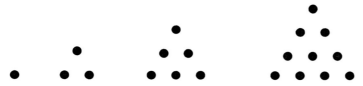

Figure 2: The Triangular Numbers 1, 3, 6, 10, ...

Notice that each triangular arrangement has one more row of dots than the arrangement to its left, and each row of dots in any one triangle has one more dot than the row above it. Besides the obvious similarity in shape between the arrangements and Pascal's Triangle, let us focus on another connection. If we count the number of dots in each arrangement, we see that there are 1, 3, 6, and 10 dots, respectively, in the triangular arrangements. Consequently, the numbers in the sequence 1, 3, 6, 10, ... are called the **triangular numbers**. From looking at the individual rows in each arrangement, we can see that the nth triangular number is the sum of the first n natural numbers. For example, the third triangular number $6 = 1 + 2 + 3$, and the fourth triangular number $10 = 1 + 2 + 3 + 4$.

> **Example 9**
> > (a) Find a recurrence relation for the sequence of triangular numbers.
> > (b) Find the fifth, sixth, and seventh triangular numbers.
>
> **Solution**
> > (a) To get the next triangular number, we add the next counting number to the previous triangular number. In other words, if t_n denotes the nth triangular number, then
> >
> > $t_n = t_{n-1} + n$, for $n \geq 2$.
> > (b) With the notation from part (a), $t_5 = t_4 + 5 = 10 + 5 = 15$, $t_6 = t_5 + 6 = 15 + 6 = 21$, and
> >
> > $t_7 = t_6 + 7 = 21 + 7 = 28$.

If we take another look at Pascal's Triangle, we find that the triangular numbers appear in a diagonal column heading down and to the left, beginning in Row 2, as highlighted in Figure 3. Of course, as a consequence of the symmetry of Pascal's Triangle, the triangular numbers also appear in a diagonal column going down and to the right.

$$
\begin{array}{c}
1 \\
1 \quad 1 \\
1 \quad 2 \quad 1 \\
1 \quad 3 \quad 3 \quad 1 \\
1 \quad 4 \quad \mathbf{6} \quad 4 \quad 1 \\
1 \quad 5 \quad \mathbf{10} \quad 10 \quad 5 \quad 1 \\
1 \quad 6 \quad \mathbf{15} \quad 20 \quad 15 \quad 6 \quad 1 \\
1 \quad 7 \quad \mathbf{21} \quad 35 \quad 35 \quad 21 \quad 7 \quad 1 \\
1 \quad 8 \quad \mathbf{28} \quad 56 \quad 70 \quad 56 \quad 28 \quad 8 \quad 1
\end{array}
$$

Figure 3: Triangular Numbers in Pascal's Triangle

From their highlighted positions in Pascal's Triangle, we see that the triangular numbers are all combination numbers of the form $C(n, 2)$, where n is an integer greater than or equal to 2. In particular, this gives us a quick way to find a nonrecursive formula for the nth triangular number:

$$
t_n = C(n+1, 2) = \frac{(n+1)!}{(n-1)!\,2!} = \frac{(n+1)n}{2} = \frac{n^2+n}{2} \; .
$$

Exploratory Exercise Set 6.3

1. Split up into pairs. As a pair, list all of the subsets of the set $\{a, b, c\}$.
   (a) Have one person construct a table as in Table 1, giving the number of r-combinations $C(3, r)$ for $r = 0, 1, 2,$ and 3.
   (b) Meanwhile, have the other person multiply out the binomial expression $(x+y)^3$ and list the coefficients of $x^3, x^2y, xy^2,$ and y^3, respectively.
   (c) Compare your results. What do you notice?
   (d) Find a row of Pascal's Triangle that matches your results. Which row is it?

2. Split up into pairs.
   (a) Have one person calculate the combination number $C(14, 4)$ using the factorial formula from the previous section:
   $$
   C(n, r) = \frac{n!}{r!(n-r)!}
   $$
   (b) Meanwhile, have the other person calculate $C(14, 4)$ by generating more rows of Pascal's Triangle. Start with Row 10, given in the solution to Example 1 at the beginning of this section, and generate through Row 14.
   (c) Compare your answers. Which method do you think is easier?

3. (a) If you have not already done the previous exercise, use Row 10 of Pascal's Triangle, given in the solution to Example 1 at the beginning of this section, to generate Row 11.
   (b) Find two adjacent numbers in Row 11 of Pascal's Triangle where one number is twice its neighbor.
   (c) In which other rows of Pascal's Triangle can you find such pairs of numbers?
   (d) On the basis of your answer to part (c), what do you think is the next row of Pascal's Triangle that will contain a pair of numbers where one number is twice its neighbor?

(e) Generate enough new rows of Pascal's Triangle to test your conjecture.

4. Split up into pairs.
   (a) Have one student compute the coefficients in the expansion of $(x+y)^{10}$ for each of the following terms: xy^9, x^3y^7, x^4y^6
   (b) Have the other student compute the coefficients in the same expansion for each of the following terms: x^6y^4, x^7y^3, x^9y
   (c) Compare your answers, and complete the following conjecture:
   In the expansion of $(x+y)^n$, the coefficient of x^ry^{n-r} is equal to the coefficient of
   _____ .
   (d) (*Proof Exercise*) Use Theorem 3 from this section and the Leftover Principle from the previous section to prove your conjecture.

5. A special coin has an x on one side and a y on the other side. The coin is tossed four times, and the resulting sequence of x's and y's is recorded.
   (a) Write down all possible sequences of x's and y's.
   (b) How many sequences have exactly three x's?
   (c) Find the coefficient of x^3y in the expansion of $(x+y)^4$.
   (d) Carefully explain the connection between the questions in parts (b) and (c).

6. When we alternately add and subtract a sequence of numbers, we call the result an **alternating sum**. For example, the alternating sum of the numbers in Row 4 of Pascal's Triangle would be $1-4+6-4+1$. Notice that the first number is positive, so we can think of adding it to 0 to start the sum. Then we subtract the second term, add the third term, and so on.
   (a) Split up into pairs. Have one person calculate the alternating sum of each of Rows 1, 3, 5, 7, and 9 of Pascal's Triangle.
   (b) Have the other person calculate the alternating sum of each of Rows 2, 4, 6, 8, and 10 of Pascal's Triangle.
   (c) Compare results, and formulate the following conjecture:
   Conjecture: The alternating sum of the numbers in Row n of Pascal's Triangle is ___ .
   (d) Examine Pascal's Triangle carefully to see whether your conjecture holds for every row. Modify your conjecture, if necessary, by placing a restriction on n.
   (e) (*Proof Exercise*) Have the person who did part (a) explain how to use symmetry to prove the conjecture when n is odd. Will this proof work when n is even?
   (f) (*Proof Exercise*) Substitute -1 for x and 1 for y into Theorem 3 to prove your conjecture whether n is even or odd. (See Example 7 from the text.)

7. (*Writing Exercise*) You have decided to include a copy of Pascal's Triangle in a space capsule that will be sent in search of extraterrestrial life. Write a paragraph describing its significance to an alien life form that might find it, assuming, of course, that the alien will be able to read your description in English.

Exercise Set 6.3

---

1. Let S be a set with 8 elements. Use Pascal's Triangle to answer the following questions.
   (a) How many subsets of S have exactly 2 elements?
   (b) How many subsets of S have exactly 3 elements?
   (c) How many subsets of S have exactly 4 elements?
   (d) How many subsets of S have exactly 5 elements?
   (e) How many subsets of S are there altogether?

2. Let S be a set with 6 elements. Use Pascal's Triangle to answer the following questions.
   (a) How many subsets of S have exactly 2 elements?

     (b) How many subsets of S have exactly 3 elements?

     (c) How many subsets of S have at least 5 elements?

     (d) How many subsets of S have at least 4 elements?

     (e) How many subsets of S are there altogether?

3.   There are ten books on Maria's reading list. Over the summer, she must read four of them. In how many ways can she choose four books to read?

4.   Mama Jane's Pizza has a special deal where you can buy a large pizza with up to four toppings for only $9.99. If there are ten toppings to choose from, how many different pizzas could you order with up to four toppings?

5.   How many arrangements of 0's and 1's are there with:

     (a) four 0's and ten 1's?

     (b) five 0's and nine 1's?

     (c) six 0's and eight 1's?

6.   A coin is tossed 14 times, and the resulting sequence of heads and tails is recorded.

     (a) How many possible sequences have exactly 4 heads?

     (b) How many possible sequences have exactly 5 heads?

     (c) How many possible sequences have exactly 6 heads?

7.   In the expansion of $(x+y)^{14}$, determine the coefficient of:

     (a) $x^4 y^{10}$

     (b) $x^5 y^9$

     (c) $x^6 y^8$

8.   In the expansion of $(x+y)^9$, determine the coefficient of:

     (a) $x^2 y^7$ (b) $x^4 y^5$

     (c) $x^7 y^2$ (d) $x^8 y$

     (e) x^9

9.   Give the complete expansion of:

     (a) $(x+y)^6$ (b) $(x-y)^6$

     (c) $(x+y)^7$ (d) $(x-y)^7$

10. Find the coefficient of $s^4 t^5$ in the binomial expansion of:

     (a) $(s+t)^9$ (b) $(2s-t)^9$

     (c) $(s+3t)^9$ (d) $(2t-s)^9$

11. Give the complete expansion of:

     (a) $(2a+b)^5$ (b) $(a-3b)^5$

     (c) $(x-2)^5$ (d) $(3s-2t)^5$

12. At Izzy's Ice Cream Igloo, you can get a sundae with one scoop of ice cream and up to three toppings.

     (a) If there are nine toppings to choose from, how many different combinations of toppings are possible?

     (b) If there are 39 flavors of ice cream, and 10 toppings to choose from, how many different sundaes could you order?

13. At Sally's Super Salad Bar, there are 4 types of lettuce, 11 different toppings, and 6 salad dressings to choose from. You get to choose one type of lettuce (you must choose one), up to four toppings, and no more than one salad dressing (you may choose to have no dressing.) How many different salad plates can you construct?

14. A coin is tossed nine times, and the resulting sequence of heads and tails is recorded.
    (a) What is the total number of sequences possible?
    (b) How many possible sequences have exactly 6 heads?
    (c) How many possible sequences have at least 7 heads?
    (d) How many possible sequences have at least 2 heads?

15. (*Proof Exercise*) Give an algebraic proof of Conjecture 2,

$$C(n, r) + C(n, r+1) = C(n+1, r+1),$$

using the factorial formula for combination numbers, $C(n, r) = \dfrac{n!}{r!\,(n-r)!}$.

16. In the song "The Twelve Days of Christmas," the singer receives a number of gifts from his or her true love. For example, on the fourth day of Christmas alone, the singer receives "four calling birds, three French hens, two turtle doves, and a partridge in a pear tree," for a total of ten gifts.
    (a) How many gifts does the singer receive on the seventh day of Christmas? Which triangular number is this?
    (b) How many gifts does the singer receive on the twelfth day of Christmas? Which triangular number is this?
    (c) How many gifts does the singer receive on the nth day of Christmas?
    (d) How many gifts does the singer receive altogether during the twelve days of Christmas?
    (e) Find your answer to part (d) in Pascal's Triangle, and describe it as a combination number $C(n, r)$.

6.4 Permutations and Combinations with Repetitions

So far in our discussion of permutations and combinations, we have assumed that the objects being permuted or chosen have all been distinct. In this section we broaden the picture to include permutations and combinations in which we are allowed to repeat one or more of the objects. We begin with the Mississippi problem, a well-known mathematical gem that quickly illustrates the difference between permutations that allow repetition and those that do not. The key difference, of course, between this problem and other permutation problems from Section 6.1 is that several letters are repeated in the word *Mississippi*. The distinguishing feature of such problems is that we attempt to count the number of permutations of a set of objects with a specified number of repetitions. We will give two solutions to the problem: one from the perspective of permutations, and another in which the problem can be viewed in terms of combinations.

MISSISSIPPI PROBLEMS

Example 1

How many distinct arrangements are there of the letters in the word MISSISSIPPI?

Solution A

First, it will be helpful to take an inventory of the letters in the word MISSISSIPPI; it has 4 I's, 4 S's, 2 P's and 1 M. Notice that since MISSISSIPPI has a total of 11 letters, there are $11! = 39,916,800$ permutations of the letters, but the problem is that not all of those arrangements are distinct, because several of the letters are identical. Given a particular arrangement, such as SIMISPIPSIS, how many ways are there to permute the 4 I's in that arrangement to give an identical arrangement? Clearly, there are $4! = 24$ of them. Similarly, there are 4! ways to permute the 4 S's, $2! = 2$ ways to permute the 2 P's, and only $1! = 1$ way to permute the 1 M. Thus the Multiplication Principle tells us that for every rearrangement of the letters of MISSISSIPPI, like SIMISPIPSIS, there are $4!\,4!\,2!\,1! = 24 \cdot 24 \cdot 2 \cdot 1 = 1152$ permutations that will result in an identical arrangement of letters. Since every distinct arrangement has 1152 permutations that will result in an identical arrangement, and there are a total of 11! permutations, the number of distinct arrangements must be

$$\frac{11!}{4!\,4!\,2!\,1!} = \frac{39,916,800}{1152} = 34,650 .$$

Solution B

Think of creating an arrangement of the letters in MISSISSIPPI by filling in 11 blanks in a 4-step process, corresponding to the 4 different types of letters. First, choose 4 blanks in which to place the 4 I's. In how many ways can this be done? Since there are 11 blanks, and we are choosing 4 of them, there must be $C(11, 4) = 330$. If, for example, we choose the 2nd, 4th, 7th, and 10th blanks for the I's, then we will have

$$\_\ I\ \_\ I\ \_\ \_\ I\ \_\ \_\ I\ \_$$

Now choose 4 of the remaining 7 blanks for the 4 S's. There are $C(7, 4) = 35$ ways we can do this. If, for example, we choose the 1st, 5th, 9th, and 11th blanks for the S's, then we will have

$$S\ I\ \_\ I\ S\ \_\ I\ \_\ S\ I\ S$$

Next, there are $C(3, 2) = 3$ ways to choose two of the remaining three blanks for the two P's, leaving just $C(1, 1) = 1$ way to choose the last remaining blank for the M. If, for example, we choose the 6th and 8th blanks for the P's, that leaves the 3rd blank for the M, which gives the arrangement

$$S\ I\ M\ I\ S\ P\ I\ P\ S\ I\ S$$

The Multiplication Principle tells us that the number of ways we can complete this four-stage process is

$$C(11, 4) \cdot C(7, 4) \cdot C(3, 2) \cdot C(1, 1) = 330 \cdot 35 \cdot 3 \cdot 1 = 34,650 .$$

Fortunately, we arrived at the same answer in both solutions. More generally, we now have two methods of counting the number of distinct arrangements of any collection of objects, where some of the objects are identical. When simplified, both lead to the same formula. We summarize the results of our two methods in Theorem 4.

Theorem 4: Suppose we have n objects of k different types, with n_1 identical objects of the first type, n_2 identical objects of the second type, ..., and n_k identical objects of the kth type. Then the number of distinct arrangements of those n objects is equal to

$$C(n, n_1) \cdot C(n - n_1, n_2) \cdot C(n - n_1 - n_2, n_3) \cdot \ \cdots \ \cdot C(n_k, n_k) = \frac{n!}{(n_1!) \cdot (n_2!) \cdot \ \cdots \ \cdot (n_k!)}.$$

Proof of Theorem 4: The methods of Solutions A and B to Example 1 show why each of the formulas gives the correct number of distinct arrangements. We will prove that the two formulas are in fact equal. Using the formula for combination numbers to expand the left-hand side of the equation, we get

$$C(n, n_1) \cdot C(n - n_1, n_2) \cdot C(n - n_1 - n_2, n_3) \cdot \ \cdots \ \cdot C(n_k, n_k)$$

$$= \frac{n!}{(n_1!) \cdot (n - n_1)!} \cdot \frac{(n - n_1)!}{(n_2!) \cdot (n - n_1 - n_2)!} \cdot \frac{(n - n_1 - n_2)!}{(n_3!) \cdot (n - n_1 - n_2 - n_3)!} \cdot \ \cdots \ \cdot \frac{n_k!}{(n_k!) \cdot 0!}$$

$$= \frac{n!}{(n_1!) \cdot (n_2!) \cdot \ \cdots \ \cdot (n_k!)}. \ \blacksquare$$

It may be a great relief that we can now count the number of distinct arrangements of the letters in any word, not just our favorite, Mississippi. Fortunately, Theorem 4 can even be used for more situations than simply scrambling letters in words. We illustrate some ways in which it can be used in the following examples.

Example 2

Marika is going off to college and decides to give her stuffed animal collection to her younger brothers and sisters. If she decides to give 5 animals to Aliya, 6 to Bob, and 3 to Carlos, and all 14 animals are different, in how many ways can she distribute the animals?

Solution

Imagine lining up the 14 animals and putting a card with the letter A, B, or C in front of each one, depending on which sibling will get that animal. In this way, the animals will spell out a 14-letter word made up of 5 A's, 6 B's, and 3 C's. Theorem 4 tells us that the number of ways in which this can be done is

$$\frac{14!}{5! \, 6! \, 3!} = \frac{87{,}178{,}291{,}200}{120 \cdot 720 \cdot 6} = 168{,}168.$$

Example 3

Ms. Abu wants to split up her class of 24 students into four groups to work on mathematics activities. She wants a group of 9 of them to work on base-10 blocks, a group of 7 to work on geoboards, a group of 4 to work on pentominos, and a group of 4 to work on tangrams. In how many ways can she assign the students to these four activity groups?

(continued)

Solution

Think of lining up the 24 students and giving each one a poster to hold with one of the letters B, G, P, or T on it, corresponding to the activity to which they will be assigned. The students will then spell out a 24-letter word made up of B's, G's, P's, and T's. Thus the assignment of students to the 4 activities is equivalent to making an arrangement of 9 B's, 7 G's, 4 P's, and 4 T's. Theorem 4 tells us that the number of such assignments is

$$\frac{24!}{9!\,7!\,4!\,4!} = \frac{620{,}448{,}401{,}733{,}239{,}439{,}360{,}000}{362{,}880 \cdot 5040 \cdot 24 \cdot 24} = 588{,}965{,}176{,}800 \,.$$

Example 4

How many distinct arrangements of the letters of DEEDEE are possible?

Solution

This is just a particularly simple example of a MISSISSIPPI problem, with only two types of letters. In particular, there are 2 D's and 4 E's. Theorem 4 tells us that there are $\dfrac{6!}{2!\,4!}$ possible arrangements. If this doesn't look familiar, our second formula confirms that there are $C(6, 2) \cdot C(4, 4) = C(6, 2) \cdot 1 = 15$ arrangements.

Notice that our answer to the problem in Example 4 is simply the combination number $C(6, 2)$, or, equivalently, $C(6, 4)$. Suppose that we had used x's and y's instead of D's and E's in our last example. Does that remind you of anything from our discussion of binomial coefficients in Section 6.3? The number of arrangements of two x's and four y's is exactly the coefficient of $x^2 y^4$ in the expansion of $(x + y)^6$, namely the binomial coefficient $C(6, 2)$. In fact, the following example shows that we can express all of the MISSISSIPPI problems we have discussed in terms of similar coefficients.

Example 5

Find the coefficient of $wx^4 y^4 z^2$ in the expansion of $(w + x + y + z)^{11}$.

Solution

We are asked to count the arrangements of *wxxxxyyyyzz*. This is exactly the MISSISSIPPI problem, except that we are using the letters w, x, y, and z instead of M, I, S, and P. The answer is 34,650, as computed in Example 1.

Thus the answer to the MISSISSIPPI problem is also a coefficient, but not a binomial coefficient. Since we have four terms that we are adding together in $(w + x + y + z)^{11}$, and four distinct letters in MISSISSIPPI, we could call the answer a quadrinomial coefficient ("quad" means four). In fact, we can think of the answers to all of the MISSISSIPPI-type problems we have considered as coefficients; they are called **multinomial coefficients**. We have in fact already proved how to compute them, as we summarize in the following corollary to Theorem 4.

Corollary to Theorem 4: If k, n, and $n_1, n_2, \ldots, n_k$ are positive integers, then the coefficient of

$$x_1^{n_1} x_2^{n_2} \cdots x_k^{n_k} \text{ in the expansion of } (x_1 + x_2 + \cdots + x_k)^n \text{ is}$$

$$\frac{n!}{(n_1!)(n_2!) \cdot \cdots \cdot (n_k!)} \,.$$

Our familiar binomial coefficients result from the special case when $k = 2$, when we have only two terms. We will return to that special case for the next type of permutations, but we will add an extra constraint. We will see that there is a nice trick we can use to recast that problem in more familiar terms.

PERMUTATIONS WITH NON-ADJACENCY CONDITIONS

We turn our attention now to the problem of counting arrangements of words with only two different types of letters, but with the added stipulation that no two of one of the types are allowed to be next to each other. We hope that the following example will illustrate exactly what we wish to count.

Example 6

How many distinct arrangements can be made from 10 H's and 7 T's if no two T's are allowed to be adjacent?

Solution

First notice that in this problem there must be a buffer of at least one H separating any two T's. The trick in our solution is to position the H's first, leaving a blank space between each pair of H's and on either end. The blank spaces may then either be left blank or filled with one T. Here is the picture for our situation:

$$\_ \ H \ \_ \ H \ \_ \ H \ \_ \ H \ \_ \ H \ \_ \ H \ \_ \ H \ \_ \ H \ \_ \ H \ \_ \ H \ \_$$

How many blanks are there? Since there are blanks on either end, there must be one more than the number of H's, that is, 11 blanks. We wish to choose 7 of those 11 blanks and fill them with T's. There are $C(11, 7) = 330$ ways to do this.

The magic of this trick is that we have turned what seemed like a difficult counting problem into a much easier and more familiar problem of counting combinations. We can generalize the method of this solution to prove the following theorem. First notice that since we have 11 blanks, we can have no more than 11 T's in our problem.

Theorem 5: If n and k are positive integers, with $k \leq n+1$, then the number of distinct arrangements of n H's and k T's with no two consecutive T's is $C(n + 1, k)$.

Example 7

There are 16 girls and 11 boys in Ms. Lopez's class. She would like for them to line up in single file to go to lunch, but she does not want any pair of boys to be next to each other. In how many ways can she line up the class?

Solution

This is not quite a direct application of Theorem 5. The difference here is that any two H's in Theorem 5 are identical, but two girls in the class are not. (Even if two students are identical twins, it would still be a different arrangement if they switched their positions in line.) We break the problem down into three steps and use the Multiplication Principle.

STEP 1: We count the number of allowable arrangements of 16 G's and 11 B's. Theorem 5 tells us that with $n = 16$ and $k = 11$, there are $C(16 + 1, 11) = C(17, 11)$ such arrangements.

STEP 2: We count the number of ways we can arrange the 16 girls into the 16 positions marked as G. This is just 16!.

STEP 3: Finally, we count the number of ways we can arrange the 11 boys: 11!.

The Multiplication Principle tells us that the answer is

$$C(17,11) \cdot 16! 11! = 12,376 \cdot 20,922,789,888,000 \cdot 39,916,800$$

$$= 10,336,074,060,910,716,518,400,000.$$

PERMUTATIONS WITH UNLIMITED REPETITIONS

In the MISSISSIPPI problem, and in the other similar problems we have considered so far in this section, we are allowed to repeat each letter only a specified number of times. In other problems, there are no such restrictions on how many repetitions are allowed. We examined such problems when we encountered license plate and password problems at the beginning of this chapter, but here we give a few brief examples here to distinguish them from the other types of problems in this section.

Example 8
How many 11-letter words can be formed using the letters I, S, P, and M?

Solution
There are four choices for the first letter, four choices for the second, and four choices for each of the 11 letters. The Multiplication Principle then tells us that there are a total of $4^{11} = 4,194,304$ such words. We can think of each word in the following example as a permutation of 11 objects, also called an 11-permutation, where each object is chosen from a set of four elements, with an unlimited number of repetitions allowed. We can express this simple principle in the following theorem. The only difficult part about the theorem is keeping straight which number is the base and which is the exponent.

Theorem 6: The number of r-letter words that can be formed from an alphabet of n letters is n^r.
Equivalently, the number of r-permutations with repetition allowed from a set of n letters is n^r.

Example 9
Passwords for the computer network at Carter Middle School must have 4, 5, or 6 characters. How many passwords are possible if each character must be:
(a) a digit (including 0 for any character)?
(b) a lowercase letter or a digit?
(c) an uppercase or lowercase letter (where the password is case-sensitive)?

Solution
(a) Theorem 6 tells us that there are $10^4, 10^5$, and 10^6 possible passwords with 4, 5, and 6 characters, respectively. Then the Addition Principle tells us that the total number of possible passwords is
$$10^4 + 10^5 + 10^6 = 1,110,000.$$
(b) Here there are $26 + 10 = 36$ choices for each letter, for a total of
$$36^4 + 36^5 + 36^6 = 2,238,928,128.$$
(c) There are 52 choices for each letter, for a total of $52^4 + 52^5 + 52^6 = 20,158,125,312$.

COMBINATIONS WITH UNLIMITED REPETITIONS

We turn now to combinations with an unlimited number of repetitions allowed. Recall that with combinations, we are choosing a subset from a larger set, without regard to the order in which the elements are chosen. In the following example, we will see that there is a nifty trick we can use to rephrase this problem into a more familiar combination problem.

Example 10

A jar of jellybeans has three different colors of jellybeans: green, red, and yellow. If you choose a handful of four jellybeans from the jar, how many different combinations of jellybeans are possible? Assume there are at least four of each color of jellybean in the jar and that any two jellybeans of the same color are identical.

Solution

We could simply list all possible 4-element combinations, with repetitions allowed, keeping in mind that the combination GYGR is the same as GGRY since the order is not important. Here is a listing:

GGGG	GGGR	GGGY	GGRR	GGRY
GGYY	GRRR	GRRY	GRYY	GYYY
RRRR	RRRY	RRYY	RYYY	YYYY

That gives us a total of 15 combinations, but this method is not very helpful in seeing how to solve the general problem.

The following method involves a nifty trick to help us turn this into a more familiar counting problem that we have already learned to solve. Consider making a tally of how many jellybeans we have of each color. We could divide a sheet into three columns, one for each color, by drawing two vertical lines. Then, for example, we could represent the combination GGRY by placing two tally marks in the leftmost column (for green), one in the middle column (for red), and one in the rightmost column (for yellow).

G	R	Y
//	/	/

As a shorthand notation for this, we can use the slash symbol (/) for each tally mark and a vertical line for the column dividers, and write // | / | / for this combination. Similarly, we can write / | | /// to represent GYYY and | | //// to represent YYYY.

Notice that in this way we can represent each combination of jellybeans with an arrangement of four (slanted) tally marks and two (vertical) column dividers. But wait; that's not all! It works in the other direction as well. Notice that every arrangement of four slanted tally marks and two vertical lines represents one of the combinations of jellybeans listed above. For example, / | / | // represents GRYY. Thus the number of combinations of four jellybeans of three colors is the same as the number of arrangements of four slash marks and two vertical lines. But how do we count arrangements like that? Just in case those / and | symbols look too frightening, suppose that we ask how many arrangements of four H's and two T's there are. Aha! That looks more familiar. Recall that we simply have to choose four of six positions for the H's, or, equivalently, two of six positions for the T's. The answer is $C(6, 4) = 15$, (or $C(6, 2) = 15$), verifying our solution to Example 10.

More important, we now have a method of solving all such combination problems with unlimited repetitions. We first note that the answer is a combination number $C(n, k)$. Now we just have to figure out what n and k represent in the problem. Let us think back and see where the numbers 6 and 4 came from in our answer $C(6, 4)$. The 4 is easy; we were choosing a handful of 4 jellybeans, that is, a 4-combination, with repetition allowed. But what about the 6? Recall that 6 is the total number of symbols in the arrangement of slashes and vertical lines. There are four tally marks, for the four jellybeans, and two vertical lines, which divided our page into three columns, for the three colors. Thus, if we want to choose a handful of r jellybeans from among n different colors, with repetitions allowed, then we would make an arrangement of r tally marks and $(n-1)$ vertical lines; the number of these is the combination number $C(r + n - 1, r)$. We have just proved the following theorem.

Theorem 7: The number of r-combinations from a set of n objects, with repetitions allowed, is $C(r + n - 1, r)$.

Example 11

If you grab five coins from a jar of change containing pennies, nickels, dimes, and quarters, how many different combinations of coins are possible?

Solution

We are counting 5-combinations from a set of four elements, with repetitions allowed. Theorem 7 tells us that there are $C(5 + 4 - 1, 5) = C(8, 5) = 56$ possible combinations.

Exploratory Exercise Set 6.4

1. Split up into pairs to consider the following problem: How many distinct arrangements are there of the letters in the word TEPEE?
   (a) Have one member list all possible arrangements. How many did you get?
   (b) Have the second member list all possible ways to fill in three of five blanks with E's. (For example, one such arrangement of E's and blanks would be E __ E E __ .)
   (c) For each arrangement of E's and blanks in part (b), how many ways are there to fill in the remaining blanks with a T and a P?
   (d) Explain why your answer to part (a) should be the product of your answers to parts (b) and (c). (If it's not, check your work in parts (a) through (c).)

2. (a) List all possible arrangements of the letters of the word DEEDED. How many did you find?
   (b) Of those arrangements listed in part (a), which do not have two D's appearing consecutively? List them.
   (c) For each of the arrangements listed in part (b), describe which of the blanks in the arrangement
       __ E __ E __ E __
       are filled with D's.
   (d) Of those arrangements listed in part (a), which do not have two E's appearing consecutively? List them.
   (e) For each of the arrangements listed in part (b), describe which of the blanks in the arrangement
       __ D __ D __ D __
       are filled with E's.

3. Split up into pairs to consider the following problem: A bag of jellybeans has three different flavors: apple, blueberry, and earwax. How many different combinations of five jellybeans are there if repetitions are allowed?
   (a) Have one member list all possible distinct combinations, keeping in mind that ABBEE is the same combination as EABEB. Within each combination, list the letters in alphabetical order. In other words, use ABBEE, rather than EABEB. How many combinations are there?
   (b) Have the second member list all possible arrangements of five slashes (/) and two vertical lines (|). How many arrangements are there?
   (c) Together, compare your answers to parts (a) and (b). Did you get the same number?
   (d) Together, match up each combination in part (a) with an arrangement of slashes and vertical lines in part (b).

4. (*Writing Exercise*) You just received a letter from your cousin, whom you have not seen in several years. She mentioned in her letter that she is really enjoying her math class, in which they are studying permutations and combinations, but not with repetitions. Write her a letter explaining how counting combinations with repetitions can be seen as equivalent to a counting problem she has already learned. Illustrate with an example not covered in the text.

Exercise Set 6.4

1. How many arrangements are there of the letters in each of the following words?
    (a) ARRANGEMENT
    (b) REORDERING
    (c) TINTINNABULATION
    (d) DEFENESTRATION

2. Three computer labs are available for 13 students to work on their computer presentations. If 4 students can work in two of the rooms and 5 students in the third, how many ways are there to assign the students to the rooms?

3. Mrs. Clark wishes to donate nine of her paintings to three local art galleries. She wants to give three paintings to the Allen Gallery, four paintings to the Babcock Gallery, and two paintings to the Carroll Gallery. In how many ways can she do this?

4. Mr. Arturo wants to divide his class of 23 students into groups to study four states. He wants 6 students to study Alaska, 4 students to study Connecticut, 7 students to study Delaware, and 6 students to study Georgia. In how many ways can he do this?

5. (a) Find the coefficient of x^2y^3z in the expansion of $(x+y+z)^6$.
    (b) How many arrangements are there of the letters in the word SEEDED?
    (c) Explain the connection between parts (a) and (b).

6. (a) Find the coefficient of $w^2x^5y^3z^4$ in the expansion of $(w+x+y+z)^{14}$.
    (b) How many arrangements are there of the letters in the word WWXXXXXYYYZZZZ?
    (c) Explain the connection between parts (a) and (b).

7. A binary string is an arrangement of 0's and 1's. (These are the fundamental building blocks of computer science.) A binary string of length 8 is called a *byte*.
    (a) How many different bytes are there?
    (b) How many different bytes are there with five 0's and three 1's.
    (c) How many different bytes are there with five 0's, three 1's, and no two consecutive 1's?

8. Suppose a coin is tossed 10 times.
    (a) How many sequences of 6 heads and 4 tails are possible?
    (b) How many sequences of 6 heads and 4 tails are possible with no two consecutive tosses being tails?
    (c) How many sequences of 6 heads and 4 tails are possible with no two consecutive tosses being heads?

9. In Ms. Smith's 14-member music class, there is a group of 5 students who are disruptive whenever any 2 of them are next to each other. In how many ways can Ms. Smith line up the students to sing a song so that no 2 of the 5 disruptive students are next to each other?

10. How many 4-letter words can be formed from an alphabet containing
    (a) 5 letters?
    (b) 10 letters?
    (c) 26 letters?

11. Computer passwords at Central Elementary School contain 8 characters. How many passwords are possible if:
    (a) each character must be a lowercase letter (a through z) or a digit?
    (b) each password must have three lowercase letters and five digits?

(c) each password must have three lowercase letters and five digits, with no two consecutive letters?

12. A package of candies comes in six flavors: apple, banana, cherry, grape, lemon, and orange. If you grab a handful of nine candies, how many combinations of flavors are possible?

13. For the candies listed in Exercise 12, write down the arrangement of slashes (/) and vertical lines (|) corresponding to the following combinations of flavors (with repetitions):
    (a) 2 apple, 3 cherry, 1 grape, 2 lemon, 1 orange
    (b) 1 apple, 2 banana, 6 lemon
    (c) 2 cherry, 4 grape, 3 orange
    (d) 1 apple, 2 banana, 3 cherry, 2 grape, 1 lemon

14. For the candies described in Exercise 12, describe the combination of flavors corresponding to the following arrangements of slashes (/) and vertical lines:
    (a) //|/||//|/|///
    (b) ||//////||//|//
    (c) //////|| ////|||
    (d) //|/|//|/|//|/

15. Leslie is packing a box with small holiday gifts to be sent to a young child. If she has four different types of gifts to choose from and she wants to put a total of ten gifts in the box, how many different combinations of gifts are possible?

6.5 The Pigeonhole Principle

> **Example 1**
>
> The 25 students in Ms. Gutierrez's class can choose from among three activities during math class, four activities during language arts class, and two activities during social studies. If each student must choose one of each type of activity, can each of the students choose a different set of activities?

Before we tackle this example, let us make a brief digression to discuss a simple childhood game. Although it might not seem related to the problem facing Ms. Gutierrez's students, it will turn out to hold the key to answering their question. Moreover, we will see that the same principle can be used to answer a wide variety of seemingly unrelated problems.

MUSICAL CHAIRS AND SOFAS

Are you familiar with the game of musical chairs? If five people are playing the game, then they walk (or dance or march) around four chairs, accompanied by a musical selection. When the music stops, the players rush to sit down in a chair. If only one person can sit in each chair, then the inevitable conclusion is that some player is left without a chair. That player is then removed from the game, ideally with a minimum of psychological trauma inflicted. One chair is then taken away, and the music begins again with four people and three chairs. This process is repeated until only one player remains; he or she is then declared the winner.

Suppose we imagine a slightly different game, called musical sofas, where we start with five players and four sofas, and each sofa is large enough to seat several players. The players walk around the sofas until the music stops, at which time they all try to sit down. What can we say about what happens next? Clearly, it is no longer the case that somebody must be left without a seat. In fact, let us assume that every player *does* find a seat. Is there anything useful that we can say about how the players are seated? What if we generalize the situation slightly and simply say that we start with more players than sofas?

If you replied that in both cases we can say that there must be at least one sofa with more than one player sitting in it, then you have put your finger on a mathematical idea that goes by several different names, but is most commonly known as the Pigeonhole Principle. A pigeonhole is simply a space, often box-shaped, for pigeons to make their nests. It may be helpful to picture a rectangular array of pigeonholes, forming a sort of pigeon apartment complex and looking much like an array of Post Office boxes. Rather than scrambling to sit in chairs or sofas, imagine a flock of pigeons descending on a collection of pigeonholes. In the following statement of the Pigeonhole Principle, the pigeonholes are analogous to the sofas in our game of musical sofas, and the pigeons correspond to the players in the game.

THE PIGEONHOLE PRINCIPLE

Theorem 8 (The Pigeonhole Principle): If m pigeons fly into n pigeonholes, where $m > n$, then there must be at least one pigeonhole containing more than one pigeon.

That's all it is? If you are wondering how such a simple idea can have such a fancy name, not to mention an entire section of a book devoted to it, let us make a few comments. The first may be welcome news: The Pigeonhole Principle really is as simple as it sounds. Later in the section we will discuss a more general version of the principle, but that will be fairly simple as well. The real utility of the Pigeonhole Principle lies in the surprising variety of ways it can be used. In many of the examples that follow, the most difficult part of the problem will not be applying the Pigeonhole Principle but, rather, figuring out how the problem is connected to the Pigeonhole Principle in the first place. This will generally involve a careful determination of what in the problem will play the role of the pigeons and what will play the role of the pigeonholes. Before exploring some of those examples, we give a proof of the Pigeonhole Principle.

> *Preparation for Theorem 8:* Review the method of proof by contradiction. In particular, recall that we want to assume that the hypotheses are true and the conclusion is false. What would the negation of the conclusion be in words?

Proof of Theorem 8: Suppose the hypotheses of the theorem are true, but the conclusion is false. Then there is no pigeonhole containing more than one pigeon. In other words, every pigeonhole contains either one pigeon or no pigeons. Then the number of pigeons, m, must be less than or equal to the number of pigeonholes, n, which contradicts the hypothesis that $m > n$. ∎

Reflect for a moment on what the Pigeonhole Principle does *not* say. Recall that in our game of musical sofas, with five players and four sofas, we cannot conclude that every sofa has at least one player on it. In a particular round of the game, this may happen, but it does not always have to happen. For example, we could have two players on each of the first two sofas, one on the third, and none on the fourth. Alternatively, we could have two players on the first sofa, three on the fourth sofa, and none on the second or third sofas. Although in these examples it is not true that every sofa has at least one player on it, it is true that there is at least one sofa with more than one person on it. It is important to keep in mind that the Pigeonhole Principle is a statement about what must happen in every possible case.

Note that although the Pigeonhole Principle can tell us that something must happen, it generally gives us no specific information about *how* it happens. In our musical sofas example with five players and four sofas, the Pigeonhole Principle tells us that there must be some sofa with more than one player on it, but it does not tell us which particular sofa that might be. In fact, it does not even tell us exactly how many sofas will have more than one player. As vague as it may seem, the information that the Pigeonhole Principle provides is often just what we need to solve the problem at hand. Finally, let us now return to Ms. Gutierrez's class and the problem we posed at the beginning of the section.

> **Solution to Example 1**
> If there are three possible math activities, four language arts activities, and two social studies activities, then the Multiplication Principle tells us that there are a total of $3 \cdot 4 \cdot 2 = 24$ possible sets of activities. With the 24 sets as pigeonholes and the 25 students as pigeons, the Pigeonhole Principle tells us there must be at least one set of activities shared by at least two students; that is, at least two students will select the same set of three activities. Thus it is not possible for each of the students to choose a different set of activities.

Example 2

Imelda has 11 pairs of shoes in her closet. Unfortunately, she was too lazy to match up the shoes when throwing them in the closet, and now they are all mixed up. How many shoes must Imelda pull out of her closet in order to be sure that she has a matching pair?

Solution

Imagine that Imelda has laid out on the floor the 11 shoeboxes that the shoes originally came in. As she pulls out a shoe, she places it in the shoebox that it came in. With the 11 shoeboxes as the pigeonholes, she must pull out 12 shoes (the pigeons), before the Pigeonhole Principle assures her that there will be a shoebox with two shoes in it. Since those shoes originally came from that shoebox, they must be a matching pair.

The next examples will require a far less obvious choice of pigeons and pigeonholes. Pay careful attention to the argument given in the solution of Example 3. With only slight modifications, the same argument can be used to solve a variety of pigeonhole problems.

Example 3

There are 11 girls and 9 boys in Ms. Logan's class. If each child in the class is assigned a different integer between 1 and 20, inclusive, show that there must be at least two girls who are assigned consecutive integers.

Solution

Order the 11 girls in some way (say, alphabetically), and designate them as Girl 1, Girl 2, ..., Girl 11. (There is no need to order the boys.) Let g_i be the number assigned to Girl i.

Consider the following two lists of numbers:

$$g_1, g_2, ..., g_{11} \quad \text{and} \quad g_1 + 1, g_2 + 1, ..., g_{11} + 1.$$

There are a total of 22 numbers in the list. These will be the pigeons. Now consider what values these 22 numbers can have. They must all be positive integers, and they must all be less than or equal to $20 + 1 = 21$, since there are 20 students in the class. The integers from 1 to 21, inclusive, will be the pigeonholes. Thus the Pigeonhole Principle guarantees us that at least two of the 22 numbers must have the same value; that is, they must be equal. Now, no two numbers from the first list $g_1, g_2, ..., g_{11}$ may be equal, since the girls were assigned different numbers.

For the same reason, no two numbers from the second list may be equal either. This means that there must be a number from the first list that is equal to a number from the second list. In other words, there must be two numbers i and j such that $g_j = g_i + 1$, where $i \neq j$, since no number is 1 more than itself. That means that Girl j's number is 1 more than Girl i's number, and hence Girl i and Girl j are assigned consecutive numbers.

Variation on Example 3

Suppose the 11 girls and 9 boys in Ms. Logan's class line up in single file to go to lunch. Explain why there must be at least two girls next to each other in line.

Solution

Notice that this is the same problem as Example 3, only slightly disguised. The assignment of numbers in Example 3 can be used to specify the order in which the students line up. In other words, if Tomika is assigned the number 14, then she will be 14th in the line. Since we know from our hard work solving Example 3 that there must be two girls with consecutive numbers, that means there must be at least two girls next to each other in line.

Example 4

An incomplete deck of playing cards with 26 black cards and 19 red cards is shuffled. Show that there must be two black cards that are exactly 6 cards apart in the deck.

(continued)
Solution

Order the 26 black cards, and let b_i represent the position in the shuffled deck of the ith black card. For example, if the 17th black card is the third card from the top in the shuffled deck, then $b_{17} = 3$. Consider the following two lists of numbers:

$$b_1, b_2, \ldots, b_{26} \text{ and } b_1 + 6, b_2 + 6, \ldots, b_{26} + 6$$

These 52 numbers will be the pigeons. The numbers in the first list must be integers between 1 and 45, inclusive, and the numbers in the second list must be integers between $1 + 6 = 7$ and $45 + 6 = 51$, inclusive. Thus each of the 52 numbers must be an integer between 1 and 51, inclusive. With the set of integers between 1 and 51, inclusive, as the pigeonholes, the Pigeonhole Principle guarantees that at least two of the 52 numbers must be equal. No two numbers from the same list can be equal, since two black cards cannot occupy the same position, so $b_i = b_j + 6$ for some two numbers i and j. That means that black cards i and j are exactly 6 cards apart.

Example 5

Suppose the Traveling Salesman from Chapter 4 visited 24 towns in 15 days. If he visited at least one town every day, show that there must have been some period of consecutive days in which he visited exactly 5 towns.

Solution

The key idea here is to keep track of the cumulative number of towns visited throughout the 15-day period. If we let t_i denote the number of towns visited during the first i days (not just on day i), then since the salesman visits at least 1 town every day, $t_1, t_2, \ldots, t_{15}$ must be an increasing sequence of distinct positive integers, with $t_{15} = 24$. Consider the two lists $t_1, t_2, \ldots, t_{15}$ and $t_1 + 5, t_2 + 5, \ldots, t_{15} + 5$. There are 30 numbers (the pigeons) in the two lists, and each number must be an integer between 1 and $24 + 5 = 29$ (the pigeonholes). Therefore, $t_i = t_j + 5$ for some two numbers i and j, and since the sequence $t_1, t_2, \ldots, t_{15}$ is increasing, i must be greater than j. That means that $t_i - t_j = 5$, so the Traveling Salesman visited a total of 5 towns on the period of consecutive days from Day $(j + 1)$ to Day i.

THE PIGEONHOLE PRINCIPLE IN GEOMETRY

Perhaps the most surprising feature of the Pigeonhole Principle is the fact that it can be applied to so many different types of problems. We give two examples below of how the Pigeonhole Principle can be applied to geometric problems. Before giving these examples, we mention a few helpful facts from the field of geometry that will come in handy.

For the first of these examples, we simply recall that a square with sides of length 1 unit has diagonals that are $\sqrt{2}$ units long. This follows from the Pythagorean Theorem, since $\sqrt{1^2 + 1^2} = \sqrt{2}$. For the second example, let us mention that a **great circle** on a sphere is a circle that divides a sphere into two hemispheres of equal size. For example, the earth's equator is a great circle, since it divides the earth into the northern and southern hemispheres. Any circle passing through the north and south poles is also a great circle. Navigators are familiar with the fact that the shortest path between two points on a sphere is along a great circle containing the two points. Fortunately, it is well known to navigators and geometers that for any two points on a sphere, there is a great circle containing them. Finally, we mention that a **closed hemisphere** is a hemisphere together with the great circle that forms its boundary.

Example 6

Suppose there are ten farmers on a square plot of land that is 3 miles long on each side. Show that there must be two farmers within $\sqrt{2}$ miles of each other.

Solution

Divide the plot of land into nine square subregions of equal size, each with sides that are 1 mile long, as in Figure 1. Since there are ten farmers (pigeons) and nine squares (pigeonholes), there must be two farmers in or on the border of the same square. The farthest apart those two farmers could be is the length of the diagonal of the square, which is $\sqrt{2}$ miles.

Figure 1: The Plot of Land in Example 6

Example 7

Given any five points on a sphere, show that there must be a closed hemisphere that contains at least four of the points.

Solution

Choose any two of the five points and draw a great circle containing them. This divides the sphere into two hemispheres. Let the remaining three points be the pigeons, and let the two closed hemispheres be the pigeonholes. The Pigeonhole Principle assures us that one of the closed hemispheres contains at least two of the other three points. Since this closed hemisphere also contains the great circle, it contains the two original points as well, so it must contain at least four of the five points.

The solutions to these examples seem so simple once you recognize how to set them up with the Pigeonhole Principle. We should caution the reader that Example 7 stumped many of the best college-age mathematical minds in the country on the 2002 W. L. Putnam exam. The Putnam exam is a notoriously difficult mathematics exam given each year to mathematics students from colleges and universities all over the United States and Canada; the top five scorers are named Putnam fellows and receive a cash prize.

THE EXTENDED PIGEONHOLE PRINCIPLE

Now that you have some experience with the variety of problems that the Pigeonhole Principle can tackle, we will introduce a souped-up version of the Principle that can give us even more information. Let us return to the musical sofas that we discussed at the beginning of the section. Suppose we have four sofas, as before, but now we have nine people playing the game. Let us also suppose, as before, that everybody finds some place to sit on some sofa. Since $9 > 4$, the Pigeonhole Principle still applies, and it is certainly still true that there must be some sofa that has more than one person on it. But can we say anything more than that?

For example, can we now say that every sofa must have at least one person on it? No; not without some additional information about how many people can fit on each sofa. Indeed, it might be the case that all nine people sit on one sofa and the other three sofas are empty. At the other extreme, imagine spreading people out as evenly as possible over the four sofas. After two people sit on each of the four sofas, we still have one more person looking for a seat. It is clear in this case, at least, that if everybody finds a seat, then some sofa must hold at least three people. Why three? Can we prove that if nine people sit on four sofas, then some sofa must hold at least three people? Perhaps more important, can we generalize this statement for an arbitrary number of people and sofas, or pigeons and pigeonholes? Fortunately, the Extended Pigeonhole Principle and its proof provide an affirmative answer to the last two questions.

Theorem 9 (The Extended Pigeonhole Principle): Suppose m pigeons fly into n pigeonholes, where $m > k \cdot n$ and k is a positive integer. Then there must be at least one pigeonhole with more than k pigeons.

Preparation for Theorem 9: Look back at the proof by contradiction used in the proof of Theorem 8.

Proof of Theorem 9: Suppose that the hypotheses are true and the conclusion is false; that is, there is no pigeonhole with more than k pigeons. In other words, every pigeonhole has k or fewer pigeons in it. Since there are n pigeonholes, the total number of pigeons, m, must be less than or equal to $k \cdot n$, contradicting the hypothesis that $m > k \cdot n$. ∎

Notice that the conclusion of Theorem 9 could also be expressed by stating that there must be at least one pigeonhole with at least $k + 1$ pigeons in it. Care must be taken in deciding which value to use for k in a given problem. Often the appropriate value for k will be 1 less than the number stated in the problem. We conclude this section with a variety of examples that take advantage of the Extended Pigeonhole Principle.

Example 8

Janet has dimes, nickels, and pennies in a coin jar. If she grabs some coins without looking at them, how many will she have to grab to be sure she has at least 10 coins of the same denomination?

Solution

Let the pigeonholes be the 3 different denominations of coins, and let $k = 9$. Then the number of pigeons would have to be greater than $9 \cdot 3 = 27$, so Janet must grab at least 28 coins to be sure that she has more than 9 of one denomination.

Example 9

Seema prepared for her math exam by working 37 problems over a period of 8 days, working a whole number of problems each day.
(a) Show that there was some day in which Seema worked at least 5 problems.
(b) Show that there was some pair of consecutive days in which Seema worked at least 10 problems.

Solution

(a) Consider the 37 problems to be the pigeons and the 8 days to be the pigeonholes. Notice that with $k = 4$, $37 > 4 \cdot 8$, so Theorem 9 assures us that there is some day in which Seema worked more than $k = 4$ problems.
(b) Now consider the 4 pigeonholes to be the 4 two-day periods (days 1 and 2, days 3 and 4, days 5 and 6, days 7 and 8). Since 37 problems must be worked over the span of these 4 two-day periods, and $37 > 9 \cdot 4$, then there must be some two-day period in which Seema worked more than 9 problems.

Example 10

Suppose there are six people in a room. Show that there are either three people in the room who are mutual acquaintances (they all know each other) or three people in the room who are mutual strangers (none of the three know either of the other two).

Solution

For convenience, let us call the six people Al, Bo, Cy, Di, Ed, and Flo. Consider the people whom Al knows. We can put the 5 people besides Al into two groups: those whom Al knows and those whom Al doesn't know. Since $5 > 2 \cdot 2$, Theorem 9 tells us that one of these two groups must have more than two people. In other words, either there are three people whom Al knows or there are three people whom Al doesn't know. We will consider each case separately.

Case 1: Suppose there are three people whom Al knows. If two of these three people know each other, then together with Al they form a threesome of mutual acquaintances. On the other hand, if no two of these three people know each other, then they form a threesome of mutual strangers.

Case 2: Suppose there are three people whom Al doesn't know. If two of these people don't know each other, then together with Al they form a threesome of mutual strangers. On the other hand, if all three of these people know each other, then they form a threesome of mutual acquaintances.

Exploratory Exercise Set 6.5

1. The Pigeonhole Principle is also known as Dirichlet's Drawer Principle, after the mathematician Lejeune Dirichlet (1805–1859), and as the Shoebox Principle.
    (a) Give a restatement of the Pigeonhole Principle involving drawers, as in Dirichlet's Drawer Principle. Are the drawers analogous to the pigeons or the pigeonholes?
    (b) Give a restatement of the Pigeonhole Principle involving shoeboxes. Are the shoeboxes analogous to the pigeons or the pigeonholes?

2. Divide into pairs to consider this problem. Recall Ms. Logan's class with 11 girls and 9 boys in Example 3 of this section. Suppose 1 girl is absent, so there are 10 girls and 9 boys in class. Is it possible for the students to line up so that no two girls are next to each other in line?
    (a) Have one member of each pair try to modify the argument in Example 3 to show that there must be at least two girls next to each other in line. Does it work? If not, what goes wrong?
    (b) Have the other member of each pair try to come up with an arrangement of the 19 students (use G for girl and B for boy) in which no two girls are next to each other in line.

3. Divide up into pairs to consider this problem. Is it possible to shuffle 16 black cards and 10 red cards so that no two black cards are exactly 5 cards apart in the shuffled deck?
    (a) Have one member of each pair try to modify the argument in Example 4 of this section to show that there must be at least two black cards that are exactly 5 cards apart. Does it work? If not, what goes wrong?
    (b) Have the other member of the pair try to come up with an arrangement of the 26 cards (use B for black and R for red) in which no two black cards are exactly 5 cards apart.
    (c) Discuss together your conclusions.
    (d) Modify the same problem so that there are 15 black cards and 10 red cards. Repeat parts (a) through (c).

4. (a) Have every student in the class answer the following question: "How many of your fellow classmates here today in this classroom were you acquainted with before this term started?" Record each student's response to this question on the board.
    (b) Are there two students that had the same number of acquaintances? Do you think this will always be the case?
    (c) Suppose your class had 100 students in it. Do you think you would still be able to find two students who were acquainted with the same number of people?
    (d) (*Proof Exercise*) Use the Pigeonhole Principle to prove that in any group of n people, there must be two people with the same number of acquaintances within the group.

5. Divide into pairs for this problem.
   (a) Have one student choose any seven integers.
   (b) Have the other student try to find some pair of those seven integers whose sum or difference is a multiple of 10.
   (c) Switch roles and repeat parts (a) and (b) three times.
   (d) As a team, formulate a conjecture based on your experiences.
   (e) (*Proof Exercise*) As a team, prove your conjecture. [*Hint*: It may be helpful to consider the following function f. If n is an integer, define $f(n)$ to be the distance between n and the nearest multiple of 10. For example, $f(42) = 42 - 40 = 2$ and $f(47) = 50 - 47 = 3$.]

6. (*Writing Exercise*) The Pigeonhole Principle Users of America (PPUA) has elected you as their national spokesperson and would like you to write a paragraph for Mathematics Awareness Month (April) that explains why the Pigeonhole Principle is so useful. Create an example not found in this text that illustrates how the Pigeonhole Principle may be used in an unexpected application.

Exercise Set 6.5

1. A standard deck of cards has 13 cards in each of 4 suits: spades, hearts, diamonds, and clubs. If you draw cards randomly from a standard deck, how many cards must you draw before you are sure to have at least two cards of the same suit?
   (a) In answering this question, what classifications would you use as the pigeonholes?
   (b) What are the pigeons?
   (c) What is the answer to the question?

2. If you have brown, blue, black, green, tan, and white socks in a drawer, how many socks must you remove from the drawer in order to be sure of having a match?
   (a) In answering this question, what classifications would you use as the pigeonholes?
   (b) What are the pigeons?
   (c) What is the answer to the question?

3. If a box contains pencils of four different colors, how many pencils must there be in the box to guarantee that there are at least 21 pencils of the same color?
   (a) In answering this question, what classifications would you use as the pigeonholes?
   (b) What are the pigeons?
   (c) What is the answer to the question?

4. (*Proof Exercise*) At a recent family reunion, Maria noticed that everybody there had a first name of David, John, Maria, or Penelope, a middle name of Anne, Lee, or Newton, and a last name of Jones, Smith, or Turner. Prove that if there were a total of 40 people at the reunion, there must have been two people there with the same full name.

5. (a) How many numbers must you choose from the set $\{1, 2, 3, ..., 15\}$ to be sure you have at least one even number?
   (b) How many numbers must you choose from the set $\{1, 2, 3, ..., 15\}$ to be sure you have at least one odd number?
   (c) If n is an odd positive integer, how many numbers must you choose from the set $\{1, 2, 3, ..., n\}$ to be sure you have at least one even number?
   (d) If n is an odd positive integer, how many numbers must you choose from the set $\{1, 2, 3, ..., n\}$ to be sure you have at least one odd number?

6. (*Proof Exercise*) An incomplete deck of playing cards with 25 red cards and 19 black cards is shuffled. Show that there must be two red cards that are exactly 4 cards apart in the deck.

7. (*Proof Exercise*) The 25 girls and 19 boys in Mr. Brown's class are going on a field trip. Their bus has 11 rows of seats for students, where four students can sit in each row. Show that some girl must sit directly behind another girl. [*Hint*: Compare to Exercise 6.]

8. (*Proof Exercise*) There are 65 women and 55 men on the faculty at a certain college. If they all line up for their commencement exercises, show that there must be at least two women who are exactly 9 people apart in the line.

9. (*Proof Exercise*) A baseball player had at least one hit in each of 34 consecutive games. Over those 34 games, he had a total of 52 hits. Show that there was some period of consecutive games in which he had exactly 15 hits.

10. A drawer has 10 C batteries and 14 D batteries. How many batteries must you pull out of the drawer to be sure you have at least four batteries of the same type?

11. Into a tool box have been spilled 22 Number Six screws, 31 Number Eight screws, 18 Number Ten screws, and 21 Number Twelve screws. How many screws must you remove to be sure that you have at least 11 of the same type?

12. (*Proof Exercise*) There are 13 farmers working on a rectangular plot of land that is 8 miles long and 6 miles wide. Show that there must be two farmers within $2\sqrt{2}$ miles of each other.

13. (*Proof Exercise*) Show that if 7 points are chosen on or inside a regular hexagon with edges of length 5 cm, then there must be two points within 5 cm of each other.

14. (*Proof Exercise*) There are 32 members in the chess club. Some members have played against each other, whereas others have not. Show that there must be two members of the club who have played against the same number of club members.

15. (a) Choose any 5 numbers from the set $\{1, 2, 3, ..., 8\}$, and label this set as S.
    (b) Find two disjoint subsets of S whose sums are equal.
    (c) Repeat (a) and (b) three more times with a different set S each time.
    (d) Complete the following conjecture:
         If S is any 5-element subset of $\{1, 2, 3, ..., 8\}$, then _____ .
    (e) (*Proof Exercise*) Prove the conjecture you made in (d). [*Hint*: How many non-empty subsets does S have?]

16. Suppose that every student enrolled at a certain university has exactly three initials. How many students would have to be enrolled at this university in order for us to be sure that there are:
    (a) at least two students with the same three initials in the same order?
    (b) at least five students with the same three initials in the same order?

6.6 The Inclusion-Exclusion Principle

Let us begin our discussion of the Inclusion-Exclusion Principle by examining the following two examples. The two examples look nearly identical at first glance, but on closer inspection they reveal an important difference that will prove to be essential. As you read these two examples, try to identify the difference between them and consider why that difference will require two fundamentally different methods of attack in solving the problems. In the discussion that follows, we will make an important mistake in one of our attempted solutions. Fortunately, like all good students, we will take the opportunity to learn from our mistake and will make an appropriate correction to arrive at the correct answer. In fact, this mistake, and the subsequent correction, will prove to be the key behind the Inclusion-Exclusion Principle.

LEARNING FROM OUR MISTAKES

Example 1
> Suppose we want to choose a number from the set $S = \{1, 2, 3, \ldots, 20\}$ that is either odd or divisible by 4. In how many ways can this choice be made?

Example 2
> Suppose we want to choose a number from the set $S = \{1, 2, 3, \ldots, 20\}$ that is either odd or divisible by 5. In how many ways can this choice be made?

Before attacking these examples, we first establish a theorem that will be helpful in solving many problems like those in Examples 1 and 2. The proof follows quickly from the definition of divisibility in Section 3.1 (see Exploratory Exercise 2).

Theorem 10: If n and d are positive integers, then the number of positive integers less than or equal to n that are multiples of d is equal to the greatest integer less than or equal to $\dfrac{n}{d}$, denoted $INT\left(\dfrac{n}{d}\right)$ or $\left\lfloor \dfrac{n}{d} \right\rfloor$.

Another important tool we will use in our analysis is the Addition Principle from Section 6.1, which we reprint below for your convenience. Recall that the Addition Principle tells us how many ways there are to choose a total of one element from two or more sets that have no elements in common. You may want to contrast this with the Multiplication Principle, in which one element is chosen from *each* of a number of sets.

THE ADDITION PRINCIPLE

(1) Suppose X_1 and X_2 are disjoint sets, where X_1 has n_1 elements and X_2 has n_2 elements. If a process is completed by choosing one element from set X_1 or by choosing one element from set X_2, then there are $n_1 + n_2$ ways of completing the process.

(2) More generally, suppose that $\{X_1, X_2, \ldots, X_k\}$ is a collection of pairwise disjoint sets (that is, no two sets have elements in common), where set X_i has n_i elements for each integer i, $1 \leq i \leq k$. If a process is completed by choosing one element from exactly one of the sets in this collection, then the number of ways to complete the process is the sum $n_1 + n_2 + \cdots + n_k$.

Solution to Example 1

We can think of X_1 as the set of odd integers in S and of the set X_2 as the set of multiples of 4 in S. We can first count the number of elements in X_1 by noticing that Theorem 10 tells us that the number of multiples of 2 in the set $\{1, 2, 3, ..., 20\}$ is INT(20/2) = 10. Since there are 10 even integers in S and a total of 20 integers in S, there must be 20 – 10 = 10 odd integers in S. Theorem 10 also tells us that the number of multiples of 4 in S is INT(20/4) = 5. (Of course, we could also simply list them all $\{4, 8, 12, 16, 20\}$ and count them by hand.) Since every multiple of 4 must be even, and hence not odd, the two sets X_1 and X_2 are disjoint. Therefore, we can use Part (1) of the Addition Principle to tell us that there are 10 + 5 = 15 ways to choose a number from S that is either odd or a multiple of 4. We can check by hand that the 15 choices are 1, 3, 4, 5, 7, 8, 9, 11, 12, 13, 15, 16, 17, 19, and 20.

WARNING! We are about to give an incorrect solution to Example 2, carefully labeled as incorrect, which will clearly give the wrong answer. Your mission is to spot the flaw in the logic. Nevertheless, we feel compelled to provide an additional disclaimer that what you are about to read is not correct, and we trust that you would never make the mistake that we are about to outline in the following erroneous argument.

INCORRECT Solution to Example 2

As in Example 1, we let X_1 be the set of odd numbers in S, and we let X_2 be the set of multiples of 5 in S. We already know that X_1 has 10 elements, and we can use Theorem 10 to see that X_2 has INT(20/5) = 4 elements. Therefore, the Addition Principle says that there should be 10 + 4 = 14 ways to choose a number in S that is either odd or a multiple of 5. Is this right?

Not so fast. A quick listing of the possibilities shows us that the only numbers in $S = \{1, 2, 3, ..., 20\}$ that are either odd or a multiple of 5 are $\{1, 3, 5, 7, 9, 10, 11, 13, 15, 17, 19, 20\}$. That's only 12 numbers, not 14. What happened? Why were we too hasty in our application of the Addition Principle? Can you spot the flaw in our argument?

It may be helpful to focus on the sets X_1 and X_2 in the two examples. In Example 1, $X_1 = \{1, 3, 5, 7, 9, 11, 13, 15, 17, 19\}$ and $X_2 = \{4, 8, 12, 16, 20\}$ are disjoint sets, as the Addition Principle specifies. Notice that the set of possible choices in Example 1 is simply the union of the sets X_1 and X_2:

$$X_1 \cup X_2 = \{1, 3, 4, 5, 7, 8, 9, 11, 12, 13, 15, 16, 17, 19, 20\}.$$

In Example 2, are the sets $X_1 = \{1, 3, 5, 7, 9, 11, 13, 15, 17, 19\}$ and $X_2 = \{5, 10, 15, 20\}$ disjoint? Clearly not, since the number 5, for example, is an element of both sets. Therefore, we made our critical flaw when we used the Addition Principle, which applies only for disjoint sets, with two sets that are not disjoint.

That explains our incorrect solution, but does it give us any insight about how to get the correct answer? Fortunately, our incorrect solution is not completely worthless. In fact, as is often the case, we can learn something quite valuable from our mistake. Our incorrect answer to Example 2 was 10 + 4 = 14, yet there are actually only 12 numbers in S that are either odd or a multiple of 5. That means that in our incorrect answer of 14, we overcounted by two. Why two?

The two sets we described in Example 2, $X_1 = \{1, 3, 5, 7, 9, 11, 13, 15, 17, 19\}$ and $X_2 = \{5, 10, 15, 20\}$, are not disjoint, but how many elements do they have in common? The answer again is two. The answer we seek in Example 2 is the number of elements in the union of X_1 and X_2. If, however, we try to count the number of elements in $X_1 \cup X_2$ by counting the number of elements in X_1 and then the number of elements in X_2, then we have counted some elements twice. Which ones? We have counted twice exactly those elements that are in both X_1 and X_2: 5 and 15.

Recalling the notation introduced in Section 2.1, where $n(S)$ represents the number of elements in a set S, we can say for the two sets in Example 2 that $n(X_1) = 10$ and $n(X_2) = 4$, but $n(X_1 \cup X_2) = 12 \neq 10 + 4$. In other words,

$$n(X_1 \cup X_2) \neq n(X_1) + n(X_2)$$

in this example. However, we can say more. The amount by which we overcounted is exactly the number of elements that are in both X_1 and X_2; that is, it is the number of elements in the intersection of the two sets, $n(X_1 \cap X_2)$. Putting all of this together, we get the first version of the Inclusion-Exclusion Principle.

Theorem 11 (Inclusion-Exclusion Principle, Version 1): If X_1 and X_2 are any two sets, then

$$n(X_1 \cup X_2) = n(X_1) + n(X_2) - n(X_1 \cap X_2).$$

Proof of Theorem 11: The name derives from the following logic in counting elements. Imagine keeping a tally of elements as you count them. If there is an element like the number 5 in Example 2 that is in both sets X_1 and X_2, then we will include 5 in our tally of elements in X_1, include 5 again in our tally of elements in X_2, but exclude 5 (erase one tally mark) because it is in the intersection of X_1 and X_2. The net effect is that we have given the element 5 two tally marks, but we have erased one, leaving us with a total of one tally mark for the element 5. If an element is in only one of the two sets X_1 and X_2, such as the number 9 in Example 2, then that number will be included once in our tally of elements of X_1, but not in our tally of the other set, and we will not exclude the number, since it is not in the intersection of the two sets. Finally, if an element is not in $X_1 \cup X_2$, such as the number 8 in Example 2, then it will not be included in either tally, nor will it be excluded, leaving it with a net tally of 0. In the same way, every element that is in the union $X_1 \cup X_2$ will have a net tally of exactly one, and any element that is not in $X_1 \cup X_2$ will have a net tally of zero. Since the right-hand side of the equation in Theorem 11 is the sum of those net tallies, it will be a sum of 1's, and the number of 1's will be the number of elements in $X_1 \cup X_2$. ∎

Example 3
   How many positive integers less than or equal to 100 are either a multiple of 3 or a multiple of 7?

Solution
   Let $S = \{1, 2, 3, \ldots, 100\}$, let X_1 be the set of multiples of 3 in S, and let X_2 be the set of multiples of 7 in S. Theorem 10 tells us that $n(X_1) = \text{INT}(100/3) = 33$, and $n(X_2) = \text{INT}(100/7) = 14$. The intersection $X_1 \cap X_2$ is the set of elements of S that are multiples of both 3 and 7; that is, they are multiples of 21. We use Theorem 10 again to see that $n(X_1 \cap X_2) = \text{INT}(100/21) = 4$. Therefore, Theorem 11 tells us that the number of elements of S that are either a multiple of 3 or a multiple of 7 is
   $$n(X_1 \cup X_2) = n(X_1) + n(X_2) - n(X_1 \cap X_2) = 33 + 14 - 4 = 43.$$

Example 4

In the fifth grade at Johnson Elementary School, 65 students are in the band, 50 students are in the chorus, and 20 students are in both the band and the chorus.
(a) How many fifth-grade students are in either the band or the chorus?
(b) If there are 100 fifth-grade students at Johnson Elementary School, how many are in neither the band nor the chorus?

Solution

(a) With X_1 as the set of fifth-graders in the band and X_2 as the set of fifth-graders in the chorus, $n(X_1) = 65$, $n(X_2) = 50$, and $n(X_1 \cap X_2) = 20$. Thus the number of fifth-graders in either the band or the chorus is $n(X_1 \cup X_2) = 65 + 50 - 20 = 95$.

(b) This is the complement of the set of students in part (a), so there are $100 - 95 = 5$ fifth-graders who are in neither the band nor the chorus.

VENN DIAGRAMS AND THE INCLUSION-EXCLUSION PRINCIPLE

Venn diagrams can often be helpful in keeping track of the information provided in a problem that requires the Inclusion-Exclusion Principle. Let us examine how a Venn diagram illustrates the method of our solution in Example 4 with the fifth-grade students at Johnson Elementary School. In Figure 1 below, we will represent the two sets X_1 and X_2 with circles and the universal set U with a rectangle enclosing the two circles, as we did in Section 2.1.

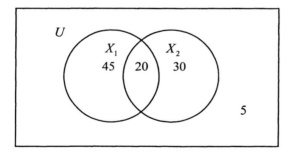

Figure 1

We will record the information from the problem on the diagram in each of four non-overlapping regions, working our way from the inside out. Since there are 20 students who are in both the band and the chorus, we will write the number 20 in the central region that is inside both of the circles. Since there are 65 students who are in the band, there must be $65 - 20 = 45$ students who are in the band but not in the chorus, so we will write the number 45 in the region that is inside circle X_1 but outside circle X_2. Similarly, there must be $50 - 20 = 30$ students in the chorus but not in the band, so we write the number 30 in the region inside circle X_2 and outside circle X_1. If we want to know the number of fifth-graders in either the band or the chorus, we can simply add these three numbers in the diagram to get $45 + 20 + 30 = 95$. Finally, we can write the number 5 in the region outside both circles, but still inside the rectangle, to represent the remaining fifth-graders, who are in neither the band nor the chorus.

Which method is better, then, for solving problems like Example 4: using the formula in Theorem 11 or using the Venn diagram? That is largely a question of personal taste, depending on whether you are more comfortable with the symbolic notation of the formula or the more graphical representation of the Venn diagram. We should, however, point out a few distinctions between the two methods. We actually have to perform more calculations in the Venn diagram than we do with the formula in Theorem 11 (two differences and two sums versus one sum and one difference to get the answer to part (a)), but the Venn diagram gives us a more comprehensive record of the data in the example. If we were asked additional questions about the number of students in one group but not another, we could simply read that information off the Venn diagram. Finally, note that the computations we make to answer part (a) using the Venn diagram are not essentially different from those in Theorem 11; they are just slightly less efficient. With the Venn diagram, we get the number 95 by computing $(65 - 20) + 20 + (50 - 20)$, which of course is equal to the $65 + 50 - 20$ that we compute with Theorem 11.

Example 5

How many two-digit positive integers (between 10 and 99, inclusive) have at least one digit that is a 7?

Solution

Let X_1 be the set of two-digit positive integers whose first digit (tens digit) is a 7, and let X_2 be the set of two-digit positive integers whose second digit (ones digit) is a 7. If the first digit is a 7, the second digit can be any digit, including 0, so $n(X_1) = 10$. But if the second digit is a 7, then the first digit cannot be 0, so $n(X_2) = 9$. There is one number, namely 77, whose first and second digits are both 7, so $n(X_1 \cap X_2) = 1$, and Theorem 11 tells us that

$$n(X_1 \cup X_2) = 10 + 9 - 1 = 18.$$

We can use the idea of complementary sets, as we did in part (b) of Example 4, to provide an alternative solution to Example 5 that does not use the Inclusion-Exclusion Principle. This alternative method cannot always be used in place of the Inclusion-Exclusion Principle, but it will provide a useful check to our calculations on problems similar to Example 5 as we develop a generalization of the principle for more than two sets.

Alternative Solution to Example 5: If U is the universal set of all two-digit positive integers, and X is the subset of two-digit positive integers whose digits contain a 7, then the complement of X, denoted $\overline{X}$, is the set of all two-digit positive integers with no 7's as digits. We can count the number of elements of $\overline{X}$ using the Multiplication Principle as follows: There are 8 choices for the first digit, since it cannot be 0 or 7, and 9 choices for the second digit, since it can be any digit except 7, so there are a total of $8 \cdot 9 = 72$ elements in the complement $\overline{X}$. Since there are $9 \cdot 10 = 90$ elements in the universal set U, there must be $90 - 72 = 18$ elements in the set X.

INCLUSION-EXCLUSION WITH MORE THAN TWO SETS

How do we count the number of elements in the union of three or more sets? We can begin with the same strategy of overcounting and correcting that we used to develop Theorem 11, but there are a few more details to keep careful track of. Let us begin by modifying the problem in Example 5 to count three-digit numbers, as follows.

Example 6

How many three-digit positive integers (between 100 and 999, inclusive) have at least one digit that is a 7?

Solution

We could begin by letting X_1 be the set of all three-digit positive integers with 7 as the first digit, letting X_2 be the set of all three-digit positive integers with 7 as the second digit, and letting X_3 be the set of all three-digit positive integers with 7 as the third digit. If the first digit is a 7, then there are ten choices for each of the remaining digits, so $n(X_1) = 10 \cdot 10 = 100$. In X_2 and X_3, however, there are only nine choices for the first digit, since it cannot be 0, and ten choices for the remaining digit, so $n(X_2) = n(X_3) = 9 \cdot 10 = 90$. After our mistake with Example 2, we are fairly certain that we will be overcounting if we simply add

(continued)

$n(X_1) + n(X_2) + n(X_3) = 100 + 90 + 90 = 280$. But by how much did we overcount? That will be the critical question as we develop a more general version of the Inclusion-Exclusion Principle.

In general, if we are trying to count the number of elements in $X_1 \cup X_2 \cup X_3$, then any element in both X_1 and X_2 will be counted twice, so we will again subtract $n(X_1 \cap X_2)$ from our total. We will also overcount any element that is in both X_1 and X_3, so we will subtract $n(X_1 \cap X_3)$ from our total as well. Elements that are in both X_2 and X_3 will be overcounted also, so we will subtract $n(X_2 \cap X_3)$ from our total. Let us take stock of our total so far. Will

$$n(X_1) + n(X_2) + n(X_3) - n(X_1 \cap X_2) - n(X_1 \cap X_3) - n(X_2 \cap X_3)$$

give us the correct total? Let us return to Example 6 to see whether this gives us an accurate count.

We have already determined that in Example 6, $n(X_1) = 100$ and $n(X_2) = n(X_3) = 90$. How many elements are in the intersection $X_1 \cap X_2$? In other words, how many three-digit positive integers have 7 as their first two digits? There are ten, since there is no restriction on the choice of the third digit. Similarly, $n(X_1 \cap X_3) = 10$, since the first and third digits must be 7, and there is no restriction on the second digit. However, $n(X_2 \cap X_3) = 9$, since the second and third digits must be 7, and the first digit cannot be 0. Therefore,

$$n(X_1) + n(X_2) + n(X_3) - n(X_1 \cap X_2) - n(X_1 \cap X_3) - n(X_2 \cap X_3) = 100 + 90 + 90 - 10 - 10 - 9$$
$$= 251.$$

How can we tell whether this is the right answer? Recall the alternative method that we used for Example 5. There are a total of $9 \cdot 10 \cdot 10 = 900$ three-digit positive integers. We can count the number of three-digit positive integers that have no 7's and subtract that number from 900. If a three-digit positive integer has no digits that are 7, then there are eight choices for the first digit, and nine choices for each of the second and third digits, for a total of $8 \cdot 9 \cdot 9 = 648$ three-digit integers that have no 7's as digits. This means that there must be $900 - 648 = 252$ three-digit numbers that have at least one digit that is a 7. How does that compare with our tally above?

Oh, dear. We are off by 1. Our strategy of overcounting and correcting gave us a total of 251, but the correct answer is 252. What happened? Despite our careful consideration of overcounting, it seems we have made another mistake. Let us focus on the number 777 for a moment to follow our tallying. Initially, we count 777 three times because it is an element of each of the sets X_1, X_2, and X_3. Then we correct for this overcounting by subtracting 1 from our total three times, because 777 is an element of each of the sets $X_1 \cap X_2, X_1 \cap X_3$, and $X_2 \cap X_3$. That leaves us with a net tally of 0 for the number 777, when we should have 1, since 777 is certainly a three-digit number with at least one digit that is a 7. The mistake that we made in our calculation is that we did not consider elements that are in *all three* of the sets X_1, X_2, and X_3. Such elements are initially counted, or *included* three times, and then *excluded* three times. To correct for our overcorrection, we will need to include such elements one more time. This leads to the second version of the Inclusion-Exclusion Principle, which we state as follows:

Theorem 12 (Inclusion-Exclusion Principle, Version 2): If X_1, X_2, and X_3 are sets, then

$$n(X_1 \cup X_2 \cup X_3) = n(X_1) + n(X_2) + n(X_3)$$
$$- n(X_1 \cap X_2) - n(X_1 \cap X_3) - n(X_2 \cap X_3)$$
$$+ n(X_1 \cap X_2 \cap X_3).$$

We will hardly be content to stop at three sets when we can generalize even further. Perhaps more important, we will see the pattern that is developing with the first two versions of the Inclusion-Exclusion Principle. Notice that there are three types of sets involved in the formula in Theorem 12, which we have separated onto three separate lines. On the top line are the individual sets we started with; on the second line are all of the sets we can form by intersecting two of the three sets; and on the third line is the only set that is the intersection of all three sets. Notice as well that we alternately add and subtract the number of elements in each of the different types of sets. This leads to our third and final version of the Inclusion-Exclusion Principle.

Theorem 13 (Inclusion-Exclusion Principle, Version 3): If $X_1, X_2, ..., X_n$ are finite sets, then the size of their union is equal to the sum of the sizes of all intersections of an odd number of those sets minus the sum of the sizes of all intersections of an even number of those sets. In symbolic notation, if A_i is the sum of the sizes of all intersections of i of the n sets, then

$$n(X_1 \cup X_2 \cup \cdots \cup X_n) = A_1 - A_2 + \cdots + (-1)^{n+1} A_n$$

Example 7

How many positive integers less than or equal to 1000 are there that are multiples of 3, 5, or 7?

Solution

Let $S = \{1, 2, 3, ..., 1000\}$. If X_1 is the set of positive integers in S that are multiples of 3, X_2 is the set of positive integers in S that are multiples of 5, and X_3 is the set of positive integers in S that are multiples of 7, then Theorem 10 tells us that $A_1 = n(X_1) + n(X_2) + n(X_3) =$ INT(1000/3) + INT(1000/5) + INT(1000/7) = 333 + 200 + 142 = 675. Since the elements of $X_1 \cap X_2$ are multiples of both 3 and 5, they must be the multiples of 15 in S, of which there are INT(1000/15) = 66. Similarly, there are $n(X_1 \cap X_3)$ = INT(1000/21) = 47 multiples of 21 and $n(X_2 \cap X_3)$ = INT(1000/35) = 28 multiples of 35 in S, so $A_2 = 66 + 47 + 28 = 141$. Finally, there are $n(X_1 \cap X_2 \cap X_3)$ = INT(1000/105) = 9 multiples of $3 \cdot 5 \cdot 7 = 105$ in S, so $A_3 = 9$. Theorem 13 tells us that the size of the union of the three sets is therefore $A_1 - A_2 + A_3 = 675 - 141 + 9 = 543$.

Example 8

How many positive integers less than or equal to 1000 are there that are multiples of 3, 5, 7, or 11?

Solution

With the same notation as above,

$$A_1 = n(X_1) + n(X_2) + n(X_3) + n(X_4) = 333 + 200 + 142 + 90 = 765,$$
$$A_2 = n(X_1 \cap X_2) + n(X_1 \cap X_3) + n(X_1 \cap X_4) + n(X_2 \cap X_3) + n(X_2 \cap X_4) + n(X_3 \cap X_4) =$$
$$66 + 47 + 30 + 28 + 18 + 12 = 201,$$
$$A_3 = n(X_1 \cap X_2 \cap X_3) + n(X_1 \cap X_2 \cap X_4) + n(X_1 \cap X_3 \cap X_4) + n(X_2 \cap X_3 \cap X_4) =$$
$$9 + 6 + 4 + 2 = 21, \text{ and}$$
$$A_4 = n(X_1 \cap X_2 \cap X_3 \cap X_4) = 0,$$

since $3 \cdot 5 \cdot 7 \cdot 11 = 1155 > 1000$. Thus Theorem 13 tells us that the number of positive integers less than or equal to 1000 that are multiples of 3, 5, 7, or 11 is 765 − 201 + 21 − 0 = 585.

We close this section with some examples in which all of the intersections of a particular number of the sets X_i have the same number of elements. This makes the computations of the numbers A_i much easier, since they are just the product of the number of intersections, $C(n, i)$, and the number of elements in each intersection.

Example 9

How many rearrangements of 12345 are there in which no digit is one more than the preceding digit?

Solution

The following pairs of digits are not allowed in the rearrangement: 12, 23, 34, and 45. Let X_1 represent the sets of rearrangements of 12345 in which the pair 12 appears. If we think of gluing the digits 1 and 2 together, as in Section 6.4, then we see that $n(X_1) = 4!$. Similarly, if $X_2, X_3,$ and X_4 represent the sets of arrangements of 12345 in which 23, 34, and 45 appear, respectively, then $A_1 = 4 \cdot 4!$, $A_2 = C(4, 2) \cdot 3!$, $A_3 = C(4, 3) \cdot 2!$, and $A_4 = 1$, so

$$n(X_1 \cup X_2 \cup X_3 \cup X_4) = A_1 - A_2 + A_3 - A_4 = 96 - 36 + 8 - 1 = 67,$$

and there are $5! - 67 = 120 - 67 = 53$ rearrangements of 12345 in which no digit is one more than the preceding digit.

Example 10

A jar of jellybeans has four flavors: cherry, grape, lemon, and orange. How many combinations of six jellybeans can be chosen if we are not allowed to choose more than two jellybeans of any flavor?

Solution

It may be helpful to review our scheme in Section 6.4 for counting combinations with repetitions. Theorem 7 tells us that there are a total of $C(6 + 4 - 1, 6) = C(9, 6) = 84$ possible combinations with no restriction on the number of each color. Here, we let C represent the set of combinations with at least three cherry jellybeans, and likewise we let G, L, and O represent the sets of combinations with at least three grape, lemon, and orange jellybeans, respectively. To compute $n(C)$, we note that if a combination has three cherry jellybeans, we need three more jellybeans, of any flavor, to complete the combination of six jellybeans. Theorem 7 from Section 6.4 tells us that there are $C(3 + 4 - 1, 3) = C(6, 3) = 20$ such combinations, so $n(C) = 20 = n(G) = n(L) = n(O)$, and $A_1 = 4 \cdot 20 = 80$. To be in the intersection of two of our sets, a combination must contain at least three jellybeans of two different flavors, and hence exactly three jellybeans of those two flavors. Thus, $n(C \cap G) = 1$, and $A_2 = C(4, 2) \cdot 1 = 6$. No 6-combination can have at least three jellybeans of more than two different flavors, so $A_3 = A_4 = 0$, and

$$n(C \cup G \cup L \cup O) = 80 - 6 = 74.$$

Finally, the number of combinations with no more than two jellybeans of any one flavor is $84 - 74 = 10$.

Exploratory Exercise Set 6.6

1. Divide up into pairs. Have each student write a subset of $\{1, 2, 3, 4, 5, 6, 7, 8, 9, 10\}$ that has at least six elements.

    (a) Have the first student name his or her set X_1 and calculate $n(X_1)$.

(b)   Have the second student name his or her set X_2 and calculate $n(X_2)$.

(c)   Together, write down the union $X_1 \cup X_2$, and calculate $n(X_1 \cup X_2)$.

(d)   Is $n(X_1 \cup X_2) = n(X_1) + n(X_2)$? If not, which is greater and by how much?

(e)   Together, write down the intersection $X_1 \cap X_2$, and calculate $n(X_1 \cap X_2)$.

(f)   What do you notice about your answers to parts (d) and (e)? Explain how this is related to Theorem 11.

2.   (*Proof Exercise*) In this exercise, we will prove Theorem 10, which says that if n and d are positive integers, then the number of positive integers less than or equal to n that are multiples of d is equal to the greatest integer less than or equal to $\dfrac{n}{d}$. Fill in the blanks in the following statements to complete the proof.

(a)   Suppose m is a positive integer less than or equal to n that is a multiple of d. This is equivalent to saying that d is a _____ of m.

(b)   From the definition of divisibility, this means that there is an integer k such that

_____ .

(c)   Since m and d are both positive integers, k must be a _____ integer as well.

(d)   Since $m = d \cdot k$ and $m \le n$, that means $d \cdot k \le$ _____ and $k \le$ _____ .

(e)   Therefore, k must be a _____ integer that is less than or equal to _____ .

(f)   The number of such integers k is the greatest integer less than or equal to _____ .

(g)   Thus the number of positive integers less than or equal to n that are multiples of d is equal to the greatest integer less than or equal to _____ .

3.   (a)   List all of the positive integers less than or equal to 100 that are multiples of both 3 and 4.

(b)   List all of the positive integers less than or equal to 100 that are multiples of both 3 and 6.

(c)   List all of the positive integers less than or equal to 100 that are multiples of both 4 and 6.

(d)   On the basis of your answers to parts (a) through (c), complete the following conjecture:

If a, b, and n are positive integers, then the number of positive integers less than or equal to n that are multiples of both a and b is _____. [*Hint*: You may want to review the discussion of greatest common divisors from Section 3.4.]

4.   (*Writing Exercise*) Write a problem in words that involves finding the number of elements in the union of four sets. Explain how to use the Inclusion-Exclusion Principle to find the answer. Then draw a Venn diagram with four sets (see Exercise 20 from Section 2.1) to solve the problem. Describe which method you prefer.

Exercise Set 6.6

1.   Find the number of positive integers less than or equal to 100 that are multiples of

(a)   6

(b)   8

(c)   13

(d)   19

(e)   20

2.   Find the number of positive integers less than or equal to 100 that are multiples of

(a)   both 5 and 8

(b)   either 5 or 8

(c)   both 6 and 8 [*Hint*: See Exploratory Exercise 3.]

(d)   either 6 or 8

3.   (a)   Find the number of positive integers less than or equal to 100 that are multiples of either 2, 3, 5, or 7.

(b) Use your answer in part (b) to determine the number of primes that are less than or equal to 100.

4. Find the number of three-digit positive integers that have either a 5 or a 7 as one of the digits.

5. There are 150 students in a particular cohort at a university. This semester, 90 of them are taking a math course, 75 are taking a history course, and 40 of them are taking both a math and a history course.
   (a) Use the Inclusion–Exclusion Principle to determine the number of students in the cohort that are taking either history or math this semester.
   (b) Draw a Venn diagram, as in Figure 1 of this section, and fill in each disjoint region with the appropriate number.
   (c) How many students are taking neither math nor history?

6. There are 400 students at the Amityville Arts Academy. There are 150 students in the band, 180 students in the chorus, 165 students in the drama club, 70 students in both the band and the chorus, 85 students in both the band and the drama club, 90 students in the chorus and the drama club, and 40 students in the band, the chorus, and the drama club.
   (a) Use the Inclusion–Exclusion Principle to determine the number of students who are in the band, the chorus, or the drama club.
   (b) Draw a Venn diagram with three circles, and fill in each region with the appropriate numbers to illustrate the problem.
   (c) How many students are in the band and the drama club but not in the chorus?
   (d) How many students are in neither the band, the chorus, nor the drama club?

7. A survey of 150 campers revealed the following information:
   60 took archery.
   70 took basketball.
   85 took canoeing.
   25 took archery and basketball.
   20 took archery and canoeing.
   40 took basketball and canoeing.
   15 took archery, basketball, and canoeing.
   (a) How many of the campers surveyed took archery but not canoeing or basketball?
   (b) How many of the campers surveyed did not take any of these three activities?

8. Find the number of rearrangements of 123456 in which no digit is two more than the preceding digit.

9. Find the number of rearrangements of 123456 in which no digit is twice as much as the preceding digit.

10. A jar of jellybeans has five flavors: apple, cherry, grape, lemon, orange. How many combinations of eight jellybeans can we select if we are allowed to choose no more than three of the same flavor?

11. Write down the ten combinations of flavors of jellybeans from Example 10 of this section that have no more than two jellybeans of any flavor.

12. Find the number of arrangements of AAABBBCCCDDD if no letter is allowed to appear three times in a row.

Probability

Chapter

7

7.1 Introduction to Probability

Many people are familiar with the language of probability from listening to weather reports. If a forecaster says, "There is a 40% chance of rain today," she is not giving a definitive yes or no answer to the question "Is it going to rain today?" However, the numerical response of 40%, or 0.40 in decimal form or $\frac{40}{100} = \frac{2}{5}$ in fractional form, might still provide valuable information that will help you decide whether or not to take an umbrella to work. What exactly does that number 40% mean? How do weather forecasters arrive at such numbers, and how should we interpret them? To answer these questions fully, we will need to introduce some of the basic ideas and vocabulary of probability.

Probability is the branch of mathematics that is concerned with assigning a numerical value, like 40%, to describe how likely it is that an event will occur. It is convenient to think of probability as a laboratory science in which we will conduct experiments. Whereas you would expect to find beakers and Bunsen burners in a chemistry lab, the indispensable equipment in a probability lab sounds a little more dubious: dice, playing cards, and coins (for tossing). Honesty compels us to admit that, historically, gambling provided much of the impetus to develop a mathematically precise theory of probability. But before you begin planning a casino junket in anticipation of the newfound knowledge you are about to acquire, let us give you a quick preview of the moral of this section as it relates to just about every casino gambling game: The more you play, the more you are likely to lose.

In probability, we will describe experiments. Some will be simple to perform, such as tossing a coin, or rolling a six-sided die. Other experiments might involve detailed medical studies or the collection of massive amounts of data, but all of them will use the same vocabulary and tools. We will rely on our knowledge of the language of sets that we developed in Chapter 2.

SAMPLE SPACES AND EVENTS

In an experiment, we are interested in the possible **outcomes**, sometimes called **sample points**. The set of all possible outcomes is called the **sample space** and is often denoted by the letter S. An **event** is defined to be some subset of the sample space. Because events are defined as sets, it makes sense to talk about unions, intersections, and complements of events. Here, the universal set is simply the sample space. Recall that two sets are disjoint if their intersection is the empty set; in other words, they have no elements in common. We say that two events are **mutually exclusive** if they are disjoint sets. Before we proceed further, let us digest these definitions with a few examples.

Example 1

Consider the experiment of rolling a standard six-sided die and recording the number of dots on the top face.
(a) What are the outcomes of this experiment?
(b) What is the sample space S for this experiment?
(c) If E is the event of rolling a number larger than 3, and F is the event of rolling an odd number, then write E and F as subsets of S.
(d) Write the union of the events E and F as a subset of S, and describe it in words.
(e) Are E and F mutually exclusive events? Explain.

Solution

(a) The outcomes are 1, 2, 3, 4, 5, and 6.
(b) The sample space is the set $S = \{1, 2, 3, 4, 5, 6\}$.
(c) $E = \{4, 5, 6\}$ and $F = \{1, 3, 5\}$.
(d) $E \cup F = \{4, 5, 6\} \cup \{1, 3, 5\} = \{1, 3, 4, 5, 6\}$. In words, the union of E and F is the event of rolling either a number larger than 3 or an odd number. You could also say that the union of E and F is the event of rolling anything other than a 2.
(e) No. $E \cap F = \{5\}$, which is not the empty set, so E and F are not mutually exclusive.

Example 2

Consider the experiment of tossing a coin twice and recording the resulting sequence of "heads" and "tails." (Write H if the coin lands with the "heads" side facing up, and record T if it lands with the "tails" side facing up.)

(a) Describe the sample space S for this experiment.

(b) If E is the event of getting heads on at least one toss, then write E as a subset of S.

(c) Describe in words the complement of the event E, and write it as a subset of S.

Solution

(a) The sample space is $S = \{HH, HT, TH, TT\}$.

(b) $E = \{HH, HT, TH\}$.

(c) The complement of E is the event of getting heads on no tosses. You could also say that it is the event of getting tails on both tosses. As a set, $\overline{E} = \{TT\}$.

Notice that the outcomes of an experiment are the building blocks with which we construct events, similar to the way molecules are built from atoms. Whereas two events may not be mutually exclusive, as in Example 1(e), two outcomes must always be mutually exclusive. In this sense, outcomes are the smallest possible building blocks; they cannot be broken down any further. Let us give two more examples to demonstrate the variety of possible sample spaces.

Example 3

Toss a coin until it comes up heads. Record the number of tosses it takes (including the last toss of heads).

(a) What is the sample space for this experiment?

(b) If E is the event of getting tails on the first two tosses, then write E as a subset of the sample space.

Solution

(a) Notice that there is no upper limit we can give to the number of tosses it will take. The sample space is the infinite set $\{1, 2, 3, \ldots\}$.

(b) We might get heads on the third toss or on any later toss. $E = \{3, 4, 5, \ldots\}$.

Example 4

Fill a cup with water from a tap, measure the temperature of the water, and record the temperature in degrees Fahrenheit.

(a) What is the sample space of this experiment?

(b) If E is the event of recording a temperature that is strictly between $60°$ F and $110°$ F, then write E as a subset of the sample space.

Solution

(a) Since water, rather than ice or vapor, comes out of the tap, it is reasonable to assume that the water temperature is strictly between $32°$ F and $212°$ F. (We may also assume that the water is relatively free from impurities that would slightly change the freezing and boiling points.) Thus the sample space is the set of all real numbers strictly between 32 and 212. In set-builder notation, $S = \{x : x \text{ is a real number and } 32 < x < 212\}$.

(b) E is the set of all real numbers between 60 and 110. In set-builder notation, $E = \{x : x \text{ is a real number and } 60 < x < 110\}$.

In Example 4, notice that it is possible to record temperatures such as $74.56°$ F, which are not whole numbers. Indeed, there are infinitely many elements in both the sample space S and the event E in both

Examples 3 and 4. In the remainder of this section, we will limit our discussion of probability to the types of experiments in Examples 1 and 2, where the sample space is a finite set.

PROBABILITY SPACES

Now that we have some examples of sample spaces and events under our belts, we are ready to define the notion of the probability of an event, which is a numerical measure of the likelihood that an event will occur. For any experiment we perform, we want to be able to assign a number, called the probability, to each event. We will use the following notation:

If E is an event, then $P(E)$ denotes the probability that the event E will occur.

In order to make this assignment in a way that makes sense, we will have to follow a few rules. The first and most important rule is that the probability must be a number between 0% and 100%, or, in decimal form, between 0 and 1. It wouldn't make any sense, for example, to say that there is a 120% chance of rain today. The more likely an event is to occur, the higher its probability will be on this scale from 0 to 1. At the two extremes of our scale, an event with probability 0 is called an **impossible event**, and an event with probability 1 is called a **certain event**. Recall that the sample space itself is an event, since it is a subset of itself. The second rule guarantees that S is a certain event. The third rule is that if an event contains a finite number of outcomes, then the probability of the event is the sum of the probabilities of the outcomes that are elements of that event. If a sample space has an assignment of probabilities that satisfies these three rules, then we call it a probability space. We summarize these rules below.

Rules for a Finite Probability Space

1. If E is an event, then $0 \le P(E) \le 1$.

2. $P(S) = 1$.

3. If $E = \{o_1, o_2, \ldots, o_k\}$, then $P(E) = P(\{o_1\}) + P(\{o_2\}) + \cdots + P(\{o_k\})$.

For convenience, we will often use the notation $P(o_1)$ rather than $P(\{o_1\})$ to denote the probability of an individual outcome. Let us now revisit our earlier examples of sample spaces and include probability assignments to turn them into probability spaces.

Example 5
Recall the experiment in Example 1 of rolling a six-sided die, in which the sample space is $\{1, 2, 3, 4, 5, 6\}$. Let us now assume that the die is "fair," which means that each of the six faces is equally likely to land face up.
(a) Determine the probability of each outcome.
(b) Determine the probabilities of E, F, and $E \cup F$, where $E = \{4, 5, 6\}$ and $F = \{1, 3, 5\}$, as in Example 1.

Solution
(a) From Rules 2 and 3, $P(1) + P(2) + P(3) + P(4) + P(5) + P(6) = P(S) = 1$. Since all 6 outcomes are equally likely, $P(1) = P(2) = P(3) = P(4) = P(5) = P(6)$, so $6P(1) = 1$, and the probability of each outcome is 1/6.
(b) $P(E) = P(\{4, 5, 6\}) = P(4) + P(5) + P(6) = 1/6 + 1/6 + 1/6 = 3/6 = 1/2$.
$P(F) = P(\{1, 3, 5\}) = P(1) + P(3) + P(5) = 1/6 + 1/6 + 1/6 = 3/6 = 1/2$.
$P(E \cup F) = \{1, 3, 4, 5, 6\} = P(1) + P(3) + P(4) + P(5) + P(6)$
$= 1/6 + 1/6 + 1/6 + 1/6 + 1/6 = 5/6$.

Example 6

Recall the experiment in Example 2 of tossing a coin twice. Assume the coin is fair.
(a) Determine the probability of each outcome.
(b) Determine the probability of getting heads at least once.

Solution

(a) Recall that the sample space for this experiment is {*HH, HT, TH, TT*}. Since each of these four outcomes is equally likely, $P(HH) = P(HT) = P(TH) = P(TT) = 1/4 = 0.25$.
(b) As a subset of the sample space, the event E of getting heads at least once is {*HH, HT, TH*}, so $P(E) = P(\{HH, HT, TH\}) = P(HH) + P(HT) + P(TH) = 1/4 + 1/4 + 1/4 = 3/4 = 0.75$.

UNIFORM SAMPLE SPACES

In Example 5 and Example 6, each of the outcomes of the experiment was equally likely to occur. A probability space in which each outcome is equally likely is called a **uniform sample space**. Uniform sample spaces are particularly nice to work with because we can easily assign probabilities to each outcome and to each event. In Example 5, notice that events E and F have the same probability because, as subsets of S, they contain the same number of elements. We can simplify our work in such cases by using Theorem 1 below. To determine the probability of an event E in a uniform probability space, we simply need to know how many elements are in E and how many elements are in the sample space, S. Recall the notation we introduced in Chapter 2, where $n(A)$ denotes the number of elements of a finite set A.

Theorem 1: Let S be a finite uniform sample space. Then:

(a) The probability of each outcome is $\dfrac{1}{n(S)}$.

(b) If E is an event (subset of S), then $P(E) = \dfrac{n(E)}{n(S)}$.

For the proof of Theorem 1, see Exploratory Exercise 5.

As we mentioned earlier, Theorem 1 greatly simplifies our work in determining probabilities for uniform sample spaces. It reduces probability problems to counting problems, and after studying Chapter 6, we are experts at counting!

Example 7

Suppose our experiment is to choose a ball at random from a box containing six red balls and four white balls. What is the probability of choosing a red ball?

Solution

Suppose we name the six red balls $B_1, B_2, \ldots, B_6$ and the four white balls B_7, B_8, B_9, and B_{10}. Since we choose one ball at random, each of the ten balls is equally likely to be chosen. If we let the sample space be $S = \{B_1, B_2, \ldots, B_{10}\}$, then S is a uniform sample space. If E is the event of choosing a red ball, then $E = \{B_1, B_2, \ldots, B_6\}$, and $P(E) = \dfrac{n(E)}{n(S)} = \dfrac{6}{10} = \dfrac{3}{5} = 0.6$. In other words, there is a 60% chance that a red ball will be chosen.

DEMONSTRATING PROBABILITIES IN THE CLASSROOM

In addition to dice and coins, spinners are also frequently used in games of chance and in elementary and middle-grades classrooms to illustrate probabilities. Typically, the interior of a circle is divided into a number of sectors, and a pointer is attached at the center of the circle and allowed to spin. (One easy method of constructing a spinner is to draw the spinner on a piece of paper, hold the tip of a pencil or pen in the center of the circle, and spin a paper clip around the pencil or pen.) If we make the assumption that the pointer is equally likely to stop in any direction, then the probability that the pointer will stop in a particular sector is equal to the proportion of the circle that the sector occupies.

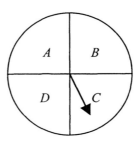

Figure 1: A Spinner

Example 8

Consider the spinner shown in Figure 1, where the four sectors are of equal size. For each of the four sectors of the circle, determine the probability that the pointer will stop in that sector.

Solution

Since each sector occupies one-fourth of the circle, the probability that the pointer will stop in that sector is 1/4, or 0.25.

Cubes are by far the most popular geometric objects to be used for dice, but there are other, less standard dice that have different numbers of faces. Among the other regular polyhedra, tetrahedral, octahedral, dodecahedral, and icosahedral dice have 4, 8, 12, and 20 faces, respectively, with each face equally likely to land face down. All of these except the regular tetrahedron have $180°$ rotational symmetry, so we generally record the face that lands face up, rather than down.

Example 9

Each of two octahedral dice has faces numbered from 1 to 8. The two dice are rolled.
(a) What is the probability that the sum of the numbers rolled is 7?
(b) What is the probability that the sum of the numbers rolled is 10?
(c) What is the most likely sum to be rolled?

Solution

There are $8 \cdot 8 = 64$ equally likely outcomes, represented by the 64 ordered pairs in the Cartesian product $A \times A$, where $A = \{1, 2, 3, 4, 5, 6, 7, 8\}$.
(a) The ordered pairs (1, 6), (2, 5), (3, 4), (4, 3), (5, 2), and (6, 1) all have a sum of 7, so the probability is $6/64 = 3/32 = 0.09375$.
(b) The ordered pairs (2, 8), (3, 7), (4, 6), (5, 5), (6, 4), (7, 3), and (8, 2) all have a sum of 10, so the probability is $7/64 = 0.109375$.
(c) There are 8 ordered pairs with a sum of 9, so the probability of rolling a sum of 8 on octahedral dice is $8/64 = 1/8 = 0.125$.

EMPIRICAL PROBABILITY SPACES

As much fun as it was to put our combinatorial skills to use in the last two examples, let us point out that we cannot use Theorem 1 unless we know that all outcomes in our experiment are equally likely. This is not always the case; in other words, many sample spaces are not uniform. Important examples can arise when we assign probabilities based on actual data. We call such a sample space an **empirical probability space**. In contrast, in our past examples we assigned probabilities on the basis of theoretical notions, such as the idea that all outcomes are equally likely to occur. Such models are called **theoretical probability spaces**. The difference between these two types of probability spaces lies not so much in the actual numbers that are assigned to each outcome as in the way those probabilities are assigned. Remember that no matter how the probabilities are assigned, it must be done in such a way that the three Rules for a Finite Probability Space still hold.

Example 10

>Let us return to our experiment of rolling a six-sided die, but suppose we have reason to believe that the die is "loaded" so that some faces are more likely to land face up than others. If we don't know which faces are more likely to land face up, how can we determine the probability of each outcome? One way to do this is to roll the die repeatedly and record how many times each face lands face up. Suppose we roll the die 1000 times and record the data shown in Table 1.

Table 1: Empirical Frequency Data for Example 10

Outcome	1	2	3	4	5	6
Frequency	110	138	82	214	165	291

Example 10 *(continued)*

>(a) How often would you expect to roll a 1 using this die?
>(b) On the basis of this empirical data, assign a probability to each of the outcomes.
>(c) What is the probability of rolling a number greater than 3 on this die?

Solution

>(a) From our observed data, 1 came up 110 out of 1000 times, so we should expect to roll a 1 on this die $\frac{110}{1000} = 0.11 = 11\%$ of the time. Compare this to a fair six-sided die, with which we should expect to roll a 1 around 1/6, or approximately 17%, of the time.
>
>(b) As in part (a), we assign a probability to each outcome by dividing the observed frequency by 1000, the total number of rolls. This gives the probabilities expressed as decimals in Table 2.

Table 2: Empirical Probabilities for Example 10

Outcome	1	2	3	4	5	6
Probability	0.11	0.138	0.082	0.214	0.165	0.291

Example 10 *(continued)*

>(c) From Rule 3, $P(\{4, 5, 6\}) = P(4) + P(5) + P(6) = 0.214 + 0.165 + 0.291 = 0.67$. Recall that the probability of this event with a fair die is $1/2 = 0.5$.

Example 11

> The fourth-grade students at Munro Elementary School were tested for reading comprehension, and the percentage of students classified in each reading level is recorded in Table 3.

<table>
<tr><td colspan="5" align="center">Table 3: Reading Level Data for Example 11</td></tr>
<tr><td>Reading Level</td><td>third</td><td>fourth</td><td>fifth</td><td>sixth</td></tr>
<tr><td>% of Students</td><td>22%</td><td>45%</td><td>29%</td><td>4%</td></tr>
</table>

Example 11 *(continued)*

> If a fourth-grade student from Munro Elementary School is chosen at random, what is the probability that the student reads:
> (a) below the fourth-grade level?
> (b) above the fourth-grade level?
> (c) at or above the fourth-grade level?

Solution

> (a) The probability is 22%, or 0.22.
> (b) The probability is $P(5\text{th}) + P(6\text{th}) = 0.29 + 0.04 = 0.33$.
> (c) The probability is $P(4\text{th}) + P(5\text{th}) + P(6\text{th}) = 0.45 + 0.29 + 0.04 = 0.78$.
>
> You may have thought of an alternative method of determining the probability in part (c). In the next section, we will discuss that alternative method, which turns out to be more efficient than the direct method we used above, particularly if we have already determined the probability in part (a). Although both methods work well in Example 11, where the sample space has only four elements, the increased efficiency of this alternative method will be more critical in examples where the sample space is considerably larger.

Exploratory Exercise Set 7.1

1. Toss a plastic cup in the air 100 times, and record whether it lands upside down, right side up, or on its side. (Give it a good spin on each toss.)
    (a) Assign probabilities to each of these outcomes on the basis of your empirical data.
    (b) Would you say that these three outcomes are equally likely?

2. Design and construct a spinner with at least five sectors, not all of the same size. (Your instructor may wish to provide you with the necessary parts; for example, paper clips or metal clasps can be used as pointers.)
    (a) Draw a sketch of your spinner on your paper, carefully illustrating the size of each sector.
    (b) On the basis of the design of your spinner, estimate the probability that the pointer will stop in each sector.
    (c) How many times would you expect the pointer to stop in each sector if you spun it 30 times?
    (d) Now spin your pointer 30 times and record how many times it stops in each sector. Compare your results with the predictions you made in part (c).
    (e) On the basis of your results in part (d), for each sector, assign an empirical probability that the pointer will stop in that sector. Compare these probabilities with those you found in part (b).

3. Dick and Jane are trying to answer the following question. **If three fair coins are flipped, what is the probability that exactly one will land heads up?** Dick says, "The number of possible heads is either 0, 1, 2, or 3, so the probability of getting heads exactly once is 1/4 = 0.25." Jane says, "There are

$2 \cdot 2 \cdot 2 = 8$ possible sequences of heads and/or tails, and three of them contain exactly 1 H (namely, *HTT*, *THT*, and *TTH*) so the probability of getting heads exactly once is $3/8 = 0.375$."

(a) Whom do you believe? Explain why you think the other one is mistaken.

(b) Test your answer by actually conducting this experiment (tossing three coins) 50 times. Record your results in Table 4.

Table 4: Empirical Data for Exploratory Exercise 3				
# of Heads	0	1	2	3
Frequency				

(c) Was the number of times you got exactly 1 "heads" closer to 25%, as Dick would predict, or to 37.5%, as Jane would predict? Would you like to change your answer to part (a)?

4. (a) Divide up into pairs. For each of the following experiments, list the elements of the sample space individually, and then compare your answers.
    (1) Toss a coin three times and record the resulting sequence of heads and/or tails.
    (2) Toss a coin three times and record the number of times it landed heads up.
    (3) Roll a six-sided die twice and record the sequence of numbers rolled.
    (4) Roll a six-sided die twice and record the sum of the numbers rolled.
    (5) Roll a six-sided die twice and record the product of the numbers rolled.
    (6) Toss a coin, roll a six-sided die, and record the sequence of results.

    (b) Suppose the coin and the die from part (a) are fair. Determine which of the sample spaces you described are uniform. Explain.

5. (*Proof Exercise*) Use the Rules for a Finite Probability Space to complete this proof of Theorem 1. We assume that S is a uniform sample space with $n(S)$ elements. It will be convenient to refer to $n(S)$ as a single letter; let's say $n(S) = N$.

    (a) In this part, we will prove that each outcome has probability $\dfrac{1}{n(S)} = \dfrac{1}{N}$.

    (1) How many outcomes are there in this experiment? Explain.

    (2) Name the outcomes $o_1, o_2, \ldots$. What will the last outcome be named? Describe S by listing its elements and using the ellipsis (...) symbol.

    (3) Fill in the blanks to complete the proof:
    From part (2) and Rule 3, we know that
    $P(S) = P(o_1, \underline{\hspace{2cm}}) = P(o_1) + \underline{\hspace{3cm}}$.
    Since S is a uniform sample space, $P(o_1) = \underline{\hspace{1cm}} = \underline{\hspace{1cm}} = \ldots = \underline{\hspace{1cm}}$.
    Thus $P(S) = P(o_1) + P(o_1) + \cdots + P(o_1) = \underline{\hspace{1cm}} \cdot P(o_1)$.
    From Rule 2, we know $P(S) = \underline{\hspace{0.5cm}}$, so $\underline{\hspace{1cm}} \cdot P(o_1) = 1$, and $P(o_1) = \underline{\hspace{1cm}}$.
    Since all outcomes are equally likely, each outcome has probability $\underline{\hspace{1cm}}$.

    (b) Complete the proof of Theorem 1 by showing that $P(E) = \dfrac{n(E)}{n(S)}$ for any event E of the uniform sample space S.

6. (*Writing Exercise*) As the weather forecaster for WPRB television station, you receive a letter from a viewer who writes, "I am tired of all your meteorological mumbo-jumbo. I have noticed that everyday it either will rain or it won't. Why don't you just say every day that there is a 50% chance that it will rain and a 50% chance that it won't?" Write a gentle response to this viewer, explaining why his logic

may not be sound. You may wish to include a brief discussion of the various factors meteorologists consider when making predictions about the weather.

Exercise Set 7.1

---

1. Write the sample space for each of the following experiments.
    (a) Toss a coin four times and record the number of times heads lands face up.
    (b) The word MISSISSIPPI is spelled on 11 scraps of paper, one letter on each scrap. Choose one scrap and record the letter on it.
    (c) Suppose the four triangular faces of a tetrahedron are numbered 1, 2, 3, and 4. Roll this tetrahedron and a standard six-sided die, and record the sequence of numbers landing face down.
    (d) Roll a pair of six-sided dice and record the number of rolls it takes until the same number lands face up on the two dice (that is, until "doubles" are rolled.)

2. For each of the experiments in Exercise 1, write the following event as a subset of the sample space.
    (a) Let E be the event of getting heads an even number of times.
    (b) Let E be the event of drawing a consonant.
    (c) Let E be the event that the sum of the numbers landing face down is even.
    (d) Let E be the event that it takes no more than ten rolls to get "doubles."

3. Write the complement of each of the events in Exercise 2 both in words and as a subset of the sample space.

4. In the MISSISSIPPI experiment in part (b) of Exercise 1, suppose that each scrap of paper is equally likely to be chosen.
    (a) Determine the probabilities of drawing each of the four distinct letters.
    (b) Determine the probability of drawing a consonant.
    (c) Determine the probability of drawing a vowel.

5. Suppose that four students line up at random. What is the probability that the students lined up in alphabetical order, from front to back?

6. Suppose that a hand of 5 cards is chosen at random from a standard deck of 52 cards. Determine the probability that the hand will contain:
    (a) all red cards
    (b) two hearts and three spades
    (c) two aces and three jacks
    (d) at least one card from each of the four suits

7. You suspect that a certain six-sided die is not fair. You roll the die 600 times and obtain the data shown in Table 5.

Table 5: Empirical Frequencies for Exercise 7						
Outcome	1	2	3	4	5	6
Frequency	105	87	82	65	125	136

(a) On the basis of these empirical data, assign a probability to each of the outcomes.
(b) How often would you expect to roll an even number on this die?
(c) How often would you expect to roll a number less than 4 on this die?

8. Suppose a fair coin is tossed eight times. What is the probability that heads came up:
    (a) not at all?
    (b) exactly twice?
    (c) at least twice?
    (d) at least once?

9. Which of the following could *not* be a probability?

    (a) 0.147   (b) $\dfrac{3}{5}$ (c) -0.75

    (d) 64   (e) $\dfrac{5}{3}$ (f) 1.05

10. Suppose you tossed a fair coin ten times and heads came up all ten times. What would the probability be of getting heads on the eleventh toss?

11. Consider the spinner illustrated in Figure 2. Assume that sector D takes up one-third of the area of the circle and that the spinner (except for the pointer) is symmetric about the vertical line through the center.
    (a) Determine the probability that the pointer will stop in each sector.
    (b) Determine the probability that the pointer will stop in a sector labeled with a vowel.
    (c) Determine the probability that the pointer will stop in either sector C or sector D.

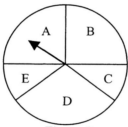

Figure 2

12. A bag contains five red marbles and seven blue marbles.
    (a) If one marble is chosen randomly from the bag, what is the probability that it is red?
    (b) Suppose you chose a red marble from the bag, and you did not replace it. If you choose a second marble from the bag, what is the probability that it will be red?
    (c) If two marbles are chosen simultaneously from the bag, what is the probability that they will both be red? (*Hint:* You may want to review combination numbers from Section 6.2.)
    (d) What connection do you see among your answers in parts (a), (b), and (c)?

13. Suppose you roll two polyhedral dice, where each die is equally likely to land on each of its n faces. If the faces on each dice are numbered from 1 to n, what is the probability that the sum of the numbers on the two faces landing face down will be:
    (a) 2?   (b) n?   (c) $n + 1$?   (d) $2n$?

7.2 Probability with Set Operations

Now that you are getting used to thinking of probability as a laboratory science, complete with experiments and lab equipment (coins, dice, playing cards, and other gambling paraphernalia), you might nonetheless be wondering why we need to drag in the language of set theory to describe events. One advantage is that we can use the structure of sets that we already have under our belts from Chapter 2 to help us describe events and to calculate their probabilities. Recall that the principal operations we use for sets are unions, intersections, and complements, rather than addition, subtraction, multiplication, and division. You may also remember that, in order to make sense of the notion of the complement of a set, we need to define a universal set. Fortunately, the structure of experiments, outcomes, and events we have described in the preceding section gives us a natural candidate for the role of a universal set: namely, the sample space S, the set of all possible outcomes.

COMPLEMENTARY EVENTS AND PROBABILITIES

Recall Example 11 from Section 7.1 about the reading comprehension of fourth-grade students at Munro Elementary School. The percentage of students classified in each reading level is recorded in Table 1.

Table 1: Reading Level Data from Munro Elementary School				
Reading Level	third	fourth	fifth	sixth
% of Students	22%	45%	29%	4%

In that example, we asked, if a fourth-grade student from Munro Elementary School is chosen at random, what the probability is that the student reads:
    (a) below the fourth-grade level?
    (b) above the fourth-grade level?
    (c) at or above the fourth-grade level?

Notice that the event in part (c) is the complement of the event in part (a). Can you see what the relationship is between the probabilities of complementary events? Let's return for a moment to the weather forecaster who got us started talking about probability. If there is a 40% probability that it will rain today, then what is the probability that it will *not* rain today? It must be 60%, right? How did we get 60%? By subtracting from 100%, or 1. We can generalize this for any two complementary events.

Theorem 2: If $\overline{E}$ is the complement of E, then $P(E) = 1 - P(\overline{E})$.

> *Preparation for Theorem 2:*
> *Example:* If E is the event in part (c), that a fourth-grade student at Munro reads at or above the fourth-grade level, then the event in part (a) is the complement of E, and $P(E) = 1 - 0.22 = 0.78$. Notice that this is the same answer that we got in the previous section when we calculated the probability directly as $0.45 + 0.29 + 0.04 = 0.78$.
> *Previous Definition:* Recall the rules for a finite probability space S:
>   1. If E is an event, then $P(E) \geq 0$.
>   2. $P(S) = 1$.
>   3. If $E = \{o_1, o_2, \ldots, o_k\}$, then $P(E) = P(o_1) + P(o_2) + \cdots + P(o_k)$.
> *Reflections:* If $\overline{E}$ is the complement of E, then $\overline{E}$ is the set of all outcomes in S that are not in E.

Proof of Theorem 2:

Suppose $\overline{E}$ is the complement of an event E in a finite sample space S. If $S = \{o_1, o_2, \ldots, o_n\}$, then we can rename the outcomes if necessary so that E consists of the first k outcomes in S, and the complement $\overline{E}$ consists of the remaining outcomes; that is, $E = \{o_1, o_2, \ldots, o_k\}$ and $\overline{E} = \{o_{k+1}, o_{k+2}, \ldots, o_n\}$. Rule 2 for a probability space tells us that $P(S) = 1$, and Rule 3 tells us that $P(S) = P(o_1) + P(o_2) + \cdots + P(o_n)$, so that $P(o_1) + P(o_2) + \cdots + P(o_n) = 1$. Notice that

$$
\begin{aligned}
P(E) + P(\overline{E}) &= \left(P(o_1) + P(o_2) + \cdots + P(o_k)\right) + \left(P(o_{k+1}) + P(o_{k+2}) + \cdots + P(o_n)\right) \\
&= P(o_1) + P(o_2) + \cdots + P(o_n) \\
&= 1,
\end{aligned}
$$

so that $P(E) = 1 - P(\overline{E})$. $\blacksquare$

We will see that Theorem 2 is particularly useful when the number of outcomes in the event E is considerably greater than the number of outcomes in the complementary event $\overline{E}$.

Example 1

A fair coin is tossed six times.
(a) Describe a sample space for this experiment.
(b) Write at least three outcomes in this sample space.
(c) Find the probability of the event E of getting heads at least twice.

Solution:

(a) If we record the sequence of heads and tails, the sample space S consists of the $2^6 = 64$ possible sequences.

(b) *HHTHTT*, *HHHHHH*, and *THTHHT* are three possible outcomes; there are many more.

(c) It is easier to consider the complementary event $\overline{E}$ of getting heads fewer than twice, that is, once or none at all. There is only one sequence with no heads (*TTTTTT*), and there are $C(6, 1) = 6$ sequences with exactly one head, so

$$
P(\overline{E}) = \frac{n(\overline{E})}{n(S)} = \frac{7}{64},
$$

and the probability of getting heads at least twice is

$$
P(E) = 1 - P(\overline{E}) = 1 - \frac{7}{64} = \frac{57}{64}.
$$

PROBABILITY OF THE UNION OF EVENTS

Suppose E and F are two events from a sample space S. If we know the probabilities of the events E and F, can we determine the probability of the union, $E \cup F$? Our first guess might be that $P(E \cup F)$ is simply equal to the sum of $P(E)$ and $P(F)$. After all, this is certainly true in the case where E and F are complementary events, as we saw in Theorem 2. However, let us reconsider Example 5 from Section 7.1, where a six-sided die is rolled. The events E and F are illustrated with a Venn diagram in Figure 1, where $E = \{4, 5, 6\}$ and $F = \{1, 3, 5\}$. We saw that with a fair die, $P(E) = 1/2$ and $P(F) = 1/2$, but $P(E \cup F) = 5/6$, rather than $1/2 + 1/2 = 1$. Why didn't we get $P(E \cup F) = P(E) + P(F)$?

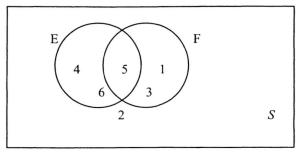

Figure 1

The reason lies in the way we define the union of two events or, indeed, of any two sets. Recall that the union of sets E and F is the set of all elements in either E or F or in both E and F. If we want to count the number of elements in $E \cup F$, we cannot simply count the number of elements in E and F (three each), and add them together, because we will have counted the element 5 twice. In general this "double counting" will happen with any element that is in both E and F. To obtain an accurate count of the number of elements in $E \cup F$, we need to correct for this double counting by subtracting the number of elements that were counted twice. But this is simply the number of elements in both E and F; that is, it is the number of elements in the intersection of E and F. We can phrase this succinctly using our mathematical notation:

$$n(E \cup F) = n(E) + n(F) - n(E \cap F).$$

You may recognize this formula as the first version of the Inclusion-Exclusion Principle from Section 6.4. In that section we generalized the formula for the union of an arbitrary number of sets.

If we have a uniform sample space, then determining probabilities is as easy as counting, so

$$P(E \cup F) = \frac{n(E \cup F)}{n(S)} = \frac{n(E) + n(F) - n(E \cap F)}{n(S)}$$

$$= \frac{n(E)}{n(S)} + \frac{n(F)}{n(S)} - \frac{n(E \cap F)}{n(S)}$$

$$= P(E) + P(F) - P(E \cap F).$$

Remarkably, this formula, summarized below in Theorem 3, holds even when the sample space is not uniform.

Theorem 3: If E and F are any two events, then

$$P(E \cup F) = P(E) + P(F) - P(E \cap F).$$

Preparation for Theorem 3: Consider the following example.
Example: Roll a fair six-sided die, and suppose E is the event of rolling a prime number and F is the event of rolling a divisor of 6. In set notation, $S = \{1, 2, 3, 4, 5, 6\}$, $E = \{2, 3, 5\}$, and $F = \{1, 2, 3, 6\}$. Then

$$P(E \cup F) = \frac{n(E \cup F)}{n(S)} = \frac{n(\{1, 2, 3, 5, 6\})}{n(\{1, 2, 3, 4, 5, 6\})} = \frac{5}{6}, \text{ and}$$

$$P(E) + P(F) - P(E \cap F) = \frac{n(E)}{n(S)} + \frac{n(F)}{n(S)} - \frac{n(E \cap F)}{n(S)}$$

$$= \frac{3}{6} + \frac{4}{6} - \frac{2}{6}$$

$$= \frac{5}{6} = P(E \cup F).$$

Previous Definitions: Recall the third property of a finite sample space with an event E: If $E = \{o_1, o_2, \ldots, o_k\}$, then $P(E) = P(o_1) + P(o_2) + \cdots + P(o_k)$.

Proof of Theorem 3:

Since S is a finite sample space, we can write $S = \{o_1, o_2, \ldots, o_n\}$. We consider four cases. If o_i is an outcome in both events E and F, then $P(o_i)$ is included in $P(E)$, $P(F)$, and $P(E \cap F)$, so it is counted a net of $1 + 1 - 1 = 1$ time in $P(E) + P(F) - P(E \cap F)$ and 1 time in $P(E \cup F)$. If o_i is an outcome in event E but not in event F, then $P(o_i)$ is counted in $P(E)$, but not in $P(F)$ or $P(E \cap F)$, so it is counted a net of $1 + 0 - 0 = 1$ time in $P(E) + P(F) - P(E \cap F)$ and 1 time in $P(E \cup F)$. The same is true if o_i is an outcome in event F but not in event E. Finally, if o_i is an outcome in neither event E nor event F, then $P(o_i)$ is not counted in $P(E)$, $P(F)$, or $P(E \cap F)$, nor is it counted in $P(E \cup F)$. Thus, in every case $P(o_i)$ appears the same number of times on both sides of the equation. Since this is true for all outcomes, the two sides must be equal. ∎

Example 2

At Georgewood Elementary School, 70% of the students ride the bus to school, 45% ride the bus home, and 35% percent ride the bus both ways. If a student from Georgewood Elementary is selected at random, what is the probability that the student:
(a) rides the bus to school or home?
(b) does not ride the bus at all?

Solution

Let E be the event that the student rides the bus to school, and let F be the event that the student rides the bus home. Then $P(E) = 0.70$, $P(F) = 0.45$, and $P(E \cap F) = 0.35$.

(a) Using Theorem 3, $P(E \cup F) = P(E) + P(F) - P(E \cap F) = 0.70 + 0.45 - 0.35 = 0.80$.

(b) The event that the student does not ride the bus at all is the complement of $E \cup F$. Using Theorem 2, $P(\overline{E \cup F}) = 1 - P(E \cup F) = 1 - 0.80 = 0.20$. In other words, 20% of the students at Georgewood Elementary do not ride the bus at all.

You may have noticed that we can simplify the formula in Theorem 3 if E and F happen to be mutually exclusive sets. In that case, $E \cap F = \emptyset$, so $P(E \cap F) = P(\emptyset) = 0$, and

$$P(E \cup F) = P(E) + P(F) - 0 = P(E) + P(F).$$

This special case of Theorem 3 is so useful that is worth labeling as a corollary, or immediate result of that theorem.

Corollary to Theorem 3: If E and F are mutually exclusive events, then

$$P(E \cup F) = P(E) + P(F).$$

We caution the reader that this simplified formula may be used only when we know that the events E and F are mutually exclusive; otherwise, we must use the more general formula in Theorem 3, where the probability of the intersection is subtracted.

> **Example 3**
>
> Of the 100 girls in the fourth grade at Carver Elementary School, 35 play on the soccer team and 25 play on the basketball team. A school policy does not allow girls to play on both teams. Suppose a fourth-grade girl from Carver Elementary School is chosen at random. What is the probability that she plays on either the soccer team or the basketball team?
>
> **Solution**
>
> If E is the event that the girl chosen is on the soccer team, and F is the event that she plays on the basketball team, then the school policy assures us that E and F are mutually exclusive, so by the Corollary to Theorem 3, $P(E \cup F) = P(E) + P(F) = 35/100 + 25/100 = 60/100 = 3/5$ or 60%.

We conclude this section with another consequence of Theorem 3. Theorem 4 brings together several of the themes discussed in this section and will be particularly useful later in this chapter.

Theorem 4: If E and F are two events, then $P(E) = P(E \cap F) + P(E \cap \overline{F})$.

> *Preparation for Theorem 4*: Recall from our work in Section 2.1 that if U is the universal set, and E and F are any subsets of U, then $E \cap U = E$, $F \cup \overline{F} = U$, and $F \cap \overline{F} = \emptyset$. Finally, recall the distributive property of intersection over union; that is, $E \cap (F \cup G) = (E \cap F) \cup (E \cap G)$.

Proof of Theorem 4: Since $E = E \cap U$ and $U = F \cup \overline{F}$, we can write $E = E \cap (F \cup \overline{F})$, which is in turn equal to $(E \cap F) \cup (E \cap \overline{F})$ by the distributive property mentioned in the preparation, so that $E = (E \cap F) \cup (E \cap \overline{F})$. Since $(E \cap F) \cap (E \cap \overline{F}) = E \cap (F \cap \overline{F}) = E \cap \emptyset = \emptyset$, the sets $E \cap F$ and $E \cap \overline{F}$ must be disjoint, which means that $P(E) = P((E \cap F) \cup (E \cap \overline{F})) = P(E \cap F) + P(E \cap \overline{F})$ by the corollary to Theorem 3.∎

Exploratory Exercise Set 7.2

1. (*Proof Exercise*) In this exploratory exercise, we will generalize Theorem 3 to determine the probability of the union of three events, E, F, and G. Specifically, we will prove that

$$P(E \cup F \cup G) = P(E) + P(F) + P(G)$$
$$-P(E \cap F) - P(E \cap G) - P(F \cap G) + P(E \cap F \cap G).$$

(a) Noting that $E \cup F \cup G = E \cup (F \cup G)$, use Theorem 3 to complete the following:
$P(E \cup F \cup G) = P(E \cup (F \cup G)) = P(E) + \underline{\hspace{2cm}} - \underline{\hspace{2cm}}$.

(b) Again using Theorem 3, $P(F \cup G) = \underline{\hspace{1.5cm}} + \underline{\hspace{1cm}} - \underline{\hspace{1.5cm}}$.

(c) Recall the Distributive Property: $E \cap (F \cup G) = (E \cap F) \cup \underline{\hspace{1.5cm}}$.

(d) Applying Theorem 3 to $(E \cap F) \cup (E \cap G)$ gives

$P((E \cap F) \cup (E \cap G)) = P(E \cap F) + \underline{\hspace{2cm}} - \underline{\hspace{2.5cm}}$.

(e) Explain why $(E \cap F) \cap (E \cap G) = E \cap F \cap G$.

(f) Putting together parts (c), (d), and (e) gives
$P(E \cap (F \cup G)) = P(E \cap F) + \underline{\hspace{2cm}} - P(E \cap F \cap G)$.

(g) Complete the proof by substituting (b) and (f) into the equation in (a).

2. Consider the following problem. At Hill High School, 50% of the students take art, 30% take badminton, 45% take chorus, 20% take both badminton and chorus, 25% percent take both art and chorus, 15% take both art and badminton, and 10% take art, badminton, and chorus. What is the probability that a Hill High School student takes at least one of art, badminton, or chorus?
   (a) Use the formula from Exploratory Exercise 1 to solve the problem.
   (b) Draw a Venn diagram with three circles to record the data and solve the problem.
   (c) What is the probability that a Hill High School student takes none of the three activities?

3. (*Writing Exercise*) The coach of your school's basketball team tells you that Tishay makes 60% of her 3-point shots and Radhika makes 40% of her 3-point shots. The coach then tells you that if Tishay and Radhika each take a 3-point shot, there is a 100% chance that at least one of them will make her shot. Write a paragraph that explains, using the language of this section, the error in the coach's reasoning.

Exercise Set 7.2

---

1. Fast Eddie's Computer Store sells floppy disks at a discount rate. Unfortunately, 10% of Fast Eddie's disks are scratched, 12% are cracked, and 5% are scratched and cracked. If you buy a floppy disk from Fast Eddie, what is the probability that the disk will be:
   (a) either scratched or cracked?
   (b) neither scratched nor cracked?

2. Suppose that a card is drawn at random from a standard deck of 52 cards. What is the probability that the card will be:
   (a) a face card (jack, queen, or king)?
   (b) either a face card or a heart?
   (c) neither a face card nor a black card?

3. A fair coin is tossed 11 times. What is the probability that heads will come up
   (a) exactly 3 times?
   (b) at least 3 times?

4. A loaded die is observed and found to land with 1 face up 20% of the time, with 2 face up 10% of the time, with 3 face up 30% of the time, with 4 face up 15% of the time, with 5 face up 5% of the time, and with 6 face up 20% of the time. If E is the event that this die lands on an even number, and F is the event that it lands on a number less than 4, then compute each of the following probabilities.
   (a) $P(\overline{E})$
   (b) $P(E \cup F)$
   (c) $P(\overline{E} \cap \overline{F})$

5. A three-digit positive integer is selected at random. What is the probability that:
   (a) none of the digits is an 8?
   (b) at least one of the digits is an 8?
   (c) at least one of the digits is greater than 7?

6. A positive integer less than or equal to 100 is chosen at random. What is the probability that it is:
   (a) either a multiple of 5 or a multiple of 7?
   (b) neither a multiple of 5 nor a multiple of 7?

7. There are 150 students in a particular cohort at a university. This semester, 90 of them are taking a math course, 75 are taking a history course, and 40 of them are taking both a math and a history course. If a student from this cohort is selected at random, what is the probability the student is taking:
   (a) either history or math this semester?
   (b) neither math nor history this semester?

8. At Logan Elementary School, 50% of the students ate a sandwich at lunch today, 60% had a salad, and 20% had both a sandwich and a salad. If a student is selected at random, what is the probability that the student:
   (a) had either a sandwich or a salad today?
   (b) had neither a sandwich nor a salad today?

9. There are 300 students at the Amityville Arts Academy. There are 150 students in the band, 180 students in the chorus, 165 students in the drama club, 70 students in both the band and the chorus, 85 students in both the band and the drama club, 90 students in the chorus and the drama club, and 40 students in the band, the chorus, and the drama club. If a student from the Amityville Arts Academy is selected at random, what is the probability that the student is:
   (a) in either the band, the chorus, or the drama club?
   (b) in the band and the drama club, but not in the chorus?
   (c) in neither the band, the chorus, nor the drama club?

10. A study of 150 campers revealed the following information:
    60 took archery.
    70 took basketball.
    85 took canoeing.
    25 took archery and basketball.
    20 took archery and canoeing.
    40 took basketball and canoeing.
    15 took archery, basketball, and canoeing.

    If a camper is selected at random, what is the probability that he or she:
    (a) took at least one of archery, basketball, or canoeing?
    (b) took archery, but not canoeing or basketball?
    (c) did not take any of these three activities?

7.3 Probability with Combinatorics

We have already seen in this chapter that a key step to calculating probabilities in uniform probability spaces is counting. Accordingly, it should come as no surprise that we can use many of the combinatorial techniques we learned in Chapter 6 to help us compute probabilities. In fact, we have already used combinations, permutations, and even the first version of the Inclusion-Exclusion Principle to help us calculate probabilities. We will expand on those ideas in this section and will use more of the combinatorial tools we have learned in Chapter 6 to address a wide variety of complex questions about probability.

PROBABILITY WITH PERMUTATIONS AND COMBINATIONS

We can now combine all of the tricks we have learned involving counting permutations and combinations to determine the probability of events where the permutations or combinations are chosen randomly. Keep in mind that we can determine probabilities by counting only if every outcome that we are counting is equally likely to occur.

Example 1

In the state of Limbo, license plates have three letters followed by three digits. If all arrangements are possible, what is the probability that a randomly assigned license plate will have:

(a) all three letters the same?

(b) all three digits the same?

(c) all three letters the same and all three digits the same?

Solution

Letting S be the sample space of all possible license plates, we can assume from the randomness of the assignment that S is a uniform probability space. The number of possible license plates is $n(S) = 26^3 \cdot 10^3 = 17,576,000$.

(a) If E is the event that all three letters are the same, then there are 26 choices for the first letter, only 1 choice for the second and third letters, and 10 choices for each of the three digits, so that $n(E) = 26 \cdot 1 \cdot 1 \cdot 10^3 = 26,000$ and

$$P(E) = \frac{n(E)}{n(S)} = \frac{26,000}{17,576,000} = \frac{1}{676} \approx 0.00148 .$$

(b) If F is the event that all three digits are the same, then there are 26 choices for each of the letters, 10 choices for the first digit, but only 1 choice for the remaining two digits, so that

$$n(F) = 26^3 \cdot 10 \cdot 1 \cdot 1 = 175,760 \text{ and } P(F) = \frac{n(F)}{n(S)} = \frac{175,760}{17,576,000} = \frac{1}{100} = 0.01 .$$

(c) If G is the event that all three letters are the same and all three digits are the same, then there are 26 choices for the first letter, 1 choice for the second and third letters, 10 choices for the first digit, and 1 choice for the second and third digits, so that $n(G) = 26 \cdot 10 = 260$

and $P(G) = \dfrac{n(G)}{n(S)} = \dfrac{260}{17,576,000} = \dfrac{1}{67,600} \approx 0.0000148 .$

Notice that in Example 1, the event G is simply the intersection of the events E and F. Do you notice a relationship among the probabilities of E, F, and G? We will explore this relationship in greater detail in Section 7.4. For now, let us return to some of our favorite paraphernalia of combinatorics (and gambling as well): coins, dice, and playing cards. It should come as no surprise that determining the probability of events can be very useful in games of chance.

Example 2

A fair coin is tossed 5 times. What is the probability that it will come up heads:
(a) exactly twice?
(b) at least twice?

Solution

Recall that there are $2^5 = 32$ sequences of heads and tails that are all equally likely. Since there are $C(5, 2) = 10$ ways to choose which 2 of the 5 tosses will come up heads, there are 10 sequences with exactly two heads, so the probability in part (a) is $10/32 = 5/16 = 0.3125$. For part (b), notice that the number of ways of getting heads once or less is $C(5, 0) + C(5, 1) = 1 + 5 = 6$, so the probability of the complementary event is $1 - 6/32 = 26/32 = 13/16 = 0.8125$.

In Chapter 6, we counted some of the possible hands we could draw from a deck of cards. We may now determine the probability of drawing particular types of hands. We should mention that in the following examples, we will focus on a relatively simple version of a card game, where a hand of five cards is dealt to each player. In other versions of the card game, for example, you are allowed to trade in some of your cards for other cards or to pick any five cards you like from a hand of seven cards.

Example 3

Five cards are dealt at random from a standard deck of 52 cards. What is the probability that:
(a) all five cards will be spades?
(b) exactly two of the cards will be aces?

Solution

For both parts, the sample space S consists of the $C(52, 5) = 2,598,960$ possible 5-card hands from a standard deck. The randomness means that each of these hands is equally likely to be dealt, so that S is a uniform sample space, and we can use Theorem 1.

(a) Let E be the event of getting all spades. There are 13 spades in the deck, so the number of elements in E is $C(13, 5) = 1287$. The probability is $\dfrac{n(E)}{n(S)} = \dfrac{1287}{2,598,960} = \dfrac{33}{66,640} \approx$ 0.000495.

(b) Let F be the event of getting exactly two aces in a 5-card hand. Let us choose the 5 cards in two steps. First choose 2 out of the 4 aces in the deck. There are $C(4, 2) = 6$ ways to do this. Then choose 3 more cards from the 48 non-aces in the deck. There are $C(48, 3) = 17,296$ ways to do this. Using the Multiplication Principle, there are a total of $6 \cdot 17,296 = 103,776$ elements in the event F, and $P(F) = \dfrac{n(F)}{n(S)} = \dfrac{103,776}{2,598,960} \approx 0.0399$.

Example 4

A five-card hand is dealt at random from a standard deck of 52 cards. What is the probability of receiving:
(a) a flush (all five cards have the same suit)?
(b) a full house (three cards of one denomination and two cards of another, such as three kings and two 8's)?

Solution

(a) If E is the event of getting a flush, we can count the outcomes in E with a two-stage process. First choose one of the four suits (spades, hearts, clubs, or diamonds); then choose 5 cards from the 13 cards of that suit. The Multiplication Principle tells us that there are $n(E) = C(4, 1) \cdot C(13, 5) = 4 \cdot 1287 = 5148$ hands that are flushes. Since there are a total of $C(52, 5) = 2,598,960$ possible hands,

(continued)

$$P(E) = \frac{5148}{2,598,960} \approx 0.00198 .$$

(b) If F is the event of getting a full house, we can count the outcomes in F with a four-stage process. First choose one of the 13 denominations (for the three-of-a-kind), and then choose three of the four cards of that denomination. Then choose one of the 12 remaining denominations (for the two-of-a-kind), and finally choose two of the four cards of that denomination. Using the Multiplication Principle and the randomness of the deal,

$$P(F) = \frac{n(F)}{n(S)} = \frac{C(13, 1) \cdot C(4, 3) \cdot C(12, 1) \cdot C(4, 2)}{C(52, 5)} = \frac{13 \cdot 4 \cdot 12 \cdot 6}{2,598,960} = \frac{3744}{2,598,960} \approx 0.00144 .$$

If two people were playing this card game, and one received a flush while the other received a full house, who do you think should win? It is no accident that in the game of poker, a hand with a full house beats a hand with a flush, since the probability of getting a full house is less than the probability of getting a flush. In fact, the hierarchy of which hands are better than others is determined by the principle that the less probable hand beats a more probable hand. Of course, we should point out that both a flush and a full house are quite rare and hence are both considered excellent hands. After all, the likelihood of even getting a flush (the more likely hand) is less than one-fifth of 1%. This means that on average, fewer than 1 in every 500 hands is a flush.

Such occurrences may seem rare, but they pale in comparison to the rarity of winning in one of the most popular games of chance: the lottery. Many lotteries have games similar to the one described in the next example. Despite slight variations, they all share the feature that winning the jackpot is extremely rare.

Example 5

In a state lottery, with each ticket you pick any set of five numbers from the set $\{1, 2, 3, \ldots, 50\}$. If those same five numbers are drawn (in any order) in the official drawing, then you win the jackpot. If you buy a ticket and pick five numbers, what is the probability that you will win the jackpot?

Solution

Since the order in which the numbers are drawn is not important, there are a total of $C(50, 5) = 2,118,760$ possible combinations. Since the winning combination is chosen at random, the probability that your combination is the winning combination is $1/2,118,760 \approx 0.000000472$. To put an event with such a small probability into perspective, it is more likely that you will be struck by lightning or killed by a tornado than that you will win the jackpot.

PROBABILITY WITH THE INCLUSION-EXCLUSION PRINCIPLE

We can combine the counting techniques we learned from the Inclusion-Exclusion Principle with probability as well. Again, we caution that there must be some indication that all outcomes we are counting are equally likely in order for us to be able to use our counting techniques to determine probability.

Example 6

A positive integer less than or equal to 100 is chosen at random. What is the probability that it will be a multiple of 2, 3, or 5?

Solution

If E_2, E_3, and E_5 are the events that the number chosen is a multiple of 2, 3, or 5, respectively, then the desired probability is $P(E_2 \cup E_3 \cup E_5)$, which is equal to $n(E_2 \cup E_3 \cup E_5)/100$. We can use the Inclusion-Exclusion Principle to determine the numerator as follows:

$$\begin{aligned} n(E_2 \cup E_3 \cup E_5) &= n(E_2) + n(E_3) + n(E_5) \\ &\quad - n(E_2 \cap E_3) - n(E_2 \cap E_5) - n(E_3 \cap E_5) \\ &\quad + n(E_2 \cap E_3 \cap E_5) \\ &= (50 + 33 + 20) - 16 - 10 - 6 + 3 \\ &= 74. \end{aligned}$$

Thus the probability in question is $74/100 = 0.74$.

You might recall from our discussion of permutations that a derangement is a special type of permutation in which no object is left in its original position. We complete this section with a discussion of how the Inclusion–Exclusion Principle can be used to determine the probability of such an event.

Example 7

Suppose a deck of cards is shuffled thoroughly. What is the probability that no card is in its original position if there are:
(a) 3 cards in the deck?
(b) 4 cards in the deck?
(c) 7 cards in the deck?
(d) k cards in the deck?

Solution

The numbers are small enough in parts (a) and (b) to handle these cases explicitly; we will then generalize for parts (c) and (d).

(a) With 3 cards there are $3! = 6$ possible arrangements of the cards. If the cards are labeled A, B, and C, then the arrangements are ABC, ACB, BAC, BCA, CAB, and CBA. If ABC is the original order then, of these 6 arrangements, only BCA and CAB are derangements, so the probability is $2/6 = 1/3$.

(b) With 4 cards in the deck, there are $4! = 24$ possible arrangements. If the cards are labeled A, B, C, and D, and the original arrangement is $ABCD$, then the derangements are $BADC$, $BCDA$, $BDAC$, $CADB$, $CDAB$, $CDBA$, $DABC$, $DCAB$, and $DCBA$. Thus the probability is $9/24 = 3/8 = 0.375$.

(c) With 7 cards in the deck, it is too difficult to actually list all of the permutations. Instead, we use the Inclusion-Exclusion Principle as follows. For values of i between 1 and 7, let E_i denote the event that the ith card is in its original place after being shuffled. As in Section 6.6, we let A_i denote the sum of the sizes of all intersections of i of the seven sets $E_1, E_2, ..., E_7$. If one card is in its original position, then there are $6!$ arrangements of the other 6 cards, so $n(E_i) = 6!$, for $i = 1, 2, 3, 4, 5, 6$, and 7. Thus $A_1 = n(E_1) + n(E_2) + \cdots + n(E_7) = 7 \cdot 6! = 7!$. Next, notice that if two cards are both in their original positions, then there are $5!$ ways to arrange the remaining 5 cards so that if i and j are any pair of integers between 1 and 7, inclusive, then $n(E_i \cap E_j) = 5!$. Since there are $C(7, 2) = 21$ possible pairs of intersecting sets,

$$A_2 = C(7, 2) \cdot 5! = \frac{7!}{2!5!} \cdot 5! = \frac{7!}{2!}.$$

(continued)

Similarly, if three cards are in their original positions, then there are $(7-3)!$ ways to arrange the remaining cards and $C(7, 3)$ possible intersections of three sets, so

$$A_3 = C(7, 3) \cdot (7-3)! = \frac{7!}{3!\,(7-3)!} \cdot (7-3)! = \frac{7!}{3!}.$$

Using the same reasoning, $A_i = \dfrac{7!}{i!}$ for $1 \le i \le 7$, so the Inclusion-Exclusion Theorem (Theorem 13 from Section 6.6) tells us that

$$n(E_1 \cup E_2 \cup \cdots \cup E_7) = A_1 - A_2 + A_3 - \cdots - A_6 + A_7$$

$$= \frac{7!}{1!} - \frac{7!}{2!} + \frac{7!}{3!} - \frac{7!}{4!} + \frac{7!}{5!} - \frac{7!}{6!} + \frac{7!}{7!}$$

$$= 5040 - 2520 + 840 - 210 + 42 - 7 + 1$$

$$= 3186$$

Thus $P(E_1 \cup E_2 \cup \cdots \cup E_7) = \dfrac{3186}{5040} = \dfrac{177}{280}$, and the probability that no card is in its original position is

$$1 - P(E_1 \cup E_2 \cup \cdots \cup E_7) = 1 - \frac{177}{280} = \frac{103}{280} \approx 0.3679.$$

(d) Finally, for a deck with k cards, we have essentially done the work in part (c). The number of elements in the intersection of any i sets, with $1 \le i \le k$, is $(k-i)!$, and the number of such intersections is $C(k, i)$. Thus,

$$A_i = C(k, i) \cdot (k-i)! = \frac{k!}{i! \cdot (k-i)!} \cdot (k-i)! = \frac{k!}{i!}, \text{ so}$$

$$n(E_1 \cup E_2 \cup \cdots \cup E_k) = k!\left(\frac{1}{1!} - \frac{1}{2!} + \frac{1}{3!} - \cdots + \frac{(-1)^{k+1}}{k!} \right), \text{ and}$$

$$P(E_1 \cup E_2 \cup \cdots \cup E_k) = \frac{k!\left(\dfrac{1}{1!} - \dfrac{1}{2!} + \dfrac{1}{3!} - \cdots + \dfrac{(-1)^{k+1}}{k!} \right)}{k!} = \frac{1}{1!} - \frac{1}{2!} + \frac{1}{3!} - \cdots + \frac{(-1)^{k+1}}{k!},$$

so that the probability that no card is in its original position is

$$1 - P(E_1 \cup E_2 \cup \cdots \cup E_k) = 1 - \left(\frac{1}{1!} - \frac{1}{2!} + \frac{1}{3!} - \cdots + \frac{(-1)^{k+1}}{k!} \right)$$

$$= \frac{1}{2!} - \frac{1}{3!} + \frac{1}{4!} - \cdots + \frac{(-1)^k}{k!}.$$

You may check (see Exploratory Exercise 3) that this formula matches the answers we got in parts (a) through (c) when k is equal to 3, 4, or 7. In that exercise you will also explore what happens to that probability as k gets larger and larger.

Exploratory Exercise Set 7.3

1. Split up into pairs and consider the following question. If four fair coins are tossed, what is the probability that exactly two coins will land heads up? Al and Bo give the following responses.

    Al: With two heads and four coins, the probability is 2/4, or 1/2.

Bo: You could get 0 heads, 1 head, 2 heads, 3 heads, or 4 heads. Since there are five possible outcomes, the probability is 1/5.

(a) Do you think Al is correct? Explain why or why not.
(b) Do you think Bo is correct? Explain why or why not.
(c) What do you think the correct answer is? Explain (if you haven't already).
(d) If you toss four coins 20 times, how many times would you expect to get exactly 2 heads? Explain.
(e) Toss four coins 20 times, and record how many times you get exactly 2 heads. Do your results match your prediction in part (d)?

2. The usual hierarchy of 5-card poker hands is given below. Determine the probability of each one and justify the order in which they are listed. Consider aces as the highest cards, so that 10-J-Q-K-A is a straight but A-2-3-4-5 is not.
(a) Straight flush (5 cards in a sequence, as in 6-7-8-9-10, all in the same suit)
(b) Four-of-a-kind (4 cards of the same denomination)
(c) Full house (3 of one denomination and 2 of another)
(d) Flush (5 cards in the same suit, not a straight flush)
(e) Straight (5 cards in a sequence, not a straight flush)
(f) Three-of-a-kind (three cards of one denomination, not a full house or four-of-a-kind)
(g) Two pair (two pairs of cards of the same denomination, not a full house or four-of-a-kind)
(h) One pair (two cards of the same denomination, and nothing else previously listed)
(i) Junk (none of the above)

3. Recall the result from part (d) of Example 7 in the text, namely that the probability that no card of a well-shuffled deck of k cards lands in its original position is $\frac{1}{2!} - \frac{1}{3!} + \frac{1}{4!} - \cdots + \frac{(-1)^k}{k!}$.
(a) Use this formula to check that the probability agrees with the answers obtained in:
    (1) part (a) of Example 7 when $k = 3$;
    (2) part (b) of Example 7 when $k = 4$;
    (3) part (c) of Example 7 when $k = 7$.
(b) In the table below, fill in the decimal approximations (accurate to four decimal places) for the aforementioned probability with a deck of k cards.

k	3	4	5	6	7
Probability					

(c) As we use more and more cards in the deck (k increases), what do you think will happen to the probability that no card will end up in its original position? Will that probability approach 1? Will it approach 0? Will it approach some other number?
(d) Take the reciprocal of each of the probabilities in the table in part (b). Do they seem to be approaching a familiar number? What is it? [*Hint*: You may have encountered this number while learning about continuously compounded interest rates or other exponential functions.]

4. (*Writing Exercise*) Pick a counting problem from Chapter 6 and explain how it can be turned into a probability problem. Be careful to state any assumptions you are making about the probability of each outcome. Then solve the probability problem that you formulated.

Exercise Set 7.3

1. Suppose a teacher arranges five books on a shelf in random order. What is the probability that the books are arranged in alphabetical order according to the author's last name? Assume that the authors have five different last names.

2. License plates in the state of Panic have three letters followed by three digits, where no repetition of letters or digits is allowed. What is the probability that a randomly selected license plate will have:
   (a) three consecutive digits in ascending order?
   (b) three consecutive digits in descending order?
   (c) three consecutive letters in alphabetical order?

3. Suppose a 4-digit positive integer is selected at random. (Recall that the first digit cannot be 0.)
   (a) What is the probability that none of its digits will be 5?
   (b) What is the probability that at least one of its digits will be 5?
   (c) What is the probability that none of its digits will be 0?
   (d) What is the probability that at least one of its digits will be 0?

4. What is the probability that a randomly dealt five-card hand will contain:
   (a) no aces?
   (b) exactly two aces?
   (c) no spades?
   (d) exactly two spades?

5. Suppose a fair coin is tossed six times. What is the probability that the coin lands heads up:
   (a) exactly twice?
   (b) at least twice?
   (c) more than it lands tails up?

6. In the game of bridge, a hand consists of 13 cards from a standard deck of cards. What is the probability of getting a hand with:
   (a) no spades?
   (b) exactly seven spades?
   (c) at least two spades?

7. A state lottery is played by choosing six positive integers less than or equal to 46.
   (a) If you buy a ticket for this lottery, what is the probability of matching all six numbers and winning the jackpot?
   (b) What is the probability of matching exactly five of the winning numbers?

8. Another state lottery is played by choosing five positive integers instead of six, as in Exercise 7. How many numbers would there have to be to choose from to make the probability of winning the jackpot in this lottery even less than in Exercise 7?

9. A positive integer less than or equal to 1000 is selected at random. What is the probability that it will be a multiple of 2, 5, 7, or 11?

10. Suppose a jar has jellybeans of five different flavors in it: apple, banana, cherry, date, and eggplant. If a sample of seven jellybeans is taken from the jar (assume that there are at least seven jellybeans of each flavor in the jar), what is the probability that it will contain:
    (a) at least three cherry-flavored jellybeans?
    (b) at least three of some flavor of jellybeans?

11. Suppose a fair coin is tossed ten times. What is the probability that no two consecutive tosses will come up heads?

7.4 Conditional Probability

Example 1

> The records of Paine Memorial Hospital indicate that during the past year 1200 patients were admitted. Of those 1200 patients, 450 were treated for cancer, 360 were smokers, and 315 were smokers who were treated for cancer. If we randomly select a patient from last year's records, what is the probability that the patient was treated for cancer?

Solution

> If we let S be the sample space of all possible patients and let E be the event that the patient selected was treated for cancer, then it is a straightforward calculation that the probability must be $n(E)/n(S) = 450/1200 = 0.375$. In other words, 37.5% of the hospital's patients were treated for cancer.

REDUCING THE SAMPLE SPACE

Suppose that we now narrow our focus to those patients at Paine Memorial Hospital who were smokers. If we select a smoker at random, what is the probability that the smoker was treated for cancer? In effect, we have now reduced our sample space to the 360 smokers at Paine. Imagine taking the records of the 360 smokers and putting them into a separate filing cabinet. This filing cabinet essentially becomes our new universal set. How many of those 360 smokers were treated for cancer? Exactly the 315 patients who both were treated for cancer and were smokers. Thus the conditional probability that a patient was treated for cancer, given that the patient was a smoker, is $315/360 = 0.875$. To put it in words, if a patient at Paine who is a smoker is chosen at random, there is an 87.5% chance that that patient is being treated for cancer. Compare this to the answer to our original question, where 37.5% of the patients at Paine were being treated for cancer.

There are several equivalent ways in which we could have worded the question in the previous paragraph. Here are a few:
  (a) What is the probability that a patient from Paine was treated for cancer, given that the patient was a smoker?
  (b) If a patient at Paine was a smoker, what is the probability that he or she was treated for cancer?
  (c) Given that a patient at Paine was a smoker, what is the probability that he or she was treated for cancer?

No matter how the question is worded, the fundamental idea is that our attention is restricted to those patients who are smokers. If we define F as the event that a randomly selected patient was a smoker, then we are restricting our attention to what happens when the event F occurs. You might say that the event F becomes a necessary condition for us to consider the questions phrased in (a) through (c) above. For that reason, the probabilities in those questions are called **conditional probabilities**.

Definition: If E and F are two events, and $P(F) \neq 0$, then the **conditional probability of E given F**, denoted $P(E \mid F)$, is given by

$$P(E \mid F) = \frac{P(E \cap F)}{P(F)}.$$

Furthermore, if E and F are events in a uniform sample space S, then

$$P(E \mid F) = \frac{n(E \cap F)}{n(F)}.$$

Notice that in our opening example, the randomness of the selection of patients assures us that we have a uniform probability space, so it was appropriate to calculate $P(E \mid F)$ by dividing $n(E \cap F)$ by $n(F)$. We could also have computed $P(E \mid F)$ by dividing $P(E \cap F)$ by $P(F)$, since

$$\frac{P(E \cap F)}{P(F)} = \frac{n(E \cap F)/n(S)}{n(F)/n(S)} = \frac{n(E \cap F)}{n(F)}.$$

Finally, notice that if we are considering the conditional probability of E given F, it is natural to assume that the given event F is not an impossible event; that is, we assume that $P(F) \neq 0$. Thus the fraction $\frac{P(E \cap F)}{P(F)}$ is always defined. If F does happen to be an impossible event, that is, $P(F) = 0$, then we say that the conditional probability $P(E \mid F)$ is undefined.

Example 2

Suppose your friend rolls a fair six-sided die and does not show you which number was rolled.
(a) What is the probability that an even number was rolled?
(b) Suppose your friend tells you (truthfully) that the number rolled was less than 4. Given that piece of information, what is the probability that an even number was rolled?

Solution

Let E be the event that an even number was rolled, and let F be the event that the number rolled was less than 4.
(a) The probability is $P(E) = 3/6 = 1/2$.

(b) The probability is $P(E \mid F) = \dfrac{n(E \cap F)}{n(F)} = \dfrac{1}{3}$, since $E \cap F = \{2\}$.

Example 3

Repeat Example 2, but with a loaded die that has the following empirical probabilities:

x	1	2	3	4	5	6
$p(x)$	0.2	0.1	0.1	0.3	0.1	0.2

Solution

Notice that in this example we do not have a uniform probability space, so we cannot simply count the number of relevant outcomes. With the events labeled as in Example 2,

(a) $P(E) = 0.1 + 0.3 + 0.2 = 0.6.$

(b) $P(E \mid F) = \dfrac{P(E \cap F)}{P(F)} = \dfrac{0.1}{0.2 + 0.1 + 0.1} = \dfrac{0.1}{0.4} = 0.25.$

INDEPENDENCE

Notice that in Examples 1 and 2, the knowledge that the number rolled was less than 4 had an effect on the probability that the number rolled was even. The probability $P(E)$ was different from the conditional probability $P(E|F)$ in each case. In some sense, we can say that the probability of the event E depends on whether or not the event F has occurred. Consider whether this is the case in Example 4.

Example 4

A fair six-sided die is rolled.
(a) What is the probability that the number rolled is even?
(b) If you are told that the number rolled is less than five, then what is the probability that the number rolled is even?
(c) What is the probability that the number rolled is less than 5?
(d) If you are told that the number rolled is even, then what is the probability that the number rolled is less than 5?

Solution

If E is the event that the number rolled is even, and F is the event that the number rolled is less than 5, then we can compute that:

(a) $P(E) = 3/6 = 1/2.$

(b) $P(E \mid F) = \dfrac{n(E \cap F)}{n(F)} = \dfrac{2}{4} = \dfrac{1}{2}.$

(c) $P(F) = 4/6 = 2/3.$

(d) $P(F \mid E) = \dfrac{n(F \cap E)}{n(E)} = \dfrac{2}{3}.$

It is worth noting that the two conditional probabilities in (b) and (d) are quite different; in general, there is no reason why $P(E|F)$ and $P(F|E)$ should be the same. From parts (a) and (b) of this example, it is clear that the knowledge that the number rolled is less than 5 does not change the likelihood that the number rolled is even. Similarly, parts (c) and (d) indicate that the likelihood that the number rolled is less than 5 does not depend on knowledge of whether the number rolled is even. In such cases where the likelihood of one event is not affected by the outcome of another event, we say that the two events are **independent**.

Definition: Two events E and F are independent if $P(E \mid F) = P(E)$.

Notice that this definition gives one of several equivalent conditions that determine whether two events are independent. Three other equivalent conditions are given in Theorem 5.

Theorem 5: If E and F are events with nonzero probabilities, then the following conditions are equivalent:

(i) $P(E \mid F) = P(E)$.

(ii) $P(F \mid E) = P(F)$.

(iii) $P(E \cap F) = P(E) \cdot P(F)$.

(iv) $P(E \mid F) = P(E \mid \overline{F})$.

Before we prove Theorem 5, notice that since the first condition defines the independence of E and F, any of the four conditions can be used to determine whether two events E and F are independent. The last condition can be particularly useful when it is difficult to calculate $P(E)$ directly.

Preparation for Theorem 5: Recall the definition of conditional probability,

$$P(E \mid F) = \frac{P(E \cap F)}{P(F)}.$$ From this, we also know that $P(F \mid E) = \dfrac{P(E \cap F)}{P(E)}$. From Section 7.2 we

recall Theorem 4, which says that $P(E) = P(E \cap F) + P(E \cap \overline{F})$.

Finally, recall that in order to prove that two statements are equivalent, we must prove that each implies the other.

Proof of Theorem 5:   We will prove that (i) is equivalent to (iii) and that (iii) is equivalent to (iv). In Exercise 6, you are asked to prove that (ii) is equivalent to (iii). Thus, all four statements must be equivalent to each other by the transitive property of equivalence.

Suppose (i) is true. Then $P(E \mid F) = P(E)$. Since $P(E \mid F) = \dfrac{P(E \cap F)}{P(F)}$, then $\dfrac{P(E \cap F)}{P(F)} = P(E)$, and

multiplying both sides by $P(F)$ gives $P(E \cap F) = P(E) \cdot P(F)$. Thus (i) implies (iii).

Now suppose that (iii) is true. Then $P(E \cap F) = P(E) \cdot P(F)$. Since $P(F) \neq 0$, we can divide both sides

by $P(F)$ to obtain $\dfrac{P(E \cap F)}{P(F)} = P(E)$. Since the left-hand side is $P(E \mid F)$, this means that

$P(E \mid F) = P(E)$. Thus (i) is equivalent to (iii).

To show that (iii) is equivalent to (iv), first suppose that (iii) is true; that is, suppose that $P(E \cap F) = P(E) \cdot P(F)$. From Theorem 4, we know that $P(E) = P(E \cap F) + P(E \cap \overline{F})$, so that $P(E \cap \overline{F}) = P(E) - P(E \cap F)$. Therefore,

$$P(E \mid \overline{F}) = \frac{P(E \cap \overline{F})}{P(\overline{F})} = \frac{P(E) - P(E \cap F)}{1 - P(F)}$$

$$= \frac{P(E) - P(E) \cdot P(F)}{1 - P(F)} = \frac{P(E)(1 - P(F))}{1 - P(F)} = P(E).$$

Since (iii) is equivalent to (i), and we are assuming that (iii) is true, then $P(E \mid F) = P(E)$, so that $P(E \mid F) = P(E \mid \overline{F})$. This proves that (iii) implies (iv). To show that (iv) implies (iii), assume that (iv) is true; that is, $P(E \mid F) = P(E \mid \overline{F})$.

Thus $\dfrac{P(E \cap F)}{P(F)} = \dfrac{P(E \cap \overline{F})}{P(\overline{F})} = \dfrac{P(E) - P(E \cap F)}{1 - P(F)}$. Cross-multiplying gives

$P(E \cap F) \cdot (1 - P(F)) = (P(E) - P(E \cap F)) \cdot P(F)$, which, after distributing, simplifies to $P(E \cap F) = P(E) \cdot P(F)$. Thus (iii) is equivalent to (iv). ■

We often make theoretical assumptions that events are independent. For example, if we roll a die and toss a coin, it is reasonable to assume that the event of rolling an odd number on the die and the event of having the coin land heads up are independent. However, as Examples 2, 3, and 4 may indicate, it is not always obvious whether two events are independent. When asked to determine whether two events are independent, it is generally best to check whether one of the equations in Theorem 5 is satisfied. If so, then the events are independent; if any of the equations is not satisfied, then the two events are not independent. Since all four equations are equivalent, it does not matter which one is used to test independence. However, the third equation is generally the easiest to use. It also gives a simple formula for the probability of the intersection of two events *if we know that the two events are independent.*

Example 5

A pair of fair dice are rolled. Let E be the event that the sum of the numbers rolled is 7, let F be the event that a 4 is rolled on the first die, and let G be the event that the sum of the numbers rolled is 6.

(a) Determine whether the events E and F are independent.

(b) Determine whether the events F and G are independent.

Solution

Recall that with a pair of fair dice, each of the 36 possible ordered pairs in the sample space $S = \{1, 2, 3, 4, 5, 6\} \times \{1, 2, 3, 4, 5, 6\}$ is equally likely, so that S is a uniform probability space.

(a) With $E = \{(1, 6), (2, 5), (3, 4), (4, 3), (5, 2), (6, 1)\}$ and

$$F = \{(4, 1), (4, 2), (4, 3), (4, 4), (4, 5), (4, 6)\}, \ P(E) \cdot P(F) = \frac{6}{36} \cdot \frac{6}{36} = \frac{1}{36}, \text{ and}$$

$$P(E \cap F) = P(\{(4,3)\}) = \frac{1}{36} = P(E) \cdot P(F), \text{ so } E \text{ and } F \text{ are independent.}$$

(b) Since $G = \{(1, 5), (2, 4), (3, 3), (4, 2), (5, 1)\}, \ P(F) \cdot P(G) = \frac{6}{36} \cdot \frac{5}{36} = \frac{5}{216}$ and

$$P(F \cap G) = P(\{(4, 2)\}) = \frac{1}{36} = \frac{6}{216} \neq P(F) \cdot P(G), \text{ so that } F \text{ and } G \text{ are not independent}$$

events.

Example 6

A bag contains seven red marbles and three blue marbles. Two marbles are randomly drawn from the bag, one at a time. Let F be the event that the first marble drawn is red, and let S be the event that the second marble is red.

(a) If the first marble is replaced after it is drawn, are the events F and S independent?

(b) If the first marble is not replaced after it is drawn, are the events F and S independent?

Solution

(a) If the first marble is replaced before the second one is drawn, then $P(S) = P(S \mid F) = 7/10$, so F and S are independent. Clearly, in this case, the color of the second marble does not depend on the color of the first marble drawn.

(b) If the first marble is not replaced, then it seems reasonable that the probability of the event S does indeed depend on the color of the first marble. We verify this by showing that $P(S \mid F) \neq P(S \mid \overline{F})$, which by Theorem 5 means that S and F are not independent. If the first marble is red, then there are six red marbles and three blue marbles remaining in the bag, so we can easily calculate the conditional probability $P(S \mid F) = 6/9 = 2/3$. If the first marble is not red, then there are seven red marbles and two blue marbles remaining in the bag, so $P(S \mid \overline{F}) = 7/9 \neq P(S \mid F)$.

Before proceeding, let us recap two formulas concerning probabilities and set operations that are always true, together with slightly simpler formulas that are true under special circumstances, in Table 1.

Table 1 : Probability Formulas		
Always True	Special Circumstances	Simpler Formula
$P(E \cup F) = P(E) + P(F) - P(E \cap F)$	E, F mutually exclusive	$P(E \cup F) = P(E) + P(F)$
$P(E \cap F) = P(E \mid F) \cdot P(F)$	E, F independent	$P(E \cap F) = P(E) \cdot P(F)$

MULTISTAGE EXPERIMENTS

In the next example we will explore how conditional probabilities can be critical tools in determining probabilities in multistage experiments ranging from picking apples out of bags to testing for various diseases.

Example 7

Suppose there are two bags of apples for snacks. The first bag has six red apples and four green apples, and the second bag has two red apples and six green apples.

(a) Suppose a student picks one of the two bags at random and then picks an apple at random from that bag. What is the probability that she picks a red apple?

(b) Suppose a student puts all of the apples into one bag and then chooses an apple at random. Now what is the probability that she picks a red apple?

(c) Under which scheme, (a) or (b), is the student more likely to get a red apple?

Solution

(a) Let B_1 represent the event that she chooses the first bag, let B_2 be the event that she chooses the second bag, and let R be the event that she chooses a red apple. Notice that if we knew she chose the first bag, then the probability of choosing a red apple would simply be $6/10$, since 6 of the 10 apples in that bag are red, and she chooses an apple randomly. This means that the conditional probability $P(R \mid B_1) = 6/10$. If we knew that the events R and B_1 were independent, then $P(R)$ would simply be equal to $P(R \mid B_1) = 6/10$.

However, we do not know that this is the case. In fact, we have a fairly good idea that the probability of choosing a red apple *does* depend on which bag is chosen. After all, there are more red apples than green in the first bag but three times as many green apples as red in the second bag. However, from the definition of conditional probability, we can say that

$P(R \mid B_1) = \dfrac{P(B_1 \cap R)}{P(B_1)}$, so that $P(B_1 \cap R) = P(B_1) \cdot P(R \mid B_1) = \dfrac{1}{2} \cdot \dfrac{6}{10} = \dfrac{3}{10}$. Similarly, if we

know that she chooses the second bag, then the probability of choosing a red apple is

$P(R \mid B_2) = 2/8 = 1/4$, and $P(B_2 \cap R) = P(B_2) \cdot P(R \mid B_2) = \dfrac{1}{2} \cdot \dfrac{2}{8} = \dfrac{1}{8}$. Finally, since she

must choose exactly one of the two bags, $B_2 = \overline{B_1}$ and

$$P(R) = P(B_1 \cap R) + P(B_2 \cap R) = \dfrac{3}{10} + \dfrac{1}{8} = \dfrac{17}{40}, \text{ by Theorem 4.}$$

(b) If she puts all the apples into one bag, then there are $6 + 2 = 8$ red apples in a bag of $10 + 8 = 18$ apples, so the probability of getting a red apple is simply $8/18 = 4/9$.

(c) Since the probability is $17/40 = 0.425$ in part (a) and is $4/9 = 0.444... > 0.425$ in part (b), she is more likely to get a red apple with the scheme in part (b).

It is often convenient to use a tree diagram to keep track of the information in multistage problems like the one in part (a) of Example 7. In the first stage of the experiment, one of the two bags is chosen. We represent this with two branches from the root. Along each of these branches we write the probabilities associated with selecting those particular branches. Since the two bags are chosen randomly, each of the initial branches is assigned a probability of 1/2.

In the second stage of the experiment, an apple is chosen from the appropriate bag. Since we are interested only in the color of the apple chosen, we draw two branches, one for red and one for green, from each of the branches in the previous stage. Along each of the new branches we write the conditional probability associated with making that particular selection. At each leaf of the tree, we calculate the probability of that sequence of choices by multiplying the probabilities along the branches leading to that leaf. For example, if we follow the topmost branch at each stage, we see that

$$P(B_1 \cap R) = P(B_1) \cdot P(R \mid B_1) = \frac{1}{2} \cdot \frac{6}{10} = \frac{3}{10}.$$

The tree diagram detailing all of this information is given in Figure 1.

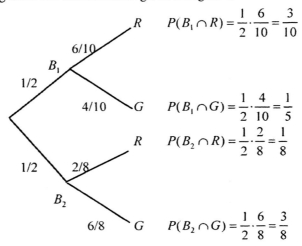

Figure 1: A Tree Diagram for Example 7

The same strategy used for the two-stage experiment in Example 7 can also be used to analyze the effectiveness of tests used to detect illness. As the next example illustrates, if the illness is particularly rare among the population tested for the illness, then the accuracy of such a test can be surprising.

Example 8

A rare blood disease occurs in 0.3% of the population. A blood test is developed for this disease that correctly detects the illness in 98% of those that actually have the disease but gives a false positive result for 1% of those that do not actually have the disease. Suppose you have a positive result from this blood test; that is, it indicates that you have the disease. What is the probability that you do *not* actually have the disease? In other words, what is the probability that you have received a false positive result?

Solution

The most difficult part of this problem can be determining which conditional probabilities the given information represents, and what information is requested. Consider the experiment to be selecting a person at random from the population. Let D be the event that the person selected actually has the disease, and let *Pos* denote the event that the person's blood test was positive.

Since the disease occurs in 0.3% of the population, $P(D) = 0.003$. (Notice that 0.3% means 3/10 of 1%.) How do we interpret the 98% figure? This is the conditional probability that the test result will be positive, given that the person tested actually has the disease. Thus $P(Pos|D) = 0.98$. Similarly, the 1% figure represents the probability that the test result will be positive,

(continued)

given that the person does not have the disease. We could write $P(Pos \mid \bar{D}) = 0.01$, where $\bar{D}$ represents the complement of the event D.

The question asks for the conditional probability that you do not have the disease, given that your test result was positive, that is, $P(\bar{D} \mid Pos)$. At first glance this might look like the probability described as 1%, but recall that $P(E|F)$ and $P(F|E)$ can be quite different, as in Example 4. The definition of conditional probability tells us that

$$P(\bar{D} \mid Pos) = \frac{P(\bar{D} \cap Pos)}{P(Pos)}$$

where the probabilities in the numerator and denominator can be determined as in Example 7.

$$P(\bar{D} \cap Pos) = P(\bar{D}) \cdot P(Pos \mid \bar{D}) = (0.997) \cdot (0.01) = 0.00997,$$

$$\begin{aligned} P(Pos) &= P(D \cap Pos) + P(\bar{D} \cap Pos) \\ &= P(D) \cdot P(Pos \mid D) + P(\bar{D} \cap Pos) \\ &= (0.003) \cdot (0.98) + 0.00997 \\ &= 0.01291. \end{aligned}$$

Thus $P(\bar{D} \mid Pos) = \dfrac{P(\bar{D} \cap Pos)}{P(Pos)} = \dfrac{0.00997}{0.01291} \approx 0.772$, so there is greater than a 77% chance that you have received a false positive test result.

The moral of the story here is that even though a test might have accuracy rates that are quite good, if the occurrence of the disease in the general population is quite small, then false positive test results are often more likely than true positives. In such cases, it is generally advisable to have those receiving positive test results be retested. Another strategy is to restrict testing to a smaller population, a high-risk group, in which particular individuals are more likely to have the disease than individuals in the wider population.

Exploratory Exercise Set 7.4

1. Split up into pairs for this exercise. Each pair will need a six-sided die.
   (a) Have Partner A roll the die and cover it so that Partner B does not see which side lands face up. (No peeking!) Have Partner B guess whether the number rolled was even or odd. Repeat this ten times, and keep track of how many times Partner B guesses correctly.
   (b) Now have Partner B roll the die and cover it so that Partner A does see which side lands face up. Have Partner B tell Partner A whether the number rolled is less than 4 or greater than 3. Now have Partner A guess whether the number is even or odd. Repeat this ten times, and keep track of how many times Partner A guesses correctly.
   (c) Who was more successful in guessing, A or B? Was this what you expected?
   (d) If E is the event that the number rolled is even, and F is the event that the number rolled is less than 4, then together compute the following probabilities, assuming that the die is fair:

   (1) $P(E \mid F)$ (2) $P(E \mid \bar{F})$

   (3) $P(\bar{E} \mid F)$ (4) $P(\bar{E} \mid \bar{F})$

2. Recall the loaded die in Example 3 of the text. Try to find different probabilities for a loaded die so that the events E and F are independent, where E is the event of rolling an even number and F is the event of rolling a number less than 4.

3. Split up into pairs. In this exercise you will simulate a curiosity from a game show that was popular in the 1970s. To prepare for the exercise, fold three index cards in half, and label them as 1, 2, and 3.
    (a) Partner B hides a coin behind one of the cards, and Partner A guesses which card it is behind. Then Partner B lifts a card without the coin that Partner A did not guess to show Partner A that it does not have the coin behind it. Partner B then offers to let Partner A switch her guess to the other remaining card. Partner A does switch her guess to the other card. Repeat this ten times and record how many times Partner A is correct *after* she switches.
    (b) Now reverse roles, but this time when Partner A offers to let Partner B switch, Partner B declines and sticks with her original guess. Repeat this ten times and record how many times Partner B's guess is correct.
    (c) Judging on the basis of your experimentation in parts (a) and (b), which partner's strategy is more successful?
    (d) Confirm your answer to part (c) by calculating the probability of a correct guess in part (a) and in part (b).

4. Suppose you received a positive test result for the disease discussed in Example 8, and you decide to be tested a second time. Assume that the results of the two tests are independent.
    (a) Draw a tree diagram to illustrate the information given.
    (b) If the second test result is positive as well, then what is the probability that you actually have the disease?

5. (*Writing Exercise*) You are a public health official in your community, and you have been given the task of explaining a new test that has just been developed for a rare but deadly disease. Write a paragraph explaining to the community that although your test is relatively accurate, you recommend that those who test positive in a random screening be tested a second time. Illustrate with specific probabilities.

Exercise Set 7.4

---

1. Notice that in the hospital example introducing this section, when we reduced our sample space from 1200 to 360, we also had to limit the number of patients who were treated for cancer. Why would it have been absurd to conclude that the probability that a smoker was treated for cancer was 450/360 = 1.25?

2. Consider an experiment in which a person from your school is selected at random. Let E be the event that the person selected has brown hair, and let F be the event that the person selected is female. Write in words (no symbols allowed) a question that asks for each of the following probabilities:
    (a) $P(E \mid F)$
    (b) $P(F \mid E)$
    (c) $P(E \cap F)$
    (d) $P(E \cup F)$
    (e) $P(E \mid \overline{F})$

3. Consider an experiment in which a registered voter is selected at random. Let E be the event that the person selected voted in the most recent election, and let F be the event that the person selected is female. For each of the following questions, write in symbolic notation (no words allowed) the probabilities that are requested.
    (a) What is the probability that the voter voted in the last election, given that the voter is female?
    (b) If the voter is female, what is the probability that she voted in the last election?

(c) What is the probability that a voter who voted in the last election is male?

(d) What is the probability that a male voter did not vote in the last election?

(e) Given that a voter voted in the last election, what is the probability that the voter is female?

4. Suppose a state legislature consists of 104 Republicans and 73 Democrats, where 12 of the Republicans are female, and 22 of the Democrats are female.

(a) What is the probability that a Republican in this legislature is female?

(b) What is the probability that a female in this legislature is a Republican?

5. At Lincoln Middle School, 63% of the students play a musical instrument, 45% play a sport, and 30% play a musical instrument and a sport.

(a) What is the probability that a student at Lincoln plays a sport, given that the student plays an instrument?

(b) What is the probability that a student at Lincoln plays an instrument, given that the student plays a sport?

6. (*Proof Exercise*) In Theorem 5, prove that conditions (ii) and (iii) are equivalent.

7. A pair of fair dice are rolled. Let E be the event that the sum of the dice is a multiple of 3, and let F be the event that the number rolled on the first die is a 2. Determine whether the events E and F are independent. Explain your reasoning.

8. A pair of fair dice are rolled. Let E be the event that the sum of the dice is a multiple of 4, and let F be the event that the number rolled on the first die is a 3. Determine whether the events E and F are independent. Explain your reasoning.

9. One bag of golf balls contains 12 white golf balls and 10 yellow golf balls. A second bag contains 7 white golf balls and 13 yellow golf balls. One of the two bags is selected at random, and then a ball is selected at random from that bag.

(a) Draw a tree diagram as in Figure 1 to illustrate this experiment, and label each branch with the appropriate probability.

(b) What is the probability that the ball selected is white?

(c) If a white ball is selected, what is the probability that it came from the first bag?

10. Rework the problem in Example 8 from the text, but with the occurrence of the disease at 10% of the population, rather than 0.3%. What now is the probability that a positive result is a false positive result?

11. Draw a tree diagram to illustrate the problem in Example 8 from the text, and fill in all of the numbers on the branches. Then give the four probabilities that correspond to the events on the four leaves of the tree diagram.

12. Using your work in the previous exercise, compute the probability that a negative test result is a false negative result with the test described in Example 8 from the text. How does that compare with the probability that a positive result is a false positive result?

13. A particular genetic disorder occurs in 0.8% of the population. A test for the disorder can accurately detect the disorder in 99.5% of those who have it, but this test gives a false positive result for 2% of those who do not have the disorder.

(a) Draw a tree diagram that illustrates the information given.

(b) If the test indicates that you have the disorder, what is the probability that the test is correct?

(c) If the test indicates that you do not have the disorder, what is the probability that the test is correct?

Discrete Applications in Political Theory

Chapter

8

8.1 Expressing Preferences in Voting

The Student Senate has a weighty task. The members have been asked to provide advice to the Vice President for Student Affairs about weekend food service in the campus cafeteria. The options that they are to evaluate are

> Hot Food (HF)
> Soup and Sandwich (SS)
> Snacks Only (SO)
> No Service (NS)

Each of the four choices has certain financial and convenience implications, and the issue is hotly debated. Finally a straw vote is taken, with these results:

> HF 10   SS 8   SO 5   NS 12

The senate leaders ponder the situation; certainly no clear mandate can be found here. They wonder, "Is there any way that a better expression of the will of the Student Senate can be found?"

The results of elections and other modes of public choice that involve three or more candidates or options are often difficult to interpret. If a single candidate or option receives more than half of the vote (a majority), most persons are comfortable with declaring the option with the majority of the vote as the winner. If no option receives a majority of the vote, however, some additional work must be done to understand the will of the voting public. One approach involves a runoff between the two options with the highest vote totals, but as we shall see, there are a number of other attractive options.

The leadership of the Student Senate asks each senator to provide not only his or her first choice but also a rank ordering of all four options for weekend food service. The resulting ballots are shown in Figure 1.

1. HF	1. HF	1. HF	1. HF	1. HF	1. HF
2. SS	2. SS	2. SS	2. SS	2. SS	2. SS
3. SO	3. SO	3. SO	3. SO	3. SO	3. SO
4. NS	4. NS	4. NS	4. NS	4. NS	4. NS
1. HF	1. HF	1. HF	1. HF	1. SS	1. SS
2. SS	2. SS	2. NS	2. NS	2. HF	2. HF
3. SO	3. SO	3. SO	3. SO	3. NS	3. NS
4. NS	4. NS	4. SS	4. SS	4. SO	4. SO
1. SS	1. SS	1. SS	1. SS	1. SS	1. SS
2. HF	2. HF	2. HF	2. HF	2. HF	2. HF
3. NS	3. NS	3. NS	3. NS	3. NS	3. NS
4. SO	4. SO	4. SO	4. SO	4. SO	4. SO
1. SO	1. SO	1. SO	1. SO	1. SO	1. NS
2. NS	2. NS	2. SS	2. SS	2. SS	2. SO
3. SS	3. SS	3. NS	3. NS	3. NS	3. SS
4. HF	4. HF	4. HF	4. HF	4. HF	4. HF
1. NS	1. NS	1. NS	1. NS	1. NS	1. NS
2. SO	2. SO	2. SO	2. SO	2. SO	2. SO
3. SS	3. SS	3. SS	3. SS	3. SS	3. SS
4. HF	4. HF	4. HF	4. HF	4. HF	4. HF
1. NS	1. NS	1. NS	1. NS	1. NS	
2. SO	2. SO	2. SO	2. SO	2. SO	
3. SS	3. SS	3. SS	3. SS	3. SS	
4. HF	4. HF	4. HF	4. HF	4. HF	

Figure 1

When the ballots are tabulated, each of the senators votes for one of the six preference lists found in Table 1. Observe that there are $4 \cdot 3 \cdot 2 \cdot 1 = 24$ possible lists of preferences, but many preference lists did not receive votes. (A table like Table 1 that summarizes and counts the preference lists of voters is called a **preference schedule**.)

Table 1: Preference Schedule for Vote on Weekend Food Service						
	List 1	List 2	List 3	List 4	List 5	List 6
First	HF	HF	SS	SO	SO	NS
Second	SS	NS	HF	NS	SS	SO
Third	SO	SO	NS	SS	NS	SS
Fourth	NS	SS	SO	HF	HF	HF
Votes	8	2	8	2	3	12

Consider the following ways in which we might evaluate these data to determine the recommendation of the Student Senate:

Method 1: Choice by plurality. Since "no service" (NS) has more first-place votes (12) than any other option, we declare NS the winner. This strategy of choosing a winner has the virtue of simplicity. However, because the winner received only 34% of the votes, a choice by plurality may well be strongly opposed by the majority of the voters. Indeed, a quick look at Table 1 shows that 19 (54%) of the senators rated NS as their third or fourth option.

Method 2: Choice by single runoff. Since the two most popular choices were "no service" (NS) with 12 votes and "hot food" (HF) with 10 votes, we conduct a runoff between the two options that have the largest numbers of votes. From our list of preferences, we can predict the results of this runoff. All of the 8 voters who voted for "soup and sandwich" (SS) will vote for HF. All of the 5 senators whose first vote went to "snacks only" (SO) will vote for NS. Therefore, in the runoff between the two options with the most votes, 18 voters will vote HF while 17 vote NS. Using choice by single runoff, the senators will recommend a hot food line (HF) on weekends.

Method 3: Choice by sequential runoffs. A criticism of choice by single runoff (Method 2) is that it ignores preferences beyond the first two preferences. Some voters may have strong feelings about their other rankings. To accommodate these concerns, we can design a sequence of runoffs to terminate when one choice receives a majority of the vote. At each step of the runoff process, we will eliminate the choice with the smallest number of first-place votes and redistribute those votes to remaining options on the basis of voter preferences. On the first vote, "snacks only" SO received the smallest number of first-place votes (5) and hence will be eliminated. In Table 2 we see the results after votes are reallocated in view of the fact that SO is no longer an option. For example, in List 1 of Table 1, 8 voters ranked "no service" (NS) fourth and "snacks only" (SO) third. Without SO as an option, these 8 voters will rank NS as their third option. More important, the votes of the voters in Lists 4 and 5 of Table 1 who ranked SO first will be tallied in new Lists 4 and 5, which reflect their choices when SO is not available. See Table 2 for the results after the first runoff.

Table 2: Preference Schedule Adjusted after "Snacks Only" (SO) Is Eliminated						
	List 1	List 2	List 3	List 4	List 5	List 6
First	HF	HF	SS	NS	SS	NS
Second	SS	NS	HF	SS	NS	SS
Third	NS	SS	NS	HF	HF	HF
Votes	8	2	8	2	3	12

Now HF has 10 first-place votes, SS has 11, and NS has 14. Still no option has a majority of the 35 votes, so we eliminate "hot foods" (HF), the option with the smallest number of first-place votes. After we eliminate this option and redistribute the votes, we have the results shown in Table 3.

Table 3: Preference Lists Adjusted after "Hot Food" HF Is Eliminated			
	List 1	List 2	
First	SS	NS	
Second	NS	SS	
Votes	19	16	

At this point, the option "soup and sandwich" (SS) has a majority with 19 votes and is chosen.

Method 4: Choice by weighted rankings. Another way to take into account the complete list of voter preferences is to give a different weight or point value to each of the different rankings and then sum the points for each option. In the straw vote of the Student Senate, we might give a weight of 4 points to a first-place ranking, a weight of 3 points to a second-place ranking, a weight of 2 points to a third-place ranking, and a weight of 1 point to a fourth-place ranking. The vote totals for each of the four options from the preferences recorded in Table 1 are computed below.

HF (hot food): $10(4) + 8(3) + 17(1)$ $= 81$
SS (soup and sandwich): $8(4) + 11(3) + 14(2) + 2(1)$ $= 95$
SO (snacks only): $5(4) + 12(3) + 10(2) + 8(1)$ $= 84$
NS (no service): $12(4) + 4(3) + 11(2) + 8(1)$ $= 90$

If we choose by weighted rankings, "soup and sandwich" (SS) will be recommended. This is the same result that we got when we chose by sequential runoff. However, as we will see in Exploratory Exercise 3, these two techniques do not always give the same result. Method 4, choice by weighted rankings, is often called a **Borda count** after its inventor, French mathematician and scholar Jean-Charles de Borda. In the remainder of this text we will refer to the weighted-rankings method as a Borda count.

Method 5: Choice by pairwise comparisons. In attempting to choose the "best" option among several, we might hope that the option we pick would beat all or most of the other options in a one-on-one vote. Hence we will use our preference lists in Table 1 to examine the six different pairings that can occur among the four options. For example, suppose the only two options being considered were "hot food" (HF) and "soup and sandwich" (SS). We see that in Lists 1 and 2 of Table 1, HF wins (10 votes) but in the remaining four lists, SS wins (25 votes). Hence, in this one-on-one competition, the winner is SS. The results of the six one-on-one competitions are shown in Table 4.

Table 4: The Results of One-on-One Competitions Between the Four Options			
	Votes: Candidate 1	Votes: Candidate 2	Winner
HF versus SS	10	25	SS
HF versus SO	18	17	HF
HF versus NS	18	17	HF
SS versus SO	16	19	SO
SS versus NS	19	16	SS
SO versus NS	13	22	NS

In pairwise competition, there is no single option that defeats all other options. Indeed, both SS and HF win two of the six pairwise competitions. Using this method, we must declare a tie between SS and HF. Hence this form of evaluation does not produce a unique winner in this example, although it does in others. This method was invented by French mathematician Marie Jean Antoine Nicholas de Caritat, Marquis de Condorcet.

Thus, we have the somewhat disturbing result that five reasonable schemes for determining voter preference produced three different winners and, in one case, failed to determine a unique winner. Closer evaluation of the five methods is even more discouraging. For example, one of the most often used methods is Method 4, the Borda count. It is used each week of the fall by various news organizations to rank the top 25 football teams. Each member of a panel of "experts" gives his or her personal ranking of the top 25 football teams. A first-place ranking is assigned 25 points, a second-place ranking is assigned 24 points, and so on. However, despite its widespread use, the Borda count has a property that has bothered many persons who study voting methods. In some cases, an option or candidate that has a majority of the vote does not win a Borda count.

Example 1

Each of 11 regulars at the Sportsman Bar and Grill are asked to rank the three best high school football teams in the county shortly before the season begins. The results are given in Table 5.

Table 5: Preference Schedule for Rankings of High School Teams			
	List 1	List 2	List 3
First	Madison	Greenhill	Jefferson
Second	Greenhill	Madison	Madison
Third	Jefferson	Jefferson	Greenhill
Votes	4	1	6

Example 1 *(continued)*

    (a) Did any team receive a majority of the 11 first-place votes?

    (b) Which team won with a Borda count, where first place was worth 3 points, second place was worth 2 points, and third place was worth 1 point?

Solution

    (a) Jefferson High School received 6 of the 11 first-place votes, so it won a majority of the votes.

    (b) The Borda counts for the three high schools are

        Jefferson: $6(3) + 5(1) = 23$

        Madison: $4(3) + 7(2) = 26$

        Greenhill: $1(3) + 4(2) + 6(1) = 17$

        Madison High School won according to the Borda count. Observe that in this example, the Borda count method did not select the team with the majority of the votes.

Methods 2 and 3, choice by single or sequential runoff, also have a disturbing property. In certain circumstances, the winner of an election using a runoff can become a loser if voters change their minds and give *more* votes to the winner. This result certainly violates our intuition about how voting schemes should work.

Example 2

Suppose Table 6 gives the preferences of 41 voters for three candidates for chairperson of the Presidential Search and Advisory Committee.

Table 6: Preference Schedule for the Election of a Chairperson					
	List 1	List 2	List 3	List 4	List 5
First	Donna	Ed	Donna	Ed	Anne
Second	Anne	Donna	Ed	Anne	Ed
Third	Ed	Anne	Anne	Donna	Donna
Votes	9	11	5	4	12

Example 2 *(continued)*

With these results, Anne has 12 first-place votes, Donna has 14 first-place votes, and Ed has 15 first-place votes. With either runoff method, we would reallocate Anne's votes to Ed, and Ed would win. What would happen if the ballots were lost, the powers-to-be decided to hold the election again, and the five voters in List 3 decided to switch their first-place votes to Ed (so that List 3 now reads Ed, Donna, Anne). Look at Table 7 and determine who the winner would be.

Solution

(See Table)

Table 7: Preference Lists for Chairperson after 5 More Persons Vote for Ed					
	List 1	List 2	List 3	List 4	List 5
First	Donna	Ed	*Ed*	Ed	Anne
Second	Anne	Donna	Donna	Anne	Ed
Third	Ed	Anne	Anne	Donna	Donna
Votes	9	11	5	4	12

Example 2 *(continued)*

With these results, Anne has 12 first-place votes, Donna has 9 first-place votes, and Ed has 20 first-place votes. With either runoff method, we would reallocate Donna's votes to Anne, and Anne would win. Isn't that strange? Ed has more first-place votes in Table 7 than in Table 6, but now he loses.

In summary, initially we examined a single election using five voting systems and found that these systems produced three different results. Further, using these examples and others, we have observed that each of the systems we have discussed has distinct drawbacks.

1. Choice by plurality may produce a winner that the majority of voters strongly oppose.
2. Choice by single runoff or by sequential runoff allows the possibility that a winner becomes a loser if the winner receives more votes.
3. Choice by pairwise comparison may fail to produce a winner.
4. Choice by Borda count may miss a winner that has a majority.

How can we construct a fair voting system? Is it even *possible* to construct a fair voting system? We will return to these questions in the next section.

Exploratory Exercise Set 8.1

1. In the examples of this section, the winner varied widely, depending on the choice of voting method. However, as you work through this exercise, you will see that this is not always the case. Choose a partner and begin.

The owner of K and W Accounting decided to take her 15 employees out for a celebratory meal on April 16. She asked each of them to rank order their preferences among restaurants *A*, *B*, *C*, and *D*. The ballots of the employees are found shown below.

1. *A*	1. *A*	1. *A*	1. *A*	1. *B*	1. *B*
2. *B*	2. *B*	2. *C*	2. *C*	2. *D*	2. *D*
3. *C*	3. *C*	3. *B*	3. *B*	3. *C*	3. *C*
4. *D*	4. *D*	4. *D*	4. *D*	4. *A*	4. *A*
1. *B*	1. *B*	1. *B*	1. *B*	1. *C*	1. *C*
2. *D*	2. *D*	2. *D*	2. *D*	2. *D*	2. *D*
3. *C*	3. *C*	3. *C*	3. *C*	3. *B*	3. *B*
4. *A*	4. *A*	4. *A*	4. *A*	4. *A*	4. *A*

1. *C*	1. *D*	1. *D*
2. *D*	2. *A*	2. *A*
3. *B*	3. *C*	3. *C*
4. *A*	4. *B*	4. *B*

(a) How many different preference lists appear?
(b) Make a preference schedule summarizing these results.
(c) Determine the winning restaurant by plurality.
(d) Determine the winning restaurant by single runoff.
(e) Determine the winning restaurant by sequential runoff.
(f) Determine the winning restaurant by a Borda count.
(g) Determine the winning restaurant by pairwise comparison.

2. In the pairwise comparison method of determining voter preference, we look at the preferences for each pair of candidates and determine the candidate who wins the most pairwise comparisons. In this exploratory exercise, we will look at the basic issues involved in implementing this method.
(a) The number of comparisons:
   (1) In an election involving three candidates *A*, *B*, and *C*, we must complete the two comparisons *A* to *B* and *A* to *C*, and then the comparison *B* to *C*. Thus the number of comparisons can be written as the sum 2 + 1. Write a sum representing the number of comparisons that must be made if there are four candidates.
   (2) Write a sum representing the number of comparisons that must be made if there are five candidates.
   (3) Write a sum representing the number of comparisons that must be made if there are six candidates.
   (4) *If you have studied Chapter 6*, use the language of combinations to describe the number of comparisons that must be made. Then show that your answers are the same as the answers given in (1), (2), and (3).
(b) When using the pairwise comparison method, we hope that one candidate defeats all others in pairwise comparisons. If not, we choose the candidate who wins the *most* pairwise comparisons. If in a given pairwise comparison the candidates tie, we will award each candidate 1/2 of a point toward their total number of pairwise wins.

Preference Schedule for Choice Involving *W*, *X*, *Y*, and *Z*					
	List 1	List 2	List 3	List 4	List 5
First	*W*	*W*	*X*	*Z*	*Z*
Second	*X*	*Y*	*Y*	*Y*	*W*
Third	*Z*	*X*	*W*	*W*	*X*
Fourth	*Y*	*Z*	*Z*	*X*	*Y*
Votes	6	6	8	10	2

    (1) *W* versus *X*: In the preference schedule we have printed in boldface type the letter representing the candidate in each column who wins the *W*-versus-*X* pairwise comparison. Who wins this comparison? Award the winner one point.

    (2) Determine the winner of the *W*-versus-*Y* pairwise comparison and award the winner one point.

    (3) Determine the winner of the *W*-versus-*Z* pairwise comparison and award the winner one point.

    (4) Explain why the *X*-versus-*Y* pairwise competition results in a tie. Award each candidate 1/2 point.

    (5) Determine the winner of the *X*-versus-*Z* comparison and award the winner one point.

    (6) Determine the winner of the *Y*-versus-*Z* comparison and award the winner one point.

    (7) After reviewing all pairwise competitions, find the winner of the election.

3. Although in many cases, all or most of the voting methods return the same winner, certainly the opposite can be true. Choose a partner and analyze the following voting situation. Determine the winner using each of the five voting methods discussed in this section.

Sponsors of the computer science luncheon/colloquium series wished to simplify their responsibility by choosing a single soft drink to serve with the lunch. They surveyed the 55 students and faculty in attendance one Wednesday and received the following preference lists for these choices: *Y* (Yello Mello), *C* (Cola Cola), *S* (Sorite), *P* (Popsi-Cola), and *RB* (Root Beer).

Preference Schedule Summarizing Student/Faculty Drink Choices						
	List 1	List 2	List 3	List 4	List 5	List 6
First	*Y*	*C*	*S*	*P*	*RB*	*RB*
Second	*P*	*RB*	*C*	*S*	*C*	*S*
Third	*RB*	*P*	*RB*	*RB*	*P*	*P*
Fourth	*S*	*S*	*P*	*C*	*S*	*C*
Fifth	*C*	*Y*	*Y*	*Y*	*Y*	*Y*
Votes	18	12	10	9	4	2

4. Imagine that your benevolent professor arranged for the class to skip a day and go out together for a good time. Have each member of the class indicate his or her preferences for such an outing by ranking the following activities: watch a movie, eat a meal, go bowling, attend a basketball game.

    First Choice: _____
    Second Choice: _____
    Third Choice: _____
    Fourth Choice: _____

    (a) Since there are four possible choices, there are $4 \cdot 3 \cdot 2 \cdot 1 = 24$ different preference lists that could possibly appear on ballots of your class members. How many different preference lists actually occurred?

    (b) Using the data from your class, create a preference schedule.

    (c) Use each of the five voting methods described in this section to choose the activity for the class.

5. (*Writing Exercise*) We observed in this section that the Borda count method is flawed because there are circumstances in which there is a candidate who receives a majority of the vote but who does not win the election using a Borda count. (See Example 1.) One reviewer of this text contended that this was not a serious flaw. Should a voting method always ensure that a candidate with a majority of the first-place votes wins? Give a carefully reasoned discussion of both sides of this question.

Exercise Set 8.1

1. Kamiya decides to take her 9 nephews "on a lark." She asks each of them to create a list that ranks his preferences for the following activities: swimming (S), ice skating (I), bowling (B), hiking (H). Here are the preference lists for all her nephews.

1. S	1. S	1. B	1. B	1. B	1. B
2. I	2. I	2. S	2. S	2. S	2. I
3. B	3. B	3. I	3. I	3. I	3. H
4. H	4. H	4. H	4. H	4. H	4. S

1. I	1. I	1. H
2. H	2. H	2. I
3. S	3. S	3. S
4. B	4. B	4. B

(a) Create a preference schedule summarizing this information.
(b) Is there a majority winner among these activities?
(c) Find the winner using plurality.
(d) Find the winner using a Borda count.

2. The members of the Executive Committee of the faculty must choose from among three nominees to determine the Teacher of the Year award. The nominees are Dr. Adams (A), Dr. Bartholomew (B), and Dr. Cook (C). The preferences of the 11 members of the Executive Committee are found in the following table.

Voter	1	2	3	4	5	6	7	8	9	10	11
First	A	A	A	C	B	C	A	C	B	A	C
Second	B	C	B	A	A	A	B	A	A	C	B
Third	C	B	C	B	C	B	C	B	C	B	A

(a) Create a preference schedule summarizing this information.
(b) Is there a majority winner?
(c) Find the winner using single runoff.
(d) Find the winner using pairwise comparison.

Exercises 3 through 7 refer to the following preference schedule. Employees have indicated which week in July they would like the factory to shut down for a company-wide holiday: A (first week in July), B (second week in July), C (third week in July), or D (fourth week in July). The results of these responses are shown in the following table.

Preference Schedule for Week of Factory Shutdown								
	List 1	List 2	List 3	List 4	List 5	List 6	List 7	List 8
First	A	A	A	B	B	B	C	D
Second	B	B	C	C	A	C	A	C
Third	C	D	B	D	C	A	D	A
Fourth	D	C	D	A	D	D	B	B
Votes	12	6	4	11	8	4	3	10

3. Determine the winner using the plurality method.

4. Determine the winner using the method of a single runoff.

5. Determine the winner using the method of sequential runoffs.

6. Determine the winner using the Borda count method.

7. Determine the winner using the pairwise comparison method.

8.

	List 1	List 2	List 3
First	*P*	*R*	*Q*
Second	*Q*	*Q*	*R*
Third	*R*	*P*	*P*
Votes	4	1	2

   (a) Is there a majority winner in the election described by this preference schedule?
   (b) What winner is found using the Borda count method?
   (c) If a candidate has a majority of the votes cast, will the Borda count method always declare that candidate the winner?

9. Make up an example of a preference schedule for an election with four candidates in which there is a candidate who receives the majority of votes but is not declared the winner by the Borda count method.

10. Does the method of pairwise comparisons yield a unique winner for the election described in the following preference schedule?

	List 1	List 2	List 3
First	*M*	*O*	*N*
Second	*N*	*M*	*O*
Third	*O*	*N*	*M*
Votes	14	12	10

11. Make up an example of a preference schedule for an election with four candidates in which the method of pairwise comparisons does not give a unique winner.

12. Explain carefully why each of the five voting methods discussed in this section will give the same winner when there are only two candidates in an election (provided the two candidates do not receive the same number of votes).

8.2 Arrow's Paradox and Voting Methods

In the previous section we examined several methods for determining a winner in an election in which voters have completed ballots expressing their preferences. However, we observed that in the same election, different methods produced different winners, and we observed flaws for each method examined. In this section we will turn to the task of developing a fair voting system that overcomes all the difficulties raised in our previous discussion. Before we do so, however, we will use the following example to understand one more flaw that can arise in a voting system.

Example 1

GROTEX Chemicals is looking for a site in the state of Georgia to build a new $160 million plant. The Georgia cities of Augusta (A), Columbus (C), Dalton (D), Macon (M), and Savannah (S) submit proposals aimed at attracting the new industry to their industrial parks. GROTEX sends a team of 20 executives, managers, and employees to the cities to make a recommendation on which site should be selected. Each member of the team ranks the cities, and the results are summarized in Table 1. Determine the winning city using the method of pairwise comparisons.

Table 1: Preference Schedule for Choice of Location for GROTEX Plant							
	List 1	List 2	List 3	List 4	List 5	List 6	List 7
First	A	S	S	C	D	D	M
Second	D	A	A	D	S	A	C
Third	C	C	D	A	C	M	D
Fourth	S	D	M	S	A	C	S
Fifth	M	M	C	M	M	S	A
Votes	2	5	4	1	1	3	4

Example 1 *(continued)*

Solution

The company needs to look at each of the ten pairs (A versus C, A versus D, and so on); determine which city wins the most pairwise comparisons; and award the new plant to that city. Should a tie occur in a particular comparison, the company will award 1/2 of a win to each of the cities involved in the tie. Since A defeats C in preference lists 1, 2, 3, and 6, A gets $2 + 5 + 4 + 3$ votes; C gets the $1 + 1 + 4$ votes from lists 4, 5, and 7. Thus A defeats C by a score of 14 to 6. The following results are easily checked:

A defeats C	14 to 6	(1 point for Augusta)
A defeats D	11 to 9	(1 point for Augusta)
A defeats M	16 to 4	(1 point for Augusta)
A loses to S	6 to 14	(1 point for Savannah)
C ties D	10 to 10	(1/2 point for Columbus; 1/2 point for Dalton)
C loses to M	9 to 11	(1 point for Macon)
C ties S	10 to 10	(1/2 point for Columbus; 1/2 point for Savannah)
D defeats M	16 to 4	(1 point for Dalton)
D defeats S	11 to 9	(1 point for Dalton)
M loses to S	7 to 13	(1 point for Savannah)

When the dust settles, Augusta has 3 points, Savannah has 2.5 points, Dalton has 2.5 points, Columbus has 1 point, and Macon has 1 point. Augusta gets the new plant!

However, before the results are announced, the Economic Development Council of Dalton withdraws its proposal because of the toxic chemicals with which GROTEX Chemicals works. Augusta hears the news but prepares to celebrate the arrival of the new $160 million plant anyway. Complete Example 2 to see what happens when we make the pairwise comparisons with Dalton eliminated.

Example 2

In Table 2 we have recorded the preference lists that remain after Dalton withdraws from consideration. Determine who is chosen for placement of the plant with the information in Table 2.

Table 2: Preference Schedule for Choice of Location for GROTEX Plant after Dalton (*D*) Is Eliminated							
	List 1	List 2	List 3	List 4	List 5	List 6	List 7
First	*A*	*S*	*S*	*C*	*S*	*A*	*M*
Second	*C*	*A*	*A*	*A*	*C*	*M*	*C*
Third	*S*	*C*	*M*	*S*	*A*	*C*	*S*
Fourth	*M*	*M*	*C*	*M*	*M*	*S*	*A*
Votes	2	5	4	1	1	3	4

Example 2 *(continued)*

Solution

Since *A* defeats *C* in preference lists 1, 2, 3, and 6, *A* gets $2 + 5 + 4 + 3 = 14$ votes; whereas *C* gets the $1 + 1 + 4 = 6$ votes from lists 4, 5, and 7. Thus *A* defeats *C* by a score of 14 to 6. The following results are easily checked:

A defeats *C*	14 to 6	(1 point for Augusta)
A loses to *S*	6 to 14	(1 point for Savannah)
A defeats *M*	16 to 4	(1 point for Augusta)
C loses to *M*	9 to 11	(1 point for Macon)
C ties *S*	10 to 10	(1/2 point for Columbus; 1/2 point for Savannah)
M loses to *S*	7 to 13	(1 point for Savannah)

Observe that with pairwise comparison Savannah now has 2.5 wins, and Augusta has only 2 wins. It looks as if it is now Savannah's turn to celebrate.

It is clear that we have identified another problem that can arise with a voting system. In this example, Augusta won using the method of pairwise comparison, but when Dalton withdrew, Augusta no longer won. Many people feel that this type of result is unfair, and that a well-designed voting system should not have such a result.

FAIRNESS CRITERIA

From this example and others in the previous section, it is clear that in order to choose a fair voting method, we must agree on what constitutes a fair voting method. Over the years, those who study voting systems have suggested four criteria that should characterize a fair system for choosing the winner of an election or making a choice in other contexts.

Fairness Criteria

1. **Majority Criterion:** If a choice receives a majority of the first-place votes, that choice should be the winner.
2. **The Condorcet Criterion:** If a choice defeats every other choice in one-on-one comparison, that choice should be the winner.

3. **The Monotonicity Criterion:** Suppose choice X wins an election. If the election is completed a second time and every voter ranks X the same or higher than in the first election (without changing the order of the other choices), then X should win the second election.
4. **The Independence-of-Irrelevant-Alternatives Criterion:** Suppose Choice X wins an election. If one or more of the other choices are removed and the ballots are recounted, then X still wins the election. (In our discussion of GROTEX Chemicals, we saw that the pairwise comparison method does not necessarily satisfy this criterion.)

Before we examine the problem of finding a fair system that satisfies all of these criteria, let us quickly look at some of our voting methods from the last section and evaluate how they fare when evaluated by these criteria.

Example 3

> Determine whether single runoff and sequential runoff satisfy the four fairness criteria for the election described by the preference schedule found in Table 3.

Table 3: Preference Schedule for Election Choosing among A, B, and C				
	List 1	List 2	List 3	List 4
First	C	C	B	A
Second	B	A	A	B
Third	A	B	C	C
Votes	7	4	8	10

Example 3 *(continued)*
Solution

> Since there is no candidate with a majority, B will be dropped; he has the smallest number of votes. His 8 votes will then be given to A, and A will win the runoff. Since the single runoff and the sequential runoff are the same when an election has three choices, A is the winner by either method.
>
> *Majority Criterion:* Since there is no majority in this election, the majority criterion is satisfied. If there had been a candidate with a majority of the vote, no runoff would have been necessary. Thus the runoff methods always satisfy the majority criterion.
>
> *The Condorcet Criterion:* In pairwise comparison B defeats A by 15 votes to 14, and B defeats C by 18 votes to 11. Hence B wins all pairwise competitions but is not the winner using runoff methods. The Condorcet criterion is not satisfied in this election.
>
> *The Monotonicity Criterion:* In order to check this criterion, we must imagine that some of the voters decided to change their votes to A. Suppose the four voters whose votes are represented in List 2 decide to change their first-place vote from C to A. This would result in the preference schedule shown in Table 4.

Table 4: Preference Schedule with New List 2				
	List 1	List 2	List 3	List 4
First	C	A	B	A
Second	B	C	A	B
Third	A	B	C	C
Votes	7	4	8	10

Example 3 *(continued)*

Now A has a total of 14 first-place votes, and C would be dropped for having the smallest number of first-place votes. Since the votes for C would be given to B, B would become the winner. Since giving A four more votes makes B the winner, the monotonicity criterion is not satisfied.

The Independence-of-Irrelevant-Alternatives Criterion: Imagine that just before the results were confirmed in this election, Candidate C dropped out. Then four of C's votes would go to A and seven of C's votes would go to B. Hence B would win. Thus in this election, the decision of Candidate C determines whether A or B is elected. The independence-of-irrelevant-alternatives criterion is not satisfied in this election.

In summary, the majority criterion is satisfied not only in this example but in all elections in which runoff methods are used. However, this example shows that there are circumstances when the runoff methods fail each of the other fairness criteria.

Example 4

Determine whether the Borda count method satisfies the four fairness criteria for the election described by the preference schedule in Table 5.

Table 5: Preference Schedule for Election Choosing among *P, Q,* and *R*.				
	List 1	List 2	List 3	List 4
First	P	P	Q	R
Second	Q	R	R	Q
Third	R	Q	P	P
Votes	5	1	3	2

Example 4 *(continued)*
Solution

Since P has 6 first-place votes and 5 third-place votes, P's Borda count is $6(3) + 5(1) = 23$. Since Q has 3 first place votes, 7 second-place votes, and 1 third-place vote, Q's Borda count is $3(3) + 7(2) + 1(1) = 24$. Similarly, since R has 2 first-place votes, 4 second-place votes, and 5 third-place votes, his Borda count is 19. Hence by a Borda count, Q wins.

Majority Criterion: In this example, P is chosen first by 6 of 11 voters and hence has a majority. This is an example in which the Borda count fails to satisfy the majority criterion.

The Condorcet Criterion: In pairwise competition, P defeats Q by 6 votes to 5, and P defeats R by a score of 6 votes to 5. However, by a Borda count, Q is the winner. Thus the Condorcet criterion is not satisfied in this election.

(continued)

The Monotonicity Criterion: In order to check this criterion, we must imagine that some additional voters decide to improve their ranking of Q while leaving the relative ranks of P and R the same. This will result in a higher Borda count for Q and a lower Borda count for the candidate or candidates whose rankings were lowered. Hence the Borda count for Q would continue to be the highest. The monotonicity criterion would hold for this example. Indeed, by similar reasoning, we can see that a Borda count will always satisfy the monotonicity criterion.

The Independence-of-Irrelevant-Alternatives Criterion: Imagine that just before the results were confirmed in this election, Candidate R dropped out. When the two votes that ranked R highest are given to Q, we get the preference schedule found in Table 6.

Table 6: Preference Schedule after R Drops Out				
	List 1	List 2	List 3	List 4
First	P	P	Q	Q
Second	Q	Q	P	P
Votes	5	1	3	2

Example 4 *(continued)*

P wins the Borda count for this preference schedule. Thus in this election, the decisions of candidate R determine whether P or Q is elected. The independence-of-irrelevant-alternatives criterion is not satisfied in this election.

In summary, the Borda count method satisfies only one of the four fairness criteria for this example.

Now we understand the four fairness criteria and hence are mindful of some of the shortcomings of the five voting methods we introduced in the last section. At this point we should be ready to roll up our sleeves and join the legions of political scientists, economists, and mathematicians who have sought to develop a fair voting system over the years. However, our efforts are doomed before we start. In 1952 a mathematical economist named Kenneth Arrow proved a very surprising result called **Arrow's Impossibility Theorem**. One consequence of Arrow's Theorem is the assertion that when there are three or more candidates and three or more voters, it is impossible to find a voting method that always satisfies all four fairness criteria.

In 1972 Dr. Arrow was awarded the Nobel Prize in economics for his investigations of voting and other forms of social choice that led to this unexpected result.

Exploratory Exercise Set 8.2

1. In this problem we will use the *plurality method* to determine the winner. Then we will ask several questions that will enable us to determine which of the four fairness criteria are satisfied by this voting method in this specific election. Work with your partner to investigate these issues.

Preference Schedule for Election Involving A, B, and C			
	List 1	List 2	List 3
First	A	B	C
Second	B	C	B
Third	C	A	A
Votes	6	5	3

(a) Determine the winner of this election using the plurality method.

(b) The Majority Criterion
   (1) Does any of the three candidates in this election obtain a majority of the vote?
   (2) Carefully explain in words why the majority criterion is always satisfied in any election in which no candidate gets a majority of the vote.

(c) The Condorcet Criterion
   (1) Is there any candidate who defeats all other candidates in pairwise comparisons?
   (2) Does the plurality method satisfy the Condorcet criterion in this election?

(d) The Monotonicity Criterion
   (1) Imagine that all or some of the voters represented in List 2 were to shift their votes to the winning Candidate A. Would this change the result of the election?
   (2) Imagine that all or some of the voters represented voters in List 3 were to shift their votes to the winning Candidate A. Would this change the result of the election?
   (3) Explain carefully in words why elections determined by the plurality method always satisfy the montonicity criterion.

(e) The Independence-of-Irrelevant-Alternatives Criterion
   (1) Suppose Candidate B were to drop out of the election and the five votes represented by List 2 were redistributed. Would this change the results of the election?
   (2) Does the plurality method satisfy the independence-of-irrelevant-alternatives criterion in this election?

2. In this problem we will use a *Borda count* to determine the winner. Then we will ask several questions that will enable us to determine which of the four fairness criteria are satisfied by this voting method in this specific election. Work with your partner to investigate these issues.

Preference Schedule for an Election Involving Candidates X, Y, and Z			
	List 1	List 2	List 3
First	Z	X	Y
Second	Y	Z	X
Third	X	Y	Z
Votes	6	15	10

(a) Determine the winner of this election using the Borda count method.

(b) The Majority Criterion
   (1) Does any of the three candidates in this election obtain a majority of the vote?
   (2) Does the Borda count method satisfy the majority criterion for this example?

(3) Review your work from Section 8.1. Does the Borda count method always satisfy the majority criterion?

(c) The Condorcet Criterion

(1) Is there any candidate that defeats all other candidates in pairwise comparisons?

(2) Does the Borda count method satisfy the Condorcet criterion in this election?

(3) Explain in words why the Condorcet criterion is always satisfied in elections in which no candidate wins all pairwise comparisons.

(d) The Monotonicity Criterion

(1) Imagine that some or all of the voters represented in List 3 were to shift their first-place votes to the winning candidate X, leaving Z in third place. How would this affect the Borda count for X? How would this affect the Borda count for Y? How would it affect the Borda count for Z?

(2) Imagine that some or all of the voters represented in List 1 were to increase their ranking of Candidate X, while leaving Candidates Y and Z in the same positions relative to one another. How would this affect the Borda count of Candidate X? How would this affect the Borda counts of Candidates Y and Z?

(3) Explain carefully in words why elections determined by the Borda count method always satisfy the montonicity criterion.

(e) The Independence-of-Irrelevant-Alternatives Criterion

(1) Suppose Candidate Y were to drop out of the election and the ten votes represented by List 3 were redistributed. Would this change the results of the election?

(2) Suppose Candidate Z were to drop out of the election and the six votes represented by List 1 were redistributed. Would this change the results of the election?

(3) Does the Borda count method satisfy the independence-of-irrelevant-alternatives criterion in this election?

3. A voting method that has gained advocates in recent years is **approval voting**. In approval voting, the voters specify on their ballots whether they approve or disapprove of each candidate. The winner is the candidate with the largest number of approval votes.

Ask the members of your class to use approval voting to recommend to the instructor the menu for the celebratory meal to follow the final examination in this class. Each class member should complete a ballot like the one that follows.

Main Course	Approve	
	Yes	No
Hot dogs		
Chili		
Pizza		
Tuna salad		

Determine the winner by finding the menu item with the largest number of approval votes.

4. Suppose Smith, Jones, and Yoblanski are three candidates competing in the local election for superintendent of education. Jones and Yablonski have similar moderate political positions, whereas Smith would like to undertake a radical revision of the school system. Suppose further that:

- 2% of the voters approve only of Jones.
- 25% of the voters prefer Jones, approve of Yablonski, do not approve of Smith.
- 3% of the voters approve only of Yablonski.
- 31% of the voters prefer Yablonski, approve of Jones, and do not approve of Smith.
- 36% of the voters approve only of Smith.
- 3% of the voters prefer Smith but also tolerate Jones.

With a partner, discuss the following three issues.
(a) Which candidate would we expect to win with the plurality method?
(b) Which candidate would we expect to win with approval voting?
(c) Identify a problem that can arise in elections where the plurality method is used that approval voting appears to address.

5. (*Writing Exercise*) Review Exploratory Exercise 3 and then use a Web search engine such as Google™ to do a Web search for "approval voting." Visit and peruse several of the appropriate Web sites. Be sure to visit the approval voting site at

http://bcn.boulder.co.us/government/approvalvote/center.html.

Write a paragraph summarizing your views on the strengths and weaknesses of approval voting.

Exercise Set 8.2

1. Majority Criterion
    (a) Explain in words why the plurality method always satisfies the majority criterion.
    (b) Explain in words why both runoff methods always satisfy the majority criterion.
    (c) Explain in words why the pairwise comparison method always satisfies the majority criterion.
    (d) Consider the following preference schedule, and determine whether the Borda count method satisfies the majority criterion.

First	A	B
Second	B	C
Third	C	A
Votes	6	4

2. Create a preference schedule for an election with at least three candidates, at least three different preference lists, and at least 15 voters in which:
    (a) a Borda count does not satisfy the majority criterion.
    (b) a Borda count does satisfy the majority criterion.

3. Condorcet Criterion: Consider the following preference schedule.

Preference Schedule for Election Involving A, B, and C			
	List 1	List 2	List 3
First	A	B	C
Second	B	C	B
Third	C	A	A
Votes	7	5	6

(a) Find the winner using the plurality method.
(b) Is there a candidate who defeats all others in pairwise comparisons?
(c) Does the plurality method always satisfy the Condorcet criterion?
(d) Find the winner using the method of sequential runoff.
(e) Does sequential runoff always satisfy the Condorcet criterion?
(f) Find the winner using a Borda count.
(g) In this specific example, does the Borda count method satisfy the Condorcet criterion?

4. Consider the following preference schedule.

Preference Schedule for Election Involving A, B, and C				
	List 1	List 2	List 3	List 4
First	A	B	C	A
Second	B	A	B	C
Third	C	C	A	B
Votes	6	5	4	2

(a) Find the winner using a Borda count.
(b) Is there a candidate who defeats all others in pairwise comparisons?
(c) Does the Borda count method satisfy the Condorcet criterion in this example?
(d) Does the Borda count ever satisfy the Condorcet criterion? (Compare with Exercise 3(g).)

5. Explain in words why the pairwise comparison method always satisfies the Condorcet criterion.

6. Devise preference schedules involving at least three candidates, at least four different preference lists, and at least 10 votes in which:
(a) the plurality method satisfies the Condorcet criterion.
(b) the plurality method does not satisfy the Condorcet criterion.
(c) the runoff methods satisfy the Condorcet criterion.
(d) the runoff methods do not satisfy the Condorcet criterion.

7. The Monotonicity Criterion: Consider Preference Schedule A.

Preference Schedule A				
	List 1	List 2	List 3	List 4
First	X	Y	X	Z
Second	Y	Z	Z	X
Third	Z	X	Y	Y
Votes	7	9	3	11

(a) Explain why Z is the winner by the method of sequential runoff.
(b) Suppose the three voters whose preferences are recorded in List 3 decide to cast their first-place votes for Z instead of X, resulting in Preference Schedule B. Who would then win by the method of sequential runoff?

Preference Schedule B				
	List 1	List 2	List 3	List 4
First	X	Y	Z	Z
Second	Y	Z	X	X
Third	Z	X	Y	Y
Votes	7	9	3	11

(c) Does the sequential runoff method satisfy the monotonicity criterion in this example?
(d) Explain in words why the plurality method always satisfies the monotonicity criterion.
(e) Explain in words why the Borda count method always satisfies the monotonicity criterion.
(f) Explain in words why the pairwise comparison method always satisfies the monotonicity criterion.

8. Devise your own preference schedule involving at least three candidates, at least four different preference lists, and at least 30 votes in which the monotonicity criterion is not satisfied for the method of sequential runoff. Explain carefully why the monotonicity criterion is not satisfied in your example.

9. Consider the following preference schedule.

Preference Schedule for Election Involving X, Y, and Z			
	List 1	List 2	List 3
First	Z	Y	X
Second	X	Z	Y
Third	Y	X	Z
Votes	5	4	3

(a) Find the winner using the plurality method.
(b) If Candidate X drops out of the race and his votes are redistributed according to the preference list, who wins the election?
(c) Does the plurality method satisfy the independence-of-irrelevant alternatives-criterion in this example?
(d) Find the winner using the sequential runoff method.
(e) If Candidate Z drops out of the race and her votes are redistributed according to the preference list, who wins the election?
(f) Does the sequential runoff method satisfy the independence-of-irrelevant-alternatives criterion in this example?
(g) Find the winner using the Borda count method.
(h) Does the Borda count method satisfy the independence-of-irrelevant-alternatives criterion in this example?

10. Consider the following preference schedule:

Preference Schedule for Candidates A, B, C, and D					
	List 1	List 2	List 3	List 4	List 5
First	A	D	B	A	C
Second	B	C	C	D	D
Third	C	B	A	B	B
Fourth	D	A	D	C	A
Votes	14	20	14	10	2

(a) Determine the winner using the pairwise comparison method.
(b) If Candidates A and C drop out of the race and their votes are redistributed according to the preference list, who wins the election?
(c) Does the pairwise comparison method always satisfy the independence-of-irrelevant-alternatives criterion?

11. Look back at the results of Exercises 1 through 10 and complete this table. Enter a "Y" in the box if the voting method always satisfies the fairness criterion and enter a "N" in the box if there are occasions when the voting method does not satisfy the fairness criterion. Remember that when there are only three candidates, single runoff and sequential runoff are the same method.

A Comparison of Voting Methods and Fairness Criteria					
Criterion	Plurality	Single runoff	Sequential runoff	Borda count	Pairwise comparison
Majority					
Condorcet					
Monotonicity					
Independence -of-irrelevant- alternatives					

8.3 Weighted Voting Systems

When Grandfather Jesse's health began to fail, he created a corporation to run his business, J & J Poultry Farms. The Board of Directors for the new corporation consisted of his brother John, with 7 votes; his daughters Juniper and Julia, each with 5 votes; and his grandchildren Jack and Jill, each with 1 vote. Further, Jesse mandated that 10 votes would be required to decide any issue. Not long after the corporation was formed, several crucial votes were taken about directions for the farms.

Should we diversify by raising turkeys as well as chickens?

Results of voting:	No votes by John and Juniper
	Yes votes by Julia, Jack, and Jill
Decision:	No, since the 12 votes of John and Juniper exceed the majority of 10 votes needed to win

Should we lease the 200 acres of river bottom to another farmer for use in raising feed corn?

Results of voting:	No votes by John and Julia
	Yes votes by Juniper, Jack, and Jill
Decision:	No, since the 12 votes of John and Julia exceed the majority of 10 votes needed to win

Should we begin to raise strawberries on some of the river bottom acreage?

Results of voting:	No votes by Juniper and Julia
	Yes votes by John, Jack, and Jill
Decision:	No, since the 10 votes of Juniper and Julia are sufficient to win

After the meeting, Jack left shaking his head. "Jill, we lost on every ballot. I know I have a vote, but I surely don't feel like I have any power!"

DUMMIES, COALITIONS, AND WEIGHTED VOTING SYSTEMS

In this section we will attempt to understand whether Jack and Jill have any power in the corporation on whose board they sit. More generally, we will think about how to measure power in weighted voting systems as they occur in corporations, local governments, and parliaments around the world. The voting system for the J & J Poultry Farms, Inc. is an example of a **weighted voting system.** Although there are five voters, there are a total of 19 votes to be cast, and each voter has a different number of votes (a different weight) for which he or she is responsible. We will consider some definitions that will help us discuss weighted voting systems.

Definitions: In a weighted voting system:

The **win point** is the number of votes required to win a ballot. The win point should be at least as large as a simple majority.

A **winning coalition** is a subset of voters who control enough votes to win any ballot. (The sum of the votes of the coalition members is greater than or equal to the win point.)

A **losing coalition** is a subset of voters who do not control enough votes to win a ballot. (The sum of the votes of the coalition members is less than the win point.)

In a winning coalition, a voter is **critical** provided that if the voter leaves the coalition, the coalition becomes a losing coalition.

A voter is a **dummy** if he is not critical in any winning coalitions.

Example 1

Consider the organization of J & J Farms Corporation.
(a) What is the win point?
(b) Give an example of a winning coalition and identify which voters are critical in the coalition.
(c) Give an example of a losing coalition.

Solution

(a) Since Grandpa Jesse declared that 10 votes would be required to decide an issue, the win point is 10.
(b) Since Juniper and John together control 12 votes, {Juniper, John} is a winning coalition, and both Juniper and John are critical in this coalition.
(c) Since Juniper and Jill together control only 6 votes, {Juniper, Jill} is a losing coalition.

As we attempt to measure the power of a specific voter within the J & J Farms leadership, we might ask how often the voter's vote is a necessary vote in achieving win point in a contested election. For example, in the first decision reviewed in the J & J example, John and Juniper voted together to determine that the farm would not raise turkeys. Further, both John's vote and Juniper's vote were important; had either of them changed their vote, the result of the balloting would have changed. In this circumstance, we say that the set {John, Juniper} is a winning coalition and that each of their votes is **critical**. (For that same decision {Julia, Jack, Jill} is a losing coalition.) In the second ballot reviewed in the J & J example, {John, Julia} was a winning coalition, and in the third ballot, {Julia, Juniper} was a winning coalition. Further, each member of these winning coalitions was critical. Had any member of a winning coalition changed his or her vote, it would have changed the result of the ballot. For example, if in the third ballot Julia had voted "yes" instead of "no," the corporation would have started raising strawberries on the river bottom.

Now, let us return to the question of whether Jack has any power. The question has become "Can we find a winning coalition including Jack, for which Jack's vote is critical?" Certainly we can find winning coalitions that contain Jack. {Jack, Julia, Juniper} is a winning coalition because it controls 11 votes, more than the win point of 10 votes. However, were Jack to abandon this coalition and change his 1 vote, the coalition would still be a winning coalition. {Jack, Jill, John, Julia} is a winning coalition because it controls 14 votes. However, again Jack can change his vote and the coalition would still win. Think about it for a minute. Can you find any winning coalitions containing Jack for which his vote is critical? *If not, then we will conclude that Jack has no power, even though he has a vote.* If a voter in a weighted voting system does not cast a critical vote as a member of any winning coalition, that voter is said to be a **dummy**. Jack and Jill are dummies in the J & J Corporation.

Using the J & J Corporation as an example, we have had a brief introduction to weighted voting systems. Weighted voting systems occur in corporations in which each shareholder has a number of votes equal to the number of shares he or she owns. Weighted voting systems also occur in parliaments where parties vote as a block and in other democratic bodies where two officials cast different numbers of votes because they represent different numbers of voters. Over the past four decades, the task of measuring the power of a voter in a weighted voting system has become quite important as we have attempted to design voting systems that reflect basic ideas of fairness and justice.

A standard way to represent a weighted voting system is first to list the win point and the number of votes (**the weights**) available to each voter. For example, the voting system for the J & J Farms Corporation can be represented as [10: 7, 5, 5, 1, 1], where 10 is the win point, 7 the number of votes for John, and so on.

Example 2

Consider a weighted voting system with voters A, B, and C represented as [5: 4, 3, 2].
(a) What is the win point for this voting system?
(b) If A has 4 votes, B has 3 votes, and C as 2 votes, list all winning coalitions.
(c) In the winning coalition $\{A, B, C\}$ which votes are critical?
(d) In the winning coalition $\{B, C\}$ which votes are critical?
(e) Is there a dummy in this weighted voting system?

Solution

(a) The first number listed in the description of a weighted voting system is the win point. Hence 5 is the win point in this system, which means that a total of at least 5 votes is required to win a ballot.
(b) $\{A, B, C\}$, $\{A, B\}$, $\{A, C\}$, $\{B, C\}$
(c) None of the votes in $\{A, B, C\}$ are critical. If C were to change his vote, the coalition would still have $4 + 3$ votes. Similarly, if B changed her vote, the coalition would still have $4 + 2$ votes, and a change by A would leave $3 + 2$ votes.
(d) Both of the votes in the winning coalition $\{B, C\}$ are critical. If C changes his vote, B does not have enough votes to win by herself, and vice versa.
(e) None of the voters is a dummy. A is critical in the winning coalition $\{A, B\}$ and $\{A, C\}$. B is critical in the winning coalitions $\{A, B\}$ and $\{B, C\}$; C is critical in $\{A, C\}$ and $\{B, C\}$. Indeed, we might conclude that A, B, and C have equal amounts of power since each is a critical vote in the same number of winning coalitions.

SOME INTERESTING EXAMPLES

In the following discussion we will examine five different weighted voting systems. In each voting system we will watch for dummies and make crude attempts to appraise the power of voters.

Example 3

Discuss each of the following weighted voting systems:
(a) [51: 55, 45]
(b) [51: 48, 48, 4]
(c) [(51: 26, 26, 26, 22]
(d) [12: 1, 1, 1, 1, 1, 1, 1, 1, 1, 1, 1, 1]

Solution

(a) In this case the win point is 51 votes and the first voter casts 55 votes. In this circumstance, the first voter can determine the outcome of all ballots. When the weight of a voter is as great as the win point, we say the voter is a **dictator**.
(b) Despite the great disparity in number of votes, these three voters have comparable amounts of power. To understand this, suppose the voters are A, B, and C with 48, 48 and 4 votes, respectively. One winning coalition is $\{A, B, C\}$ and no voters are critical to this coalition. The other three winning coalitions are $\{A, B\}$, $\{A, C\}$, and $\{B, C\}$. Note that each voter is critical in each of these coalitions. Hence each of A, B, and C is critical in each of two coalitions.

(continued)

   (c)   Despite the fact that the fourth voter has nearly the same number of votes as the first three, he is a dummy. Verify this by trying to find a winning coalition for which the fourth voter is critical.

   (d)   The only winning coalition is a coalition consisting of all voters, because the win point is equal to the sum of the votes. This is the voting system found in the American jury system, in which all jurors must agree in order to convict. In this weighted voting system, any voter can prevent a motion from passing. A voter who alone can keep any motion from passing is said to have **veto power**. All 12 voters in this weighted system have veto power.

Example 4

One of the first weighted voting systems to be studied carefully was New York's Nassau County Board of Supervisors. In 1965 a lawyer, John Banzhaf, wrote a paper in which he analyzed the power of the six persons who made up this board. One large district (Hempstead) elected two representatives in an at-large election, and each of four other districts elected a single representative. The number of votes allowed to each representative were roughly based on the population of the district. A simple majority vote of 58 was required to make a decision. Table 1 gives the population of each municipality and the distribution of the votes.

Table 1: Nassau County Board of Supervisors 1960		
Municipality	1960 Population	Number of Votes
Hempstead 1	728,625	31
Hempstead 2		31
Oyster Bay	285,545	28
North Hempstead	213,225	21
Glen Cove	22,752	2
Long Beach	25,654	2
Totals	1,275,801	115

Example 4 *(continued)*

   (a)   Describe this weighted voting system by recording the win point and the weight of each voter.

   (b)   Determine whether any of the supervisors is a dummy.

Solution

   (a)   There were a total of 115 votes, and a simple majority of 58 would win any ballot. Hence this voting system can be represented by [58: 31, 31, 28, 21, 2, 2].

   (b)   In thinking about the second question, we discover that several things are immediately apparent:

   • Any set of two representatives chosen from the three representatives of Hempstead and Oyster Bay could form a winning coalition in which both voters were critical.

   • On the other hand, there were no winning coalitions in which the representative of North Hempstead, Glen Cove, or Long Beach was critical. Together their votes totaled $21 + 2 + 2 = 25$. Hence, if they formed a coalition with one of the representatives of the big three (Hempstead I and II and Oyster Bay), the coalition was not a winning coalition. If they formed a coalition with two or more of the big three, the representative of none of the smaller municipalities was critical. The representatives of North Hempstead, Glen Cove, and Long Beach were dummies.

Think about what we learned in Example 4 for a minute. The representatives from North Hempstead, Glen Cove, and Long Beach may have attended meetings and entered into discussion. However, they never cast a vote that mattered. This is not quite our idea of representative government.

THE BANZHAF POWER INDEX

As attorney Banzhaf from Example 4 studied voting power in Nassau County government, he discussed a measure of power that we will call the Banzhaf power index.

> **Definition:** The **Banzhaf power index** for a voter is the quotient
>
> $$\frac{\text{The number of winning coalitions in which the voter is critical}}{\text{The sum over all voters of the number of winning coalitions in which the voter is critical}}$$

To put it another way, suppose we have n voters. Suppose that the first voter is critical in k_1 winning coalitions, the second voter is critical in k_2 winning coalitions, and so on. Then the Banzhaf power index of

$$\text{Voter } j \text{ is } \frac{k_j}{k_1 + k_2 + ... + k_n}$$

Example 5

Compute the Banzhaf power index for each voter in the weighted voting system described by [51: 48, 48, 4].

Solution

Suppose we let the voters be A with 48 votes, B with 48 votes, and C with 4 votes. We will list all winning coalitions in Table 2, together with the critical voters for each coalition.

Table 2	
Winning Coalitions	Critical Voters
$\{A, B, C\}$	No voter is critical.
$\{A, B\}$	A, B
$\{A, C\}$	A, C
$\{B, C\}$	B, C

Example 5 (continued)

A is critical in 2 winning coalitions.
B is critical in 2 winning coalitions.
C is critical in 2 winning coalitions.
The sum over all voters of the number of winning coalitions in which the voter is critical is $2 + 2 + 2 = 6$.

Hence the Banzhaf power index for A is $\frac{2}{2+2+2} = \frac{1}{3}$. Similarly, the Banzhaf indices of B and

C are both $\frac{1}{3}$. The Banzhaf indices suggest that all of the three voters have equal power. This

confirms the observations we made in Example 3(b).

The major difficulty in computing the Banzhaf indices stems from the fact that we need to find all winning coalitions first. Remember that when looking for coalitions, we are just looking for subsets of the set of voters, and we can call upon knowledge we have acquired earlier in identifying characteristics of those subsets. At the most basic level, we know that a set of n voters has 2^n subsets, and one of these subsets is the empty set. Hence, when looking for winning coalitions in a set of voters, we know an upper bound on the number of coalitions we might have to examine. At a slightly more sophisticated level, those who have completed Chapter 6 should remember that there are $C(n, k)$ subsets with k elements in a set of n elements. This kind of information can help us ensure that we do not miss any winning coalitions.

Example 6

Find the Banzhaf power index of each voter in the weighted voting system $\{51: 40, 30, 20, 10\}$.

Solution

Let us name the voters P, Q, R, and S and suppose they have 40, 30, 20, and 10 votes, respectively. We need to find all winning coalitions and then determine which voters are critical in each winning coalition. By remembering that there are $C(4, 4) = 1$ coalition with four voters, $C(4, 3) = 4$ coalitions with 3 voters, and so on, we can gather the information found in Table 3.

Table 3	
Winning Coalitions	Critical Voters
$\{P, Q, R, S\}$	No voter is critical.
$\{P, Q, R\}$	P
$\{P, Q, S\}$	P, Q
$\{P, R, S\}$	P, R
$\{Q, R, S\}$	Q, R, S
$\{P, Q\}$	P, Q
$\{P, R\}$	P, R

(continued)

P is critical in 5 winning coalitions.
Q is critical in 3 winning coalitions.
R is critical in 3 winning coalitions.
S is critical in 1 winning coalition.
The sum over all voters of the number of winning coalitions in which the voter is critical is 12. Hence the Banzhaf indices are $\frac{5}{12}, \frac{3}{12}, \frac{3}{12}$, and $\frac{1}{12}$, respectively. These indices suggest that P is the most powerful voter, that Q and R have equal amounts of power, and that S is the least powerful voter but is not a dummy.

Exploratory Exercise Set 8.3

1. Work in pairs. Complete each part of the following exercises independently; then compare answers and discuss the results. Remember that you have described a weighted voting system when you have identified the win point and the number of votes for each voter. The win point should be at least as great as a simple majority.
   (a) Describe a weighted voting system with four voters in which no voter is a dummy.
   (b) Describe a weighted voting system with four voters in which the first voter is a dictator.
   (c) Describe a weighted voting system with four voters in which two of the voters are dummies yet each has at least one vote.
   (d) Describe a weighted voting system with four voters in which each voter has veto power.

2. Work in pairs. Have each member of the pair describe a weighted voting system with four voters. Swap voting systems with your partner, and compute the Banzhaf power index for each voter in your partner's voting system. When finished with your computations, discuss your work with your partner.

3. We suggested in the text that our ability to count the number of subsets of a specified size could be useful in computing Banzhaf power indices. Let us use such knowledge to help us compute the Banzhaf power indices for the J & J Corporation discussed in this section. Remember that the J & J weighted voting system could be described as [10: 7, 5, 5, 1, 1]. Remember also that in computing the Banzhaf power indices, we need to find all winning coalitions (subsets) in which some voter is critical. Now answer the following questions.

(a) Since there are five voters in this voting system, there are only _____ non-empty subsets that we need to consider when looking for winning coalitions.

(b) If all three of John (with 7 votes), Jennifer (with 5 votes), and Julia (with 5 votes) are in a winning coalition, then no voter in the coalition can be critical. Why?

(c) If only one of John, Jennifer, or Julia is in a coalition, it cannot be a winning coalition. Why?

(d) Hence, to find winning coalitions in which a voter is critical, we need only look at coalitions that contain exactly two of John, Jennifer, and Julia. How many different ways can we choose exactly two members from the set {John, Jennifer, Julia}?

(e) If a winning coalition contains exactly two members from {John, Jennifer, Julia}, then it must contain 0, 1, or 2 members from {Jack, Jill}. How many winning coalitions must we examine in search of voters who are critical in a winning coalition?

(f) Complete the following table, which lists all winning coalitions in which some voter is critical. Then compute the Banzhaf power index for each of the voters.

Winning Coalitions	Critical Voters
{John, Julia}	John, Julia
{John, Julia, Jack}	
{John, Julia, Jill}	

4. A primary obstacle to computing Banzhaf power indices by hand when the number of voters is large is the difficulty in tabulating all winning coalitions. On some occasions, however, special voting rules make it particularly easy to find the winning coalitions. Consider a medical practice with two senior partners, Dr. Saw and Dr. Bones, and four junior partners Dr. W, Dr. X, Dr. Y, and Dr. Z. Suppose that in order to change policy for the medical practice, both Dr. Saw and Dr. Bones must agree on the change, as well as least two of the junior partners.

(a) If we were to naively list all non-empty coalitions (subsets) of voters, how many would we need to list?

(b) In searching for winning coalitions, why do we need to list only coalitions that contain Drs. Saw and Bones?

(c) A winning coalition would then consist of Drs. Saw and Bones and either 2 or 3 or 4 of the junior partners. How do we know that there are $C(4, 2) + C(4, 3) + C(4, 4)$ winning coalitions?

(d) Complete this table, which lists all winning coalitions and the critical voters in each coalition.

	Winning Coalitions	Critical Voters
1	{S, B, W, X}	S, B, W, X
2	{S, B, W, Y}	
3		
4		
5		
6		
7		
8		
9		
10		
11		

(e) Compute the Banzhaf power index for each of the six doctors.

(f) Choose an appropriate win point w so that the voting system of the Saw and Bones practice can be described as [w: 10, 10, 2, 2, 2, 2].

5. (*Writing Exercise*) Use the concepts and terminology of this section to write a careful discussion of political power in these contexts.

(a) Imagine a parliament with 61 members in which each of the two major parties has 29 representatives and a small splinter party has 3 members. Further, the parties vote as a bloc on major issues. Use the concepts of this section as well as your own observations to discuss the assertion "All three parties have equal power."

(b) The country of Israel has two major political parties and several parties that are much smaller. Research the current alignment of political parties in Israel, and use the concepts of this section to discuss the relative power of those parties.

Exercise Set 8.3

In the weighted voting systems described in Exercises 1 through 4, assume that the voters are named A, B, C, and so on. In each voting system:

(a) Determine the win point for the weighted voting system.
(b) Determine the weight of each voter in the system.
(c) Find one winning coalition for the system.
(d) Find any dictators.
(e) Find any voters with veto power.

1. [51: 52, 48]   2. [4: 2, 2, 1]

3. [3: 2, 1, 1]   4. [12: 4, 3, 2, 2, 1]

In the weighted voting systems described in Exercises 5 through 8 assume that the voters are named A, B, C, and so on.

(a) In each voting system, list all winning coalitions.
(b) In each winning coalition, identify which voters are critical in the coalition.
(c) In each voting system, identify all dummies.

5. [4: 3, 3, 1]   6. [5: 5, 3, 1]

7. [10 : 6, 6, 2, 1]   8. [9: 7, 5, 3, 2]

9. Compute the Banzhaf power index for each voter in the weighted voting system of:
(a) Exercise 5   (b) Exercise 6
(c) Exercise 7   (d) Exercise 8

10. The Human Resources Advisory Committee of Excell University consists of (A)dministrators, (F)aculty, and (S)taff. The members of each group tend to vote as a bloc. Analyze the Advisory Committee as a weighted voting system if the committee consists of 4 administrators, 5 faculty members, and 2 staff members. Assume that 6 votes are required to win a ballot.
    (a) Find all winning coalitions of the three groups {A, S, F}.
    (b) Compute the Banzhaf index for each of the three groups.
    (c) What do the Banzhaf indices suggest about the relative power of the groups?

Review Exploratory Exercise 4 about the medical practice of Drs. Saw and Bones and then complete Exercises 11 and 12, in which changes in the voting system of that medical practice are proposed. In each case:
    (a) First make a guess about how you expect the voting power of each of the six doctors to be modified by the change.
    (b) Then compute the Banzhaf power indices and compare the result with your guess.

11. Suppose that the practice of Drs. Saw and Bones adds another junior partner, Dr. P, and continue to require concurrence by Saw and Bones and at least two of the associates to change policy.

12. Suppose Dr. Gristle joins the practice as a senior partner so that there are three senior partners and four junior partners. Suppose further that a change in policy now requires concurrence by at least two of {Saw, Bones, Gristle} and at least two of the junior partners.

13. The weighted voting system [3: 1, 1, 1, 1, 1] is an example of a "one-person, one vote" voting system that we usually think of when considering democratic decision making.
    (a) Before you compute anything, how do you think power should be distributed in this system?
    (b) Compute the Banzhaf power index of each of these voters, and compare the result with your intuition about how the power should be distributed.

14. In the weighted voting system [9: 9, 4, 2, 2] the first voter, Voter A, is a dictator. Compute the Banzhaf power indices of all of the voters, and discuss how they confirm (or fail to confirm) your intuition about distribution of power when there is a dictator.

15. Compute the Banzhaf power index for each of the voters in the Nassau County government described in Example 4.

16. In 1970 the county government of Nassau County described in Example 4 changed the win point from 58 to 63 but left the weights of the six voters the same. Compute the Banzhaf power indices for all of the voters in that government, which can be described as [63: 31, 31, 28, 21, 2, 2].

8.4 The Problem That Began in 1790

In 1790 the first census was completed, and the founding fathers of the United States had to look back at the Constitution they had finally agreed upon in 1787. This is what they saw.

> ARTICLE I, SECTION 2 OF THE CONSTITUTION
> Representatives and direct Taxes shall be apportioned among the several States which may be included within this Union, according to their respective Numbers.

In order to put together the House of Representatives, our founding fathers had two tasks:
   (1) Determine how many members would be in the House of Representatives.
   (2) Determine how many representatives would be seated from each state.

Politicians quickly realized that the answers to these questions were very important to their health, welfare, and chance for re-election, and they got to work immediately. A group working in the House of Representatives under the leadership of Thomas Jefferson met and decided on one set of answers to the question. The progress of their deliberations alarmed the Federalists, and so a second group working in concert with Alexander Hamilton proposed a second apportionment. The Hamilton proposal narrowly passed both houses of Congress and found its way to the desk of President Washington. We will examine first Hamilton's solution and then Jefferson's solution. Next we will review the resulting conversation over the subsequent half-century. Then we will see, in the exercise sets, that this seemingly narrow question has broad applications in unexpected areas.

The Hamilton proposal that was passed to the president was based on a House of Representatives with 120 members, whereas Jefferson proposed a House of 105. However, the real differences were not about the size of the House but about how these plans distributed, or apportioned, the members of the House among the 13 states. In order to see the basis of their disagreement, let us examine the circumstances of the state of Connecticut.

Example 1

In 1790 the total population of the country was 3,615,920. Say we desire a House of Representatives with 120 members.
(a) How many persons should each member of the House represent?
(b) If Connecticut has a population of 236,841, how many representatives should be elected from Connecticut?

Solution

(a) In a House of 120, each representative should represent approximately $\dfrac{3,615,920}{120}$ or about 30,133 persons.

(b) Since Connecticut had a population of 236,841, Connecticut should be represented by $\dfrac{236,841}{30,133}$ or (approximately) 7.86 representatives.

Now we see the problem. Although we might find 7 persons willing to go to Congress from Connecticut, we would probably have difficulty persuading someone to send along only 0.86 of himself or herself to the capitol. The problem is the fractions.

HAMILTON'S METHOD

The method recommended by the congress of 1790 is sometimes called Hamilton's method but is also called the Method of Largest Fractions. In order to understand Hamilton's method, we need the following definitions.

Definitions: Suppose the House of Representatives has H members, State k has a population of P_k and the country has population P.

- The **ideal divisor** $= \dfrac{\text{population of country}}{\text{number of representatives}} = \dfrac{P}{H}$. [From Example 1, the ideal divisor in 1790 was 30,133.]

- The **quota** for state k is defined by

  Quota $= \dfrac{\text{population of state } k}{\text{ideal divisor}} = \dfrac{P_k}{\text{ideal divisor}}$. [From Example 1, the quota for Connecticut in 1790 was 7.86.]

- The **lower quota** for a state is the largest integer that is less than or equal to the quota for that state. [The lower quota for Connecticut in 1790 was 7.]

- The **upper quota** for a state is the smallest integer that is greater than or equal to the quota for that state. [The upper quota for Connecticut in 1790 was 8.]

Example 2

In 1790 Georgia had a population of 70,835. Find the quota, lower quota, and upper quota for Georgia.

Solution

(a) As we saw earlier, the ideal divisor in 1790 was $\dfrac{3,615,920}{120}$ or approximately 30,133.

Hence Quota $= \dfrac{70,835}{30,133} = 2.35$.

(b) Lower quota = 2 (the largest integer less than or equal to 2.35).

(c) Upper quota = 3 (the smallest integer greater than or equal to 2.35).

Those who designed Hamilton's method of apportionment concluded that the number of representatives for each state should at least be equal to the lower quota of the state. Hence Connecticut would have at least 7 representatives, and Georgia would have at least 2 representatives. After apportioning to each state its lower quota, these congressmen advocated distributing the additional representatives to the states on the basis of which states had the largest fractions remaining when the lower quota was subtracted from the quota. Hence Hamilton's method can be described as follows:

HAMILTON'S METHOD

1. Choose the size of the House.
2. Find the quota of each state, and assign to each state its lower quota.
3. Assign the remaining seats to those states whose quotas exceed the lower quotas by the largest fractions

Note: Early politicians understood that an assignment must result in each state receiving at least one representative. If in Step 2 the lower quota is zero, the state is assigned one representative, but is not eligible to receive an additional representative in Step 3.

In Table 1 you will find the 1790 apportionment advocated by Hamilton.

Example 3

Explain how it was determined using Hamilton's method that New Jersey, with a population of 179,570, received 6 seats.

Solution

As we have already observed, the ideal divisor for this Congress was 30,133. Hence the quota for New Jersey was $\dfrac{179{,}570}{30{,}133} = 5.96$. Thus the lower quota for New Jersey is 5. After all states received their lower quotas, 111 seats had been distributed and 9 seats remained to be distributed. We then note that the fraction $5.96 - 5 = 0.96$ was the largest of the fractions remaining when lower quotas were subtracted from quotas. Thus New Jersey got another, sixth seat.

In addition to the extra seat awarded to New Jersey, the remaining 8 seats went to Virginia, Connecticut, Delaware, South Carolina, Vermont, Massachusetts, North Carolina, and New Hampshire. These states are marked with an asterisk in Table 1. Observe that the method of Hamilton will always assign a number equal to either the lower quota or the upper quota.

Table 1: Hamilton Apportionment: 1790					
State	Population in 1790	Quota	Lower Quota	Hamilton Apportionment	Upper Quota
Connecticut	236,841	7.86*	7	8	8
Delaware	55,540	1.84*	1	2	2
Georgia	70,835	2.35	2	2	3
Kentucky	68,705	2.28	2	2	3
Maryland	278,514	9.24	9	9	10
Massachusetts	475,327	15.77*	15	16	16
New Hampshire	141,822	4.71*	4	5	5
New Jersey	179,570	5.96*	5	6	6
New York	331,589	11.00	11	11	12
North Carolina	353,523	11.73*	11	12	12
Pennsylvania	432,879	14.37	14	14	15
Rhode Island	68,446	2.27	2	2	3
South Carolina	206,236	6.84*	6	7	7
Vermont	85,533	2.84*	2	3	3
Virginia	630,560	20.93*	20	21	21
Totals	3,615,920		111	120	

Example 4

Table 2 gives the population distribution for the country Consternation, which has five states, a population of 27,000, and a legislature of 30 persons.
(a) Find the ideal divisor for the country of Consternation.
(b) Find the quota for the state of Dismay.
(c) Apportion this legislature using Hamilton's method.

| Table 2: Population of States of Consternation ||
State	Population
Dismay	9461
Anxiety	6869
Distress	5359
Loneliness	3780
Distraction	1531
Total	27,000

Example 4 *(continued)*
Solution

   (a) The ideal divisor is the number of persons ideally represented by each legislator. Hence, in this case, the ideal divisor is $\dfrac{27000}{30} = 900$.

   (b) The quota for the state of Dismay $= \dfrac{9461}{900} = 10.51$. Hence the lower quota for Dismay is 10 and the upper quota is 11.

   (c) Table 3 gives the lower quotas and quotas for all of the states. The sum of the lower quotas is 27, so we have 3 additional seats to distribute. The 3 additional seats go to Distress, Distraction, and Anxiety because 0.95, 0.70, and 0.63 are the largest three fractions remaining when lower quotas are subtracted from quotas. Dismay receives 10 seats, Anxiety 8, Distress 6, Loneliness 4, and Distraction 2.

| Table 3: Population of States of Consternation ||||
State	Population	Quota	Lower Quota
Dismay	9461	10.51	10
Anxiety	6869	7.63	7
Distress	5359	5.95	5
Loneliness	3780	4.20	4
Distraction	1531	1.70	1
Total	27,000		27

Hamilton's method sounds like a thoroughly reasonable solution to the problem, does it not? Indeed, this was the solution sent by Congress to President Washington. But here is a question to ask your American history professor: "What was the first bill that suffered a presidential veto?" The answer, of course, is the Hamilton apportionment bill.

JEFFERSON'S METHOD OF APPORTIONMENT

Why was the Hamilton apportionment bill vetoed? Perhaps fellow Virginian Thomas Jefferson was whispering in President Washington's ear, "Mr. President, it seems unfair to recognize some fractions and ignore others. I have a better idea. Let's ignore all of the fractions." Whatever the conversation, Jefferson's method of apportionment requires that we adjust the ideal divisor to arrive at a divisor for which the sum of the largest integers that are less than or equal to the quotients is 120.

The function that accepts a number x and returns "the largest integer less than or equal to x" is called the greatest integer function and is sometimes denoted by $INT(x)$. Hence $INT(3.1) = 3$, $INT(2.7) = 2$, $INT(5.3) = 5$, and $INT(4) = 4$. With this notation we can describe the method advocated by Jefferson and his colleagues. Although there are more direct algorithms available, we will describe this method using the time-honored problem-solving strategy of guess-and-test.

JEFFERSON'S METHOD
1. Choose the size of the House, H.
2. Suppose the populations of the states are $P_1, P_2, \ldots, P_n$. Experiment as follows with several divisors D (which will be smaller than the ideal divisor):

   Compute the integers $INT(\dfrac{P_k}{D})$ for each State k.

   Form the sum of the integers, $INT(\dfrac{P_1}{D}) + INT(\dfrac{P_2}{D}) + INT(\dfrac{P_3}{D}) + \ldots + INT(\dfrac{P_n}{D})$.

3. When we find a divisor D for which the sum equals H, we allocate $INT(\dfrac{P_k}{D})$ representatives to

   State k for each of the states.

Example 5

Investigate a Jefferson apportionment for the country Consternation from Example 4. In order to complete a Jefferson apportionment, we need to try several different divisors. As we saw in Example 4, the ideal divisor is 900, so we will experiment with divisors smaller than this.
(a) Use the divisor $D = 880$ to attempt a Jefferson apportionment of Consternation.
(b) Complete a Jefferson apportionment of Consternation.

Solution

(a) The populations for the states of Consternation are found in the first column of Table 4.

   When we use the divisor $D = 880$, the quotient for Dismay is $\dfrac{9461}{880} = 10.75$, so

   $INT(\dfrac{9461}{880}) = 10$. Similarly, for Anxiety, $INT(\dfrac{P_k}{D}) = INT(\dfrac{6869}{880}) = INT(7.81) = 7$. Verify

   the rest of the entries in the column for $D = 880$ in Table 4. We are trying to build a legislature with 30 members. When we add the apportionments suggested by the column for $D = 880$, we find that we have allocated only 28 seats. We will need to experiment with a smaller divisor.

(b) In the third and fourth columns of Table 4 we find the computations of $INT(\dfrac{P_k}{D})$ for $D =$

   860 and $D = 840$. With $D = 860$ we distribute only 29 legislators. Hence we need to try a smaller divisor. With $D = 840$ we distribute 30 legislators. The results in the column labeled $D = 840$ constitute the Jefferson apportionment for Consternation.

State	Population P_k	$D = 880$ $INT(\dfrac{P_k}{D})$	$D = 860$ $INT(\dfrac{P_k}{D})$	$D = 840$ $INT(\dfrac{P_k}{D})$
Dismay	9461	10	11	11
Anxiety	6869	7	7	8
Distress	5359	6	6	6
Loneliness	3780	4	4	4
Distraction	1531	1	1	1
Total	27,000	28	29	30

Table 4: Jefferson Apportionment for Consternation

Table 5 presents a comparison of the Hamilton apportionment and the Jefferson apportionment for a House with 120 members based on the census of 1790.

Table 5 Comparison of Hamilton and Jefferson Apportionments			
State	Population in 1790	Hamilton Method	Jefferson Method
Connecticut	236,841	8	8
Delaware	55,540	2	1
Georgia	70,835	2	2
Kentucky	68,705	2	2
Maryland	278,514	9	9
Massachusetts	475,327	16	16
New Hampshire	141,822	5	4
New Jersey	179,570	6	6
New York	331,589	11	11
North Carolina	353,523	12	12
Pennsylvania	432,879	14	15
Rhode Island	68,446	2	2
South Carolina	206,236	7	7
Vermont	85,533	3	3
Virginia	630,560	21	22
Total		120	120

If we examine carefully the two apportionments, Hamilton's and Jefferson's, we understand what motivates Jefferson in addition to his concern for fair treatment of fractions. Observe who gains seats and who loses seats as one moves from Hamilton's apportionment to Jefferson's Apportionment. Little Delaware and New Hampshire lose seats; big Pennsylvania and Virginia gain seats. If we doubt the political nature of the apportionment controversy, we should remember what state Mr. Jefferson called home.

After President Washington vetoed the Hamilton apportionment, the matter returned to Congress, where Jefferson's method prevailed. (For historical accuracy, we should note that the Jefferson method was first implemented with a House of 105 members, not 120 members.) Indeed, the Jefferson method of apportionment was also used after the censuses of 1800, 1810, 1820, and 1830. However, the method was the target of increasingly heavy criticism over time for two reasons. In each apportionment, Jefferson's method favored large states over small states. More painfully, the Jefferson method sometimes awarded large states more representatives than their upper quota. For example, in 1830 the state of New York was allocated 40 seats in the House of Representatives via Jefferson's method. Compare this to the results in Example 6.

Example 6
> In 1830 the national population was 11,931,000, the population of New York was 1,918,578, and the House of Representatives had 240 members.
> (a) Find the ideal divisor for this apportionment.
> (b) Find the quota for the state of New York.
> (c) Find the upper quota and the lower quota for New York.

Solution
> (a) The ideal divisor is $\dfrac{11,931,000}{240} = 49,713$.
>
> (b) The quota for New York is $\dfrac{1,918,578}{49,713} = 38.59$.
>
> (c) The lower quota is 38 and the upper quota is 39, yet Jefferson's method allocated 40 seats to New York.

> **Definition:** An apportionment method **satisfies quota** if, for all the states, the number of representatives allocated to the state is either the lower quota or the upper quota for the state. If for some state this is not true, we say the apportionment method **violates quota**.

In some instances Jefferson's method violates quota, whereas the Hamilton method always satisfies quota. Eventually, the concerns with the tilt toward the larger states and the concerns about violating quota led to a reconsideration of the method of apportionment.

WEBSTER'S METHOD OF APPORTIONMENT

Following the census of 1840, the Senate became involved in a heated discussion of apportionment. On a single day, over 50 motions were made in the Senate on how to use Jefferson's method to complete the new apportionment. Each motion, of course, utilized a different divisor, ranging from 30,000 to 140,000. Finally, the politicians tired and, in their weariness, adopted a new method of apportionment forcefully advocated by the distinguished senator from Massachusetts, Daniel Webster.

Webster recommended that after finding the quota for each state, we round the quota up if it is greater than or equal to 0.5 and we round the quota down if it is less than 0.5. If the resulting distribution of seats does not provide the desired number of representatives, then we try another divisor (near the ideal divisor) until such rounding produces the correct number of representatives. Again, we will describe the method using guess-and-test methods since these methods help us understand the connection between the process and the desired outcome.

Webster's Method
1. Choose the size of the House, H.
2. Suppose the populations of the states are $P_1, P_2, \ldots, P_n$. Experiment as follows with several divisors D (which will be near the ideal divisor):

    Compute the quotients $\dfrac{P_k}{D}$ for each State k.

    Round each quotient up if the fractional part of the quotient is greater than or equal to 0.5 and round it down if the fractional part is less than 0.5. Let R_k denote the rounded quotient for State k. Form the sum of the rounded quotients, $R_1 + R_2 + R_3 + \ldots + R_n$.
3. When we find a divisor D for which $R_1 + R_2 + R_3 + \ldots + R_n$ equals H, we allocate R_k representatives to State k for each of the states.

As with the Jefferson method of apportionment, the Webster method often requires that we experiment with several different divisors until we find one with the desired properties.

> **Example 7**
>   Investigate a Webster apportionment for the country Consternation from Example 4.
>   (a) Determine whether we can complete a Webster apportionment using the ideal divisor.
>   (b) If we are not able to use the ideal divisor, experiment with other divisors in order to complete the Webster apportionment.
> **Solution**
>   (a) The third column of Table 6 shows the quotas, the quotients formed when we divide by the ideal divisor, 900. If we round those quotients, we get the results in the fourth column. Notice that they sum to 31. Since the legislature is to have 30 seats, we will need to divide by a slightly larger divisor.
>   (b) In the fifth column are the quotients that result when we divide by the divisor 910. Notice that the rounded quotients now sum to 30. Hence the Webster apportionment assigns 10 representatives to Dismay, 8 to Anxiety, 6 to Distress, 4 to Loneliness, and 2 to Distraction.

Table 6: A Webster Apportionment for Consternation					
State	Population	Quota (Quotient with $D = 900$)	Rounded Quotient	Quotient with $D = 910$	Rounded Quotient
Dismay	9461	10.51	11	10.40	10
Anxiety	6869	7.63	8	7.55	8
Distress	5359	5.95	6	5.89	6
Loneliness	3780	4.20	4	4.15	4
Distraction	1531	1.70	2	1.68	2
Total	27,000		31		30

The Webster method of apportionment was used only one time, in the apportionment that followed the census of 1840. In the next decade, however, a somewhat surprising event occurred. Before the results of the census were even known, Samuel Vinton of Ohio presented a bill to avoid the "disreputable contest about unrepresented fractions," and the so-called Vinton method of 1850 was adopted. However, close readers observed that Vinton's method was nothing but Hamilton's method, the one that Washington vetoed. Nonetheless, the Hamilton, alias Vinton, method was used through the census of 1900. We will examine some interesting consequences of this decision in Section 8.5.

The methods used to allocate legislative seats to states can be used in many other common problem situations. Consider these problems:

- A middle school principal has resources to teach 10 sections of mathematics. This year 33 students need geometry, 101 students need algebra, 79 students need pre-algebra, and 74 students need general math. How many sections of each course should be taught?

- The College of Education wishes to form a committee of 9 persons that should fairly represent its three departments. The departments have 13, 8, and 7 faculty members. How many representatives should be chosen from each group?

- A statewide committee of 12 persons is being formed to consider education from pre-kindergarten to college. It should fairly represent a pool of participants formed by 233 elementary teachers, 162 middle and secondary educators, 97 college and university personnel, and 46 representatives of business and industry. How many representatives should be chosen from each group?

- Organizers of a boat show must allocate 26 booths in an exhibition hall in a way that will reflect the priorities of the public. A survey suggests that local boat owners are distributed as follows: 654 own power boats, 477 own personal water craft, 246 own sailboats, and 155 own sea kayaks. How should the 26 booths be allocated among the vendors?

Very shortly you will be involved in solving all of these problems.

Exploratory Exercise Set 8.4

Apportionment methods can be used in many contexts other than elections. Hillsborough Middle School has two mathematics teachers to teach all sections of eighth-grade mathematics. Together the two teachers are able to teach 10 sections and, with those 10 sections, must meet the needs of 33 honors geometry students, 101 algebra students, 79 pre-algebra students, and 74 students in general math. In Exploratory Exercises 1 through 3 we will use the Hamilton, Jefferson, and Webster methods to determine the number of sections of each course that should be taught. Work in pairs as you increase your understanding of how to use these apportionment methods.

1. Hamilton's Method
    (a) The ideal divisor in this case will represent the average number of students per section. Compute the ideal divisor.

    (b) The quota for the number of sections of algebra is $\dfrac{101}{ideal\ divisor} = $ ____.

    (c) Complete Table 7, giving the quota, lower quota, and upper quota for the number of sections of each of the four subjects to be taught.

Table 7

Subject	Enrollment	Quota	Lower Quota	Upper Quota
Geometry	33			
Algebra	101	3.52	3	
Pre-algebra	79			
General math	74			
Totals	287		8	

    (d) Judging by the fractions (Quota – Lower Quota), which two subjects have the largest remaining fractions?
    (e) How would the ten sections be distributed if we used Hamilton's method to distribute them?

2. Jefferson's Method: Observe that when we used the ideal divisor 28.7 as the divisor in Exploratory Exercise 1 and ignored fractions, we distributed only 8 sections. Hence we will need to use a divisor smaller than the ideal divisor. In our work below, we will compute the quotient $\dfrac{Enrollment}{D}$ for each of the subjects, with $D = 27$, $D = 26$, and $D = 25$. Our goal will be to find a divisor such that the sum of the greatest integers less than or equal to the quotients will be 10.

    (a) When we use the divisor $D = 27$ to find the quotient for geometry, we compute $\dfrac{33}{27} = 1.22$. The greatest integer less than or equal to 1.22, $INT(1.22)$, is 1. Continue to use the divisor $D = 27$ and complete columns 2 and 3 of Table 8.

Table 8: $D = 27$			
Subject	Enrollment	Quotient $D = 27$	INT(Quotient) $D = 27$
Geometry	33	1.22	1
Algebra	101		
Pre-algebra	79		
General math	74		
Totals	287	XXXX	8

(b) When we review Table 8, we observe that we still have distributed only 8 sections. Complete Table 9 with $D = 26$ to see if we can get closer to the desired distribution of 10 sections.

Table 9: $D = 26$			
Subject	Enrollment	Quotient $D = 26$	INT(Quotient) $D = 26$
Geometry	33		
Algebra	101		
Pre-algebra	79		
General math	74		
Totals	287	XXXX	

(c) When we review Table 9, what suggests that we need to try yet a smaller divisor?

(d) Use the divisor $D = 25$ to complete Table 10.

Table 10: $D = 25$			
Subject	Enrollment	Quotient $D = 25$	INT(Quotient) $D = 25$
Geometry	33		
Algebra	101		
Pre-algebra	79		
General math	74		
Totals	287	XXXX	

(e) How many sections does the Jefferson method allocate to each subject?

3. Webster's Method: Observe that when we use the ideal divisor, 28.7, as the divisor in Exploratory Exercise 1 and round the quotients, the sum of the rounded quotients is 11 whereas we have only 10 sections available. Hence we will need to use a divisor larger than the ideal divisor. In our work

below, we will compute the quotients $\dfrac{Enrollment}{D}$ for larger divisors such as 29, 30, and 31. Our goal will be to find a divisor such that the sum of the rounded quotients is 10.

(a) When we use the divisor $D = 29$ to find the quotient for geometry, we compute $\dfrac{33}{29} = 1.14$. Use the divisor $D = 29$ to find the quotient for algebra.

(b) Continue to use the divisor $D = 29$ and complete columns 2 and 3 of Table 11. If this completes the Webster apportionment, then you are done. Indicate the number of sections allocated to each subject. Otherwise, create a table using $D = 30$ and other divisors near the ideal divisor until you find one that produces a sum of 10.

Table 11: $D = 29$			
Subjects	Enrollment	Quotient $D = 29$	Rounded Quotient
Geometry	33	1.14	1
Algebra	101	3.48	3
Pre-algebra	79		
General math	74		
Totals	287	XXXX	

4. Spread Sheets and Apportionment: Spreadsheets are an ideal technological tool to use in completing apportionments, particularly when completing a Jefferson or Webster apportionment for which we need to experiment with a number of divisors.

(a) This spreadsheet can be used to help complete the Hamilton apportionment found in Exploratory Exercise 1. Using this spreadsheet as a model, create a spreadsheet to find the quota and lower quota for each state in Table 1 of this section. Use this information to verify the Hamilton apportionment of 1790 found in that table. Observe that in Column E we use the function *INT*, which returns the largest integer less than or equal to the number to compute the lower quota. This function is provided in Excel® and similar spreadsheets.

	A	B	C	D	E
1	Subject	Enrollment	Ideal Divisor	Quota	Lower Quota
2			28.7		
3	Geometry	33		=(B3/C2)	=INT(D3)
4	Algebra	101		=(B4/C2)	=INT(D4)
5	Pre-algebra	79		=(B5/C2)	=INT(D5)
6	General math	74		=(B6/C2)	=INT(D6)
7					
8	Totals	=SUM(B3..B6)			=SUM(E3..E6)

(b) Create a spreadsheet that will help find the Jefferson apportionment for the census of 1790 that is found in Table 5 for a House with 120 members. You will need to experiment with several divisors smaller than the ideal divisor of 30,133. You might try divisors such as 30,000, 29,500, 29,000, 28,500, …. You will continue until the sum of the integer parts of the quotients is equal to 120. As a model you might use the following spreadsheet, which can be used to complete the

Jefferson apportionment that we completed in Exploratory Exercise 2. (*Note*: To use this spreadsheet in completing Exploratory Exercise 2, you will need to enter several different divisors smaller than the ideal divisor of 28.7 into C2 until your sum in column E equals 10.)

	A	B	C	D	E
1	Subject	Enrollment	Divisor	Quotient	INT(Quotient)
2			27		
3	Geometry	33		=(B3/C2)	=INT(D3)
4	Algebra	101		=(B4/C2)	=INT(D4)
5	Pre-algebra	79		=(B5/C2)	=INT(D5)
6	General math	74		=(B6/C2)	=INT(D6)
7					
8	Totals	=SUM(B3..B6)			=SUM(E3..E6)

(c) Consider these questions:
  (1) Although the Webster method of apportionment was not under discussion in 1790, look at Table 1 in this section and explain why the Webster apportionment would have been the same as the Hamilton apportionment for these data.
  (2) Create a spreadsheet that will help compute the Webster apportionment for the country of Consternation that is found in Example 4. As a model you might use the following spreadsheet, which helps compute the Webster apportionment that we completed in Exploratory Exercise 3. Note that we use the function ROUND(n,0), which rounds the decimal number n to the nearest integer. This function is provided in Excel® and similar spreadsheets. (In Exploratory Exercise 3 we were able to find a useful divisor with one try; more often we will need to try several divisors.)

	A	B	C	D	E
1	Subject	Enrollment	Divisor	Quotient	Rounded Quotient
2			29		
3	Geometry	33		=(B3/C2)	=ROUND(D3,0)
4	Algebra	101		=(B4/C2)	=ROUND(D4,0)
5	Pre-algebra	79		=(B5/C2)	=ROUND(D5,0)
6	General math	74		=(B6/C2)	=ROUND(D6,0)
7					
8	Totals	=SUM(B3..B6)			=SUM(E3..E6)

5. (*Writing Exercise*) Read the first three chapters (pages 1 through 22) of *Fair Representation: Meeting the Ideal of One Man, One Vote* by M. L. Balinsky and H. P. Young (Yale Unversity Press, 1982). Compare the extent to which the choices made by early Congresses relative to apportionment methods were influenced by political considerations and by mathematical considerations.

Exercise Set 8.4

In each of Exercises 1 through 3 we will use the Hamilton method of apportionment to solve problems. Before completing these exercises, you should review Exploratory Exercise 1.

1. The College of Education wishes to form a committee to determine a new core curriculum. The committee will have 9 members with representatives from each of the three departments in the college. The number of faculty members in each of the departments is found in the table.
    (a) Find the ideal divisor for this apportionment.
    (b) Complete the table by computing the quota, the lower quota, and the upper quota for each department.
    (c) Use a Hamilton apportionment to determine the number of committee members that should be chosen from each department.

College of Education				
Department	Faculty Members	Quota	Lower Quota	Upper Quota
Early Childhood Ed.	13			
Secondary Ed.	8			
Special Ed.	7			

2. The P-16 Initiative of the state of Georgia is directed at ensuring that there is a seamless transition from pre-kindergarten education through college education. The membership of regional councils across the state consists of 223 early childhood educators, 162 middle and secondary educators, 97 college and university faculty, and 46 representatives of business and industry. A special select committee of 12 persons is to be formed with representatives from each of the four groups.
    (a) What is the total number of members on the regional councils?
    (b) Find the ideal divisor for this apportionment situation.
    (c) Find the quota for each of the four groups.
    (d) Determine how the Hamilton method will allocate the 12 seats on the select committee.

3. Organizers of the Savannah Boat Show have space for 26 booths in the exhibition hall. They decide to allocate those booths to vendors on the basis of the kinds of boats purchased in Savannah. The organizers learn that during the past year new boat sales were as follows: 654 powerboats, 477 personal water craft, 246 sailboats and 155 sea kayaks. Use a Hamilton apportionment to allocate the 26 booths to vendors of powerboats, vendors of personal water craft, vendors of sailboats, and vendors of sea kayaks.

4. Compute the greatest integer function, *INT(x)*, for each of the following numbers.
    (a) 5.23   (b) 6   (c) 4.65   (d) 0.98   (e) 2.1

5. Round each of the numbers in Exercise 4 to the nearest integer.

In Exercises 6 and 7 we will use the Jefferson method of apportionment to solve problems. Before completing these exercises, you might like to review Exploratory Exercise 2.

6. An apartment complex has three buildings. Building 1 has 44 units, Building 2 has 35 units, and Building 3 has 24 units. A ten-person resident council will make policies for the residents in the complex. We desire that membership on the council be proportional to the number of units in each building. Use Jefferson's method of apportionment to determine how many persons on the council will come from each building.
    (a) What is the ideal divisor?
    (b) What is the sum of the lower quotas when the ideal divisor is used as the divisor.

(c) Complete the following table, in which we experiment with several divisors smaller than the ideal divisor. Which divisor allows us to complete the Jefferson apportionment? How many persons on the council will come from each building?

	Number of Units	INT(Quotient) $D = 10$	INT(Quotient) $D = 9$	INT(Quotient) $D = 8$	INT(Quotient) $D = 8.8$
Building 1	44				
Building 2	35				
Building 3	24				
Sum					

7. Solve Exercise 3 using a Jefferson apportionment.

In Exercises 8 and 9 we will use the Webster method of apportionment to solve problems. Before completing these exercises, you might like to review Exploratory Exercise 3.

8. Hourly employees for the S & S Cafeteria Corporation are negotiating a new contract and wish to select a 12-person negotiating committee. There are 46 cashiers, 118 servers, and 44 cooks that will be affected by this negotiation. Membership on the negotiating committee is to be chosen from these three groups proportionally to membership in the groups. Use the Webster method of apportionment to determine the membership of the committee.
(a) What is the ideal divisor?
(b) What is the sum of the rounded quotients when the ideal divisor is used as the divisor?
(c) Complete the following table, in which we experiment with several divisors larger than the ideal divisor. Which divisor allows us to complete the Webster apportionment? How many persons on the negotiating committee will come from each group of workers?

	Enrollment	Quotient $D = 17.3$	Rounded Quotient	Quotient $D = 18$	Rounded Quotient	Quotient $D = 18.5$	Rounded Quotient
Cashiers	46						
Servers	118						
Cooks	44						
Totals	208	XXXX		XXXX		XXXX	

9. Solve Exercise 3 using a Webster apportionment.

10. Let P represent the population of the nation and H the number of members of the House of Representatives. Let P_k be the population of State k. We have defined the quota for State k in the following way:

$$\text{The ideal divisor is } \frac{P}{H}.$$

$$\text{The quota } q_k \text{ for state } k \text{ is } \frac{P_k}{ideal\ divisor}.$$

(a) Show that a formula for the quota of State k is given by $q_k = \frac{P_k H}{P}$.

(b) Show that this is equivalent to solving this proportion for q_k.

$$\frac{q_k}{H} = \frac{P_k}{P}$$

The politicians of New England were incensed by the use of the Jefferson method. The portion of the American population located in New England had fallen as time passed, and the little New England states had borne the brunt of the inequities of Jefferson's method. As the apportionment following the 1830 census drew near, John Quincy Adams, the former president, reported that

> *I passed an entirely sleepless night. The iniquity of the Apportionment Bill, and the disreputable means by which so partial and unjust a distribution had been effected agitated me till I could not close my eyes.*

Evidently desperation bred inspiration, for when he arose, Mr. Adams was ready to propose his own method of apportionment.

> ***Adams' Method:*** *Proceed exactly as in the Jefferson method, except for each divisor; form the sum of the least integers **greater** than or equal to the quotients. When the sum is equal to the desired number of legislators, the apportionment is completed.*

The Adams' method was never used, but let's investigate how it works.

11. (a) Complete the following table, which will use the Adams' method to complete the apportionment from Exercise 6.
    (b) Which divisor produces the Adams' apportionment?
    (c) How many committee members does the Adams' method of apportionment allocate to each building?

	Number of Units	Least Integer ≥ Quotient $D = 11$	Least Integer ≥ Quotient $D = 12$	Least Integer ≥ Quotient $D = 11.7$
Building 1	44	4		
Building 2	35	4		
Building 3	24	3		
Sum		11		

12. The states of Arizona, California, Nevada, Oregon, and Washington wish to create an Energy Advisory Council of 20 members, with membership on the council proportional to the populations of the states. The population of Arizona is 5.1 million; California, 33.9 million; Nevada, 2.0 million; Oregon, 3.4 million; and Washington, 5.9 million. *Note: There is no requirement that each state have a representative on the council.*
    (a) Use Hamilton's method to allocate membership on this council.
    (b) Use Jefferson's method to allocate membership on this council.
    (c) Use Webster's method to allocate membership on this council.
    (d) Use Adams' method (see Exercise 11) to allocate membership on this council.
    (e) From your work do you see evidence that Jefferson's method favors large states, whereas Adams' method favors small states?

8.5 Apportionment Revisited

For the last half of the nineteenth century, the Hamilton method was the mathematical basis for the apportionment of the House of Representatives. Because the number of members of the House of Representatives was not fixed by law, prior to each apportionment eager politicians examined Houses of various sizes to determine which gave them the greatest political advantage. In 1881 C. W. Seaton, the chief clerk of the Census Office, was given the responsibility of providing Congress with apportionments based on the number of representatives being some integer between 275 and 350. In a letter to Congress dated October 25, 1881, he reported,

> While making these calculations I met with the so-called "Alabama" paradox where Alabama was allotted 8 Representatives out of a total of 299, receiving but 7 when the total became 300.

Can you imagine the uproar this caused? When the total number of house members was increased, the number of representatives of a state declined. In order to see how this outrageous thing could happen, examine the work in Example 1.

Example 1

> Consider the country with five states described in Table 1. Suppose they discuss whether to have a House with 30 members or 31 members. Use the Hamilton method of apportionment to allocate the membership in both cases.

Table 1: Hamilton Apportionment for House of 30 Members and 31 Members					
State	Population	Quota with 30 Seats	Hamilton Method with 30 Seats	Quota with 31 Seats	Hamilton Method with 31 Seats
South Car	9061	10.54	11	10.89	11
Flaroda	7188	8.36	8	8.64	9
Georgea	5469	6.36	6	6.57	7
Tennessaw	2910	3.38	4	3.50	3
North Car	1172	1.36	1	1.41	1
Total	25,800		30		31

Example 1 *(continued)*
Solution

> For a House of Representatives with 30 members, the ideal divisor is 860, whereas with a House of 31 members, the ideal divisor is near 832. Columns 4 and 6 of Table 1 show a Hamilton apportionment for these two Houses. Observe that when the size of the House of Representatives is *increased* from 30 to 31 seats, the representation of the state of Tennessaw *decreases* from 4 members to 3. This is an example of the Alabama paradox.

Various manifestations of the Alabama paradox led to rancorous debates following the census of 1900. Finally, in 1910, the furor was calmed when Congress resumed using Webster's method.

MEASURING INEQUITY IN REPRESENTATION

At about the same time, the Washington statistician Joseph Hill suggested another point of departure for the apportionment questions. Hill suggested that in allocating additional seats, we ask the question "How can we allocate this additional seat in such a way that we will minimize inequity in representation between the states?" This seems like a reasonable suggestion, but it is not entirely clear what Hill meant by "inequity of representation." In order to understand the meaning Hill attached to the phrase, consider Example 2 and Example 3.

Example 2

Suppose that State A has a population of 22,000 and 6 representatives, and State B has a population of 15,000 and 4 representatives. Suppose the decision is made to give an additional representative to one of these two states. We will first examine what happens when we give the additional representative to State A.
(a) What is the average number of constituents represented by each of the 7 representatives of State A?
(b) What is the average number of constituents represented by each of the 4 representatives of State B?
(c) Compute the ratio

$$\frac{Average\ number\ of\ constituents\ for\ State\ B}{Average\ number\ of\ constituents\ for\ State\ A}.$$

Solution

(a) Since there is a population of 22,000 and 7 representatives, the average number of constituents is $\dfrac{22,000}{7} = 3143$.

(b) 3750

(c) $\dfrac{3750}{3143} = 1.19$

Example 3

Observe that the result in (c) of Example 2 indicates that if State A is given the extra representative, there are 19% more constituents per representative for State B than for State A. Now suppose that for the states in Example 2, State B is given the extra representative.
(a) What is the average number of constituents represented by each of the 6 representatives in State A?
(b) What is the average number of constituents represented by each of the 5 representatives in State B?
(c) Compute the ratio

$$\frac{Average\ number\ of\ constituents\ for\ State\ A}{Average\ number\ of\ constituents\ for\ State\ B}$$

(d) Determine which of the two assignments of the extra representative produces the greater inequity.
(e) To which state would Hill assign the additional representative?

Solution

(a) $\dfrac{22,000}{6} = 3667$

(b) $\dfrac{15,000}{5} = 3000$

(c) $\dfrac{3667}{3000} = 1.22$

(d) When the new representative is given to State A as in Example 2, each representative for State B represents 19% more citizens than representatives for State A. When the new representative is given to State B, each representative in State A represents 22% more

(continued)

citizens than the representatives of State B. Assigning the new representative to State B produces the greater inequity.

(e) Hill wished to minimize the inequity observed in (d), so he would assign the new representative to State A.

Hill's Principle: Suppose that State A has N_A representatives and a population P_A and that State B has N_B representatives and population P_B. State A has priority for the next representative if

$$\frac{Average\ constituency\ of\ B\ with\ N_B\ reps}{Average\ constituency\ of\ A\ with\ N_A+1\ reps} < \frac{Average\ constituency\ of\ A\ with\ N_A\ reps}{Average\ constituency\ of\ B\ with\ N_B+1\ reps}$$

If we translate this inequality into symbols, it becomes

$$\frac{\dfrac{P_B}{N_B}}{\dfrac{P_A}{N_A+1}} < \frac{\dfrac{P_A}{N_A}}{\dfrac{P_B}{N_B+1}}$$

By multiplying both sides of the inequality by the denominators, we see that Hill will allocate the representative to State A when

$$\frac{(P_B)^2}{(N_B)(N_B+1)} < \frac{(P_A)^2}{(N_A)(N_A+1)}$$

Attend closely to the two quotients that appear in this last inequality. They will be important as we finish the apportionment story.

Although the principle that Hill advocated is clear, it is not so clear how to complete an apportionment using Hill's reasoning. Indeed, Hill himself did not develop an adequate algorithm for completing the apportionment. However, an eminent mathematician who was a friend from Hill's college years, E. V. Huntington, found algorithms that would accomplish the apportionment and became a vigorous proponent of the method before the United States Congress. In 1940, because of a fortuitous combination of political expediency and Huntington's persuasive powers, the Hill-Huntington method became the law of the land. It continues to be the method of apportionment for the House of Representatives to this day.

THE HILL-HUNTINGTON METHOD OF APPORTIONMENT

Several algorithms can be described for completing the apportionment on the basis of Hill's reasoning, including an iterative process that uses the quotients we derived when we discussed the basis of that reasoning.

Hill-Huntington Quotient: If a state has a population of P and has been allocated N representatives, then the Hill-Huntington Quotient for the state is

$$\frac{P^2}{N(N+1)}$$

The Hill-Huntington Method
1. Choose the size of the House.
2. Assign one representative to each state.

3. Repeat until all representatives have been assigned:
    Compute the Hill-Huntington quotient of each state.
    Assign the next representative to the state with the largest Hill-Huntington quotient.

Example 4

Use the Hill-Huntington method to apportion a House of six representatives for the country with three states described in Table 2.

Table 2		
State	Population	Representatives
Kentuck	360	1
Arkansee	850	1
Indianola	460	1

Example 4 *(continued)*
Solution

Table 2 gives not only the population information for the three states but also the results after Step 1 of the Hill-Huntington method. Since three of the six representatives have been assigned, we now must assign the fourth. To do so, we need to compute the Hill-Huntington quotients for the three states:

$$\frac{360^2}{(1)(2)} = 64,800 \qquad \frac{850^2}{(1)(2)} = 361,250 \qquad \frac{460^2}{(1)(2)} = 105,800$$

Clearly, Arkansee is the state with the largest Hill-Huntington quotient, so we assign the fourth representative to this state.

Table 3			
State	Population	Representatives	Hill-Huntington Quotient
Kentuck	360	1	64,800
Arkansee	850	2	120,417
Indianola	460	1	105,800

Example 4 *(continued)*

Table 3 describes how the first four representatives are allocated. In addition, in the fourth column are computed the new Hill-Huntington quotients (rounded to the nearest integer). Since Arkansee still has the largest Hill-Huntington quotient, Arkansee is allocated the fifth representative. Table 4 gives the Hill-Huntington quotients to use in allocating the last (sixth) representative.

Table 4			
State	Population	Representatives	Hill-Huntington Quotient
Kentuck	360	1	64,800
Arkansee	850	3	60,208
Indianola	460	1	105,800

Example 4 *(continued)*

Since Indianola has the largest Hill-Huntington quotient, Indianola is allocated the next representative. Hence the apportionment of the legislature by the Hill-Huntington method gives 1 representative to Kentuck, 3 representatives to Arkansee, and 2 representatives to Indianola.

In describing the Hill-Huntington method of apportionment, we have used an iterative description. Surprisingly, it can also be described very simply using a divisor method similar to the ones used for the Jefferson and Webster methods. We will examine that description in Exploratory Exercise 3.

WRAP-UP

During the 1970s a pair of mathematicians, Michel Balinski and H. Peyton Young, did a careful of analysis of methods of apportioning the House of Representatives. They identified three desirable characteristics of such a method.
1. The method should satisfy quota. That is, the number of representatives assigned to each state should be either the upper quota or the lower quota.
2. The method should avoid the Alabama paradox. That is, as the size of the House increases, the number of representatives for a state should grow or stay the same (the number should not decrease).
3. The method should avoid what they called the **population paradox**. That is, as the population of State A increased relative to State B, State A should not lose representatives to State B.

In our investigations, we have seen that the Hamilton method satisfies quota but suffers from the Alabama paradox. We shall see in Exercises 9 through 11 that the population paradox can also occur with Hamilton's method. We have also seen that the Jefferson method sometimes fails to satisfy quota. Indeed, all divisor methods such as the Jefferson method, the Webster method, and the Hill-Huntington method violate quota in certain instances. The question, then, is whether we can find any method that exhibits all three desirable properties. In 1982 Balinsky and Young published a proof of the following theorem.

Theorem 1: There is no apportionment method that always satisfies quota and also avoids the Alabama paradox and the population paradox.

Much like Arrow's Impossibility Theorem, which we mentioned in Section 8.2, Theorem 1 tells us that there is no perfect solution to the apportionment problem. We can stop looking for perfection and focus on what is the best possible. Although the Hill-Huntington method continues to be the law of the land, Balinski and Young advocate the Webster method quite enthusiastically. Both Hill-Huntington and Webster avoid the Alabama paradox and the Population paradox, but the Webster method is more likely to satisfy quota than Hill-Huntington and is completely unbiased relative to the large-state/ small-state issue. (Hill-Huntington favors small states.) Perhaps you should write a letter to your member of congress.

Exploratory Exercise Set 8.5

1. In this exercise we will review Hamilton's method and seek to find an example of the Alabama paradox. Work with a partner to complete the following problem.

    Three corporations, Triax, Acme, and Umax, form a consortium to build a jointly owned "nano-technology" laboratory. Triax invests 52 million dollars, Acme contributes 34 million dollars, and Umax contributes 15 million dollars. A governing council for the facility is to be appointed, with memberships on the governing council proportional to the contributions made by the corporations.

    (a) The initial plan was to choose a governing council consisting of 9 persons. Use Hamilton's method to determine how many representatives to the governing council should be chosen from each corporation.

(1) With total investments of 101 million dollars and a council of 9 persons, what is the ideal divisor?

(2) Complete Table 5 giving the solution to the problem using Hamilton's method.

Table 5			
Corporation	Investment	Quota with 9 Persons	Number of Members
Triax	52		
Acme	34		
Umax	15		

(b) In the midst of negotiations about the project, the CEOs of the corporations decide that they need a somewhat larger governing council. Use Table 6 to complete a Hamilton apportionment for councils consisting of 10 and 11 members.

Table 6					
Corporation	Investment	Quota with 10 Persons	Number of Members	Quota with 11 Persons	Number of Members
Triax	52				
Acme	34				
Umax	15				

(c) Review your results from (b) carefully, and explain why the representation of Umax shows that Hamilton's method suffers from the Alabama paradox.

2. Work with your partner to use the Hill-Huntington method to apportion the 9-person governing committee from Exploratory Exercise 1.

(a). Table 7 shows the results after the first step of the Hill-Huntington method.

Table 7			
Corporation	Investment	Representatives	Hill-Huntington Quotient
Triax	52	1	1352
Acme	34	1	
Umax	15	1	

(1) The Hill-Huntington quotient for Triax is $(52)^2/(1)(2) = 1352$. Compute the Hill-Huntington quotient for Acme.

(2) Compute the Hill-Huntington quotient for Umax.

(3) Which of the three corporations gets the next representative on the committee?

(b) Table 8 shows how the first 4 of the 9 council members have been allocated. The Hill-Huntington quotient for Triax is now $(52)^2/(2)(3) = 451$ when rounded to the nearest integer. Compare with the other Hill-Huntington quotients to determine who gets the next member.

Table 8			
Corporation	Investment	Representatives	Hill-Huntington Quotient
Triax	52	2	451
Acme	34	1	
Umax	15	1	

(c) Proceeding similarly, complete the Hill-Huntington apportionment for a governing council of 9 members.

3. The Hill-Huntington method of apportionment can be described as a divisor method very much like the Webster method. The difference between the two methods lies in the following statements:

$$\text{Suppose } A = INT(\frac{population}{divisor}).$$

In the Webster method, we round the quotient $\frac{population}{divisor}$ up if the quotient is greater than or equal to the arithmetic mean of A and $A + 1$ (that is, if the quotient $\geq A + 0.5$).

In the Hill-Huntington method, we round the quotient $\frac{population}{divisor}$ up if the quotient is greater than or equal to the geometric mean of A and $A + 1$ (that is, if the quotient $\geq SQRT(A(A + 1)))$

(a) The geometric mean of numbers X and Y is defined by $SQRT(X \cdot Y)$. Compute the geometric means of the following pairs of numbers.
   (1) 5, 10   (2) 1, 2   (3) 3, 4
(b) A divisor method for computing the Hill-Huntington apportionment can be described as follows:
   (i) Choose the size of the House, H.

   (ii) Suppose the population of the states is $P_1, P_2, ..., P_n$. Experiment as follows with several divisors D (which will be near the ideal divisor):
   Compute the quotients $\frac{P_k}{D}$ for each State k and find $A_k = INT(\frac{P_k}{D})$.
   Round each quotient up if the quotient $\geq$ the geometric mean. $SQRT(A_k(A_k + 1))$ and round it down otherwise [let R_k denote the rounded quotient for State k] .
   Form the sum of the rounded quotients, $R_1 + R_2 + R_3 + ...+ R_n$.
   (iii) When we find a divisor D for which $R_1 + R_2 + R_3 + ... + R_n$ equals H, we allocate R_k representatives to State k for each of the states.

Complete Table 9 to determine whether $D =$ "the ideal divisor" will yield a Hill-Huntington apportionment for the governing council of 9 discussed in Exploratory Exercise 1(a). Note that since $4.63 \geq$ the geometric mean of 4 and 5, we round up to 5 when allocating representatives to TRIAX.

Table 9				
Corporation	Investment	Quotient with $D = 11.22$	Geometric Mean	Number of Members
Triax	52	4.63	SQRT($4 \cdot 5$) = 4.47	5
Acme	34	3.03	SQRT($3 \cdot 4$) = 3.46	?
Umax	15	1.34	SQRT($1 \cdot 2$) = ?	?

(c) Use the divisor algorithm for the Hill-Huntington method to apportion a governing committee of 10 members.

(d) Use the divisor algorithm for the Hill-Huntington method to apportion governing committees of 11 members. (*Hint*: You will need to experiment with divisors greater than the ideal divisor to complete this apportionment.)

(e) Compare your work in (c) and (d) with the Hamilton method applied in Exploratory Exercise 1(b). Does the Hill-Huntington method show any sign of the Alabama paradox in these examples?

4. Create a spreadsheet that will assist in computing the Hill-Huntington apportionment for the country with three states that is found in Example 4 of this section. As a model, you might like to use this spreadsheet, which assists in the Hill-Huntington apportionment that we completed in Exploratory Exercise 3(b). Note that we use the function SQRT(n) that computes the square root of the number n; we also use $\cdot$ to denote multiplication. This notation is available in Excel® and similar spreadsheets. (In Exploratory Exercise 3 we were able to find a useful divisor with one try; more often we will need to try several divisors.)

	A	B	C	D	E	F
1	Corporation	Investment	Divisor	Quotient	INT(Quotient)	Geometric Mean
2			11.22			
3	Triax	52		=(B3/C2)	=INT(D3)	=SQRT(E3$\cdot$(E3+1))
4	Acme	34		=(B4/C2)	=INT(D4)	=SQRT(E4$\cdot$(E4+1))
5	Umax	15		=(B5/C2)	=INT(D5)	=SQRT(E5$\cdot$(E5+1))

5. (*Writing Exercise*) Write a letter to your congressperson advocating a change of our present system of apportionment to one of the historical systems we have discussed. You may advocate for any of the historical systems for any reason (including selfish interests). If you feel you need more background before writing the letter, read Chapters 1 through 4 and 10 (pages 1 through 35 and pages 79 through 86) of *Fair Representation: Meeting the Ideal of One Man, One Vote* by M. L. Balinsky and H. P. Young (Yale University Press, 1982).

Exercise Set 8.5

Suppose State A has a population of 2532 and $N_A = 3$ representatives and State B has a population of 6423 and $N_B = 8$ representatives. We wish to give a new representative to one of the two states. In Exercises 1 and 2 we will examine the inequity that would result in each case and compare the two.

1. Suppose that the new representative is given to State A.
    (a) What is the average number of constituents represented by each of the 4 representatives of State A?

(b) What is the average number of constituents represented by each of the 8 representatives of State B?

(c) Compute the ratio $\dfrac{Average\ constituency\ of\ B\ with\ N_B\ reps}{Average\ constituency\ of\ A\ with\ N_A+1\ reps}$.

(d) In this case, each representative of State B represents what percent more constituents than each representative of State A?

2. Suppose that the new representative is given to State B.

(a) What is the average number of constituents represented by each of the 3 representatives of State A?

(b) What is the average number of constituents represented by each of the 9 representatives of State B?

(c) Compute the ratio $\dfrac{Average\ constituency\ of\ A\ with\ N_A\ reps}{Average\ constituency\ of\ B\ with\ N_B+1\ reps}$.

(d) In this case, each representative of State A represents what percent more constituents than each representative of State B?

(e) Compare your work in 1(d) and 2(d). To which state would Hill's principle assign the new representative.

3. The Chatham Area Transit System (CATS) uses four buses on Route 14, and they carry 2300 passengers each day. CATS uses five buses on Route 24, and they carry 3200 passengers per day. Suppose CATS has the resources to add a bus to one of these two routes. In view of Hill's principle, which route should get the additional bus? [Hint: See Exercises 1 and 2.]

4. Maine is represented by 2 congresspersons and has a population of 1.3 million. Massachusetts is represented by 10 congresspersons and has a population of 6.4 million. According to Hill's principle, which state is more deserving of a new representative?

After reviewing Exploratory Exercise 2, solve Exercises 5 through 7 using the Hill-Huntington method of apportionment. [Alternatively, solve these exercises after reviewing Exploratory Exercise 3.]

5. The members of three unions that work at the Kimika chemical plant decide to form a labor council. The pipefitters' local has 45 members, the electricians' local has 32 members, and the carpenters' local has 28 members. The labor council is to consist of 7 representatives with at least one representative from each union. Determine the number of representatives that should come from each union.

6. Eight seats on the Youth Performing Arts Board are to be apportioned. In the city 410 young people participate in music, 350 young people participate in theater, and 110 young people participate in dance. How many young people from each performance area should be placed on the Board?

7. The honors committee at Armstrong University is able to award nine full scholarships to students who are finishing their junior year. The committee members want to apportion the scholarships on the basis of the number of students in the three colleges. If the College of Arts and Sciences has 354 juniors, the College of Education has 89 juniors, and the College of Health Professions has 119 juniors, how should the nine scholarships be awarded?

8. (a) Use the Hamilton method to apportion a legislature of 20 members for the country described in Table 10.

Table 10	
State	Population
New Yark	7,480,000
Delaska	2,520,000
West Carolina	4,510,000
Merryland	5,490,000

(b) Use the Hamilton method to apportion a legislature of 21 members for the country in Table 10.

(c) Review your work from (a) and (b) to find evidence of the Alabama paradox.

One undesirable property that plagues the Hamilton method and some other apportionment methods that satisfy quota is the so-called population paradox. The population paradox occurs when State A increases in population relative to State B yet loses a representative to State B in a subsequent apportionment. In Exercises 9 through 11 we will discover an instance when the Hamilton method suffers from the population paradox.

9. Use the Hamilton method to apportion a House of 30 members for the country with five states found in Table 11. The population figures come from the census of 1990.

Table 11			
State	Population	Quota	Hamilton Apportionment
Dismay	10,000		
Anxiety	7,100		
Distress	5,359		
Loneliness	3,780		
Distraction	1,531		

10. The results of the census of 2000 are found Table 12. Use the Hamilton method to apportion a House of 30 members in view of the new population figures.

Table 12			
State	Population	Quota	Hamilton Apportionment
Dismay	10,600		
Anxiety	7,400		
Distress	5,800		
Loneliness	4,100		
Distraction	1,525		

11. Carefully review your work from Exercises 9 and 10.
    (a) Between 1990 and 2000, what happened to the population of Anxiety?
    (b) Between 1990 and 2000, what happened to the number of representatives allocated to Anxiety?
    (c) Between 1990 and 2000, what happened to the population of Distraction?
    (d) Between 1990 and 2000, what happened to the number of representatives allocated to Distraction?
    (e) Do you see evidence of the population paradox?

Answers to Odd Questions

Exercise Set 1.1

1. (a) Statement—False (b) Not a statement   (c) Statement—True
   (d) Not a statement   (e) Not a statement   (f) Statement—False
   (g) Not a statement   (h) Not a statement
3. (a) False   (b) False   (c) False
   (d) False   (e)True   (f) False
5. (a) November has 30 days and Thanksgiving is always on November 25. False
       November has 30 days or Thanksgiving is always on November 25. True
   (b) The smallest counting number is 2, and 10 is not a multiple of 5. False
       The smallest counting number is 2, or 10 is not a multiple of 5. False
   (c) $2 + 3 = 4 + 1$ and $8 \cdot 6 = 4 \cdot 12$. True
       $2 + 3 = 4 + 1$ or $8 \cdot 6 = 4 \cdot 12$. True
   (d) Triangles are not squares and 3 is smaller than 5. True
       Triangles are not squares or 3 is smaller than 5. True
7. (a) It is snowing and the roofs are not white.
   (b) The roofs are not white or the streets are slick.
   (c) It is snowing, and either the roofs are white or the streets are not slick.
   (d) It is snowing or the streets are slick, and the trees are green.
   (e) It is not true that it is snowing and the trees are not green.
   (f) It is not true that it is snowing and the streets are slick.
9. (a) $p \wedge q$ (b) $\sim r$ (c) $\sim q$ (d) $(q \vee r) \wedge \sim p$
11. (a) $(A \vee D) \wedge W$
    (b) Yes. A is true. D is false. W is true. It follows that $A \vee D$ is true so the statement is true.
    (c) No. A is false. D is true. It follows that $A \vee D$ is true. However, W is false so the statement is false.
13. (a) Husband is a Liar; wife is a Truthteller.
    (b) Husband is a Liar; wife is a Truthteller.
    (c) The first giant is a Liar; the second is a Truthteller; and the third is a Liar.
15. (a)

p	q	$\sim q$	$p \vee \sim q$
T	T	F	T
T	F	T	T
F	T	F	F
F	F	T	T

(b)

p	$\sim p$	$p \wedge \sim p$
T	F	F
F	T	F

17. (a) The stock market is not bullish and the Dow average is increasing.
    (b) The stock market is bullish or the price of utilities is not decreasing.
    (c) The stock market is bullish or the price of utilities is decreasing, and the Dow average is increasing.
    (d) The price of utilities is not decreasing or the stock market is not bullish, and the Dow average is increasing.
19. (a) $p \wedge q$ (b) $p \vee q$ (c) $\sim p \wedge q$ (d) $\sim p \vee \sim q$
21 (a) False   (b) False   (c) False   (d) False
23. (a)

p	q	r	$\sim q$	$(p \vee \sim q)$	$(p \vee \sim q) \wedge r$
T	T	T	F	T	T
T	T	F	F	T	F
T	F	T	T	T	T
T	F	F	T	T	F

F	T	T	F	F	F
F	T	F	F	F	F
F	F	T	T	T	T
F	F	F	T	T	F

(b)

p	q	r	$(q \vee r)$	$p \wedge (q \vee r)$
T	T	T	T	T
T	T	F	T	T
T	F	T	T	T
T	F	F	F	F
F	T	T	T	F
F	T	F	T	F
F	F	T	T	F
F	F	F	F	F

(c)

p	q	$\sim p$	$(p \vee \sim p)$	$(p \vee \sim p) \wedge q$
T	T	F	T	T
T	F	F	T	F
F	T	T	T	T
F	F	T	T	F

(d)

p	q	r	$(p \vee q)$	$\sim (p \vee q)$	$\sim (p \vee q) \wedge r$
T	T	T	T	F	F
T	T	F	T	F	F
T	F	T	T	F	F
T	F	F	T	F	F
F	T	T	T	F	F
F	T	F	T	F	F
F	F	T	F	T	T
F	F	F	F	T	F

Exercise Set 1.2

1. (a) $A \rightarrow D$ (b) $\sim A \rightarrow \sim B$ (c) $\sim C \rightarrow \sim A$ (d) $C \rightarrow \sim D$
3. (a) Converse: If one angle of a triangle measures 90°, then the triangle is a right triangle.
   Inverse: If a triangle is not a right triangle, then no angle has a measure of 90°.
   Contrapositive: If no angle of a triangle measures 90°, then the triangle is not a right triangle.
   (b) Converse: If a number is odd, then it is prime.
   Inverse: If a number is not prime, then it is not odd.
   Contrapositive: If a number is not odd, then it is not prime.
   (c) Converse: If alternate interior angles are equal, then two lines are parallel.
   Inverse: If two lines are not parallel, then alternate interior angles are not equal.
   Contrapositive: If alternate interior angles are not equal, then two lines are not parallel.
   (d) Converse: If Joyce is happy, then she is smiling.
   Inverse: If Joyce is not smiling, then she is not happy.
   Contrapositive: If Joyce is not happy, then she is not smiling.
5. (a) Converse: If $2 + 2 = 5$, then the moon orbits the earth.
   Inverse: If the moon does not orbit the earth, then $2 + 2 \neq 5$.
   Contrapositive: If $2 + 2 \neq 5$, then the moon does not orbit the earth.
   (b) Converse: If $2 + 2 = 5$, then Lincoln was the first U.S. president.
   Inverse: If Lincoln was not the first U.S. president, then $2 + 2 \neq 5$.

Contrapositive: If $2 + 2 \neq 5$, then Lincoln was not the first U.S. president.
(c) Converse: If $3 + 3 = 6$, then Mexico is the southern neighbor of the United States.
Inverse: If Mexico is not the southern neighbor of the United States, then $3 + 3 \neq 6$.
Contrapositive: If $3 + 3 \neq 6$, then Mexico is not the southern neighbor of the United States.
(d) Converse: If $3 + 3 = 6$, then Canada is in Asia.
Inverse: If Canada is not in Asia, then $3 + 3 \neq 6$.
Contrapositive: If $3 + 3 \neq 6$, then Canada is not in Asia.

7. The car is behind Door 1; the skateboard is behind Door 2. [Explanation: If the sign on the second door is true, then the signs on both doors are true. This is not possible, so the sign on the first door is true and the sign on the second door is false.]

9. (a) Converse: If you have fewer cavities, then you brush your teeth with White-as-Snow.
Inverse: If you do not brush your teeth with White-as-Snow, then you do not have fewer cavities.
Contrapositive: If you do not have fewer cavities, then you do not brush your teeth with White-as-Snow.
(b) Converse: If you love mathematics, then you like this book.
Inverse: If you do not like this book, then you do not love mathematics.
Contrapositive: If you do not love mathematics, then you do not like this book.
(c) Converse: If you eat Barlies for breakfast, then you are strong.
Inverse: If you are not strong, then you do not eat Barlies for breakfast.
Contrapositive: If you do not eat Barlies for breakfast, then you are not strong.
(d) Converse: If your clothes are bright and colorful, then you use Wave.
Inverse: If you do not use Wave, then your clothes are not bright and colorful.
Contrapositive: If your clothes are not bright and colorful, then you do not use Wave.

11. (a) The minimum penalty is $75.
(b) The policy statement provides no guidance in this circumstance.

13. (a)

p	q	$\sim q$	$p \rightarrow q$	$\sim (p \rightarrow q)$	$p \wedge \sim q$
T	T	F	T	F	F
T	F	T	F	T	T
F	T	F	T	F	F
F	F	T	T	F	F

Since the last two columns have the same truth values, these statements are equivalent.
(b) It is not true that if Nakisha is a music major, then she sings well.
(c) My dog has fleas and he does not scratch often.

15. (a) It is not true that if Ed drives a truck, then he wears socks.
(b) It is not true that if Ana lives in Memphis, then she does not ride a bus to work.
(c) The black dog sits and he does not want to eat.
(d) The phone rings and trouble does not follow.

17. (a)

p	q	$\sim p$	$\sim q$	$p \vee q$	$\sim (p \vee q)$	$\sim p \wedge \sim q$
T	T	F	F	T	F	F
T	F	F	T	T	F	F
F	T	T	F	T	F	F
F	F	T	T	F	T	T

Since the last two columns have the same truth values, these statements are equivalent.

(b)

p	q	$\sim p$	$\sim q$	$p \wedge q$	$\sim (p \wedge q)$	$\sim p \vee \sim q$
T	T	F	F	T	F	F
T	F	F	T	F	T	T
F	T	T	F	F	T	T
F	F	T	T	F	T	T

Since the last two columns have the same truth values, these statements are equivalent.

    (c) I will not take history and I will not take physics.

    (d) I am not on the baseball team or I am not on the basketball team.

19. (a) If a geometric figure is a triangle, then it is not a square.

    (b) If they are birds of a feather, then they flock together.

    (c) If a politician is honest, then she/he does not accept bribes.

21. (a)

p	q	$\sim p$	$p \vee q$	$\sim p \to q$
T	T	F	T	T
T	F	F	T	T
F	T	T	T	T
F	F	T	F	F

    Since the last two columns have the same truth values, these statements are equivalent.

    (b)

p	q	$\sim p$	$\sim q$	$p \wedge q$	$\sim(p \wedge q)$	$\sim p \wedge \sim q$
T	T	F	F	T	F	F
T	F	F	T	F	T	F
F	T	T	F	F	T	F
F	F	T	T	F	T	T

    Since these two statements have different truth values in two cases, they are not equivalent.

    (c)

p	q	$\sim p$	$\sim q$	$p \vee q$	$\sim(p \vee q)$	$\sim p \vee \sim q$
T	T	F	F	T	F	F
T	F	F	T	T	F	T
F	T	T	F	T	F	T
F	F	T	T	F	T	T

    Since these two statements have different truth values in two cases, they are not equivalent.

    (d)

p	q	$\sim p$	$\sim q$	$p \to q$	$\sim p \to \sim q$
T	T	F	F	T	T
T	F	F	T	F	T
F	T	T	F	T	F
F	F	T	T	T	T

    Since these two statements have different truth values in two cases, they are not equivalent.

23. (a) If two lines are not parallel and not skew, then they intersect.

    (b) If the conjunction $p \wedge q$ is false, then either p is false or q is false.

25. (a) This statement is a tautology.

p	$p \to p$
T	T
F	T

(b) This statement is not a tautology.

p	q	$\sim q$	$p \to q$	$p \wedge (p \to q)$	$[p \wedge (p \to q)] \to \sim q$
T	T	F	T	T	F
T	F	T	F	F	T
F	T	F	T	F	T
F	F	T	T	F	T

(c) This statement is not a tautology.

p	q	$p \to q$	$(p \to q) \to p$
T	T	T	T
T	F	F	T
F	T	T	F
F	F	T	F

(d) This statement is a tautology.

p	q	$\sim p$	$\sim p \vee q$	$(\sim p \vee q) \wedge p$	$[(\sim p \vee q) \wedge p] \rightarrow q$
T	T	F	T	T	T
T	F	F	F	F	T
F	T	T	T	F	T
F	F	T	T	F	T

Exercise Set 1.3

1. (a)

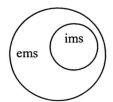

(b)

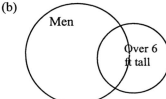

(c)

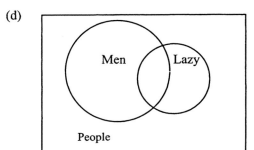

(d)

3. (a) $x = 3$ (b) $x = 11$ (c) $x = 4$ or -4 (d) Let x be any number.
5. (a) For some number x, $x + 4 = 7$.
    (b) For some number x, $x - 7 = 4$.
    (c) For some number x, $x^2 = 16$.
    (d) For all numbers x, $x + 3 = 3 + x$.
7. (a) Some athletes over 7 feet tall do not play basketball.
    (b) No students work hard at their studies.
    (c) Some men do not use Bob-Bob hair oil.
    (d) All professors are intelligent.
    (e) Some men weigh more than 500 pounds. (There exists a man that weighs more than 500 pounds.)
    (f) Some Martian either is not green or does not have three eyes, and my name is Captain Midnight.
    (g) Some mystery novel involves neither a murder nor a bungled robbery.
9. (a) Premises: All donors are generous. Some young persons are donors.
        Conclusion: Some young persons are generous.
    (b) Premises: All diligent students succeed. All education majors are diligent.
        Conclusion: All education majors succeed.
11 (a) 3 is a counting number, and 3 is odd.
    (b) Jane Swift was the governor of Massachusetts, and she is a woman.
13. (a) Some figure is a square and is not a rectangle.
    (b) Some number is divisible by 4 and is not divisible by 2.
    (c) Some squares are not rectangles.
    (d) Some numbers are divisible by 4 and not divisible by 2.
15. Dr. X is a professor. Dr. X is not a woman professor. Dr. X is neither dull nor boring.

17. Invalid reasoning

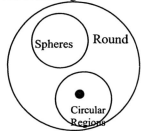

19. Invalid reasoning

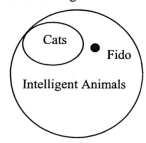

21. Invalid reasoning

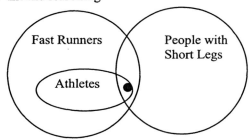

23. Conclusion (c)

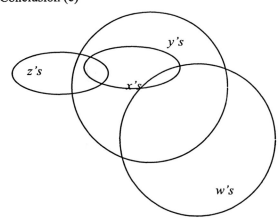

25. Conclusion (b)

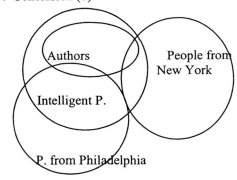

27. (a) $\exists x,\ x^2 = 4.$ (b) $\forall x,\ x^2 = 4.$ (c) $\exists$ triangles, such that the sum of the measures of the interior angles is equal to 180°. (d) $\forall$ triangles, the sum of the measures of the interior angles is equal to 180°.

Exercise Set 1.4

1. $s \to p$

   s

   $\therefore p$

3. $p \to c$

   $c \to m$

   $\therefore p \to m$

5. (a) Rule of contraposition   (b) Chain rule   (c) Rule of detachment
   (d) Rule of contraposition   (e) Chain rule

7. (a) Susan takes mathematics. (rule of detachment)   (b) You studied. (rule of contraposition)
   (c) You cry. (rule of detachment)   (d) You did not cut the grass. (rule of contraposition)

9. (a)

p	q	$\sim q$	$p \to q$	$(p \to q) \wedge \sim q$	$\sim p$
T	T	F	T	F	F
T	F	T	F	F	F
F	T	F	T	F	T
F	F	T	T	T	T

(b) In every case in which $(p \to q) \wedge \sim q$ is true, $\sim p$ is true.

11.

p	q	$p \vee q$	$(p \vee q) \wedge p$	$\sim q$
T	T	T	T	F
T	F	T	T	T
F	T	T	F	F
F	F	F	F	T

This argument is invalid, because there is an instance in which the conjunction of the premises is true but the conclusion is not true.

13.

p	q	r	$\sim p$	$\sim q$	$(p \vee \sim q)$	$\sim p \vee r$	$(p \vee \sim q) \wedge$ $(\sim p \vee r)$	$q \rightarrow r$
T	T	T	F	F	T	T	T	T
T	T	F	F	F	T	F	F	F
T	F	T	F	T	T	T	T	T
T	F	F	F	T	T	F	F	T
F	T	T	T	F	F	T	F	T
F	T	F	T	F	F	T	F	F
F	F	T	T	T	T	T	T	T
F	F	F	T	T	T	T	T	T

This argument is valid.

15.

p	q	r	$\sim p$	$q \rightarrow r$	$\sim p \vee q$	$(q \rightarrow r) \wedge (\sim p \vee q) \wedge$ p	r
T	T	T	F	T	T	T	T
T	T	F	F	F	T	F	F
T	F	T	F	T	F	F	T
T	F	F	F	T	F	F	F
F	T	T	T	T	T	F	T
F	T	F	T	F	T	F	F
F	F	T	T	T	T	F	T
F	F	F	T	T	T	F	F

This argument is valid.

17. Invalid as shown by the truth table. Let p be the statement "I can go to town," and let q be the statement "I will go to the shopping center." The argument then reads $(\sim p \rightarrow q) \wedge p; \therefore \sim q$.

p	q	$\sim p$	$\sim p \rightarrow q$	$(\sim p \rightarrow q) \wedge p$	$\sim q$
T	T	F	T	T	F
T	F	F	T	T	T
F	T	T	T	F	F
F	F	T	F	F	T

19. Valid by the rule of detachment. Argument reads $(\sim s \rightarrow f) \wedge \sim s; \therefore f$.

21. Invalid as shown by the truth table in Example 7. Argument reads $(p \rightarrow o) \wedge o; \therefore p$.

23. Invalid as shown by the truth table in Exploratory Exercise 2(c). Argument reads $(p \rightarrow q) \wedge \sim p; \therefore \sim q$.

25. The questions place you on the defensive by implying that something is true that may or may not be true. The first question implies that you are beating your wife (which may or may not be true). The second question assumes that you will be buying a suit in which you look good.

27. In this fallacy we assume that because two things occur at the same time or in the same location, one caused the other. The presence of a black cat at the time of a catastrophe in no way suggests that the cat was connected to the catastrophe.

Exercise Set 2.1

1. (a) T   (b) F   (c) F   (d) F   (e) T   (f) F   (g) T   (h) T   (i) F   (j) F
3. {2, 4, 6, 8}
5. $E = \{-3, -2, -1, 0, 1, 2, 3\}$
7. (a) {x | x is a single-digit odd counting number}
    (b) {x | x is a living former U.S. president}
    (c) {x | x is a U.S. state capital named for a U.S. president}
    (d) {x | x is a square number and x < 50}
    (e) {x | x is an even counting number}
9. (a) {1, 3, 5, 7, 9}   (b) {−3, 3}   (c) {11, 69, 88, 96}
    (d) {Alabama, Alaska, Arizona, Arkansas}
11. The sets A, C, and D are equal.
13. (a) {a, b, c, d, e, f}   (b) {b, e}   (c) { }   (d) {e}   (e) {a, c, e}
    (f) {a, b, c, e}   (g) { }   (h) {a, c}   (i) {d, f}   (j) {a, b, c, e}
15. (a) {January, April, June, July, September, November} (b) {January}
    (c) {May, June, July, August} (d) {January, May, June, July, August}
    (e) {February, March, April, September, October, November, December}
17. (a) $\overline{B \cup C}$, $\overline{B} \cap \overline{C}$ is also correct.   (b) $A \cup C$
19.

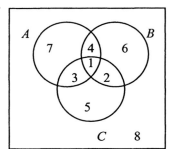

Exercise Set 2.2

1. (a) {(1, 2), (1, 3), (3, 2), (3, 3), (5, 2), (5, 3)}
    (b) {(1, 1), (1, 3), (1, 5), (3, 1), (3, 3), (3, 5), (5, 1), (5, 3), (5, 5)}
    (c) {(2, 2),(2, 3),(3, 2),(3, 3)}
3. (a) 20   (b) 130
5. (a) The pairs (a, b) satisfy the relation a "is a major league baseball team in" b.
    (b) The pairs (a, b) satisfy the relation a "is a state with capital" b.
    (c) The pairs (a, b) satisfy the relation a "is a pen name for" b.
7.

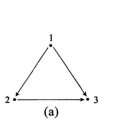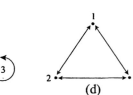

(a)   (b)   (c)   (d)

9.

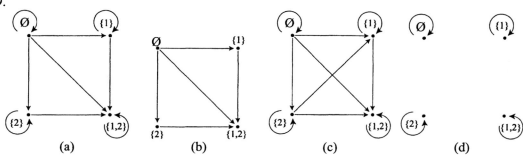

    (a)   (b)   (c)   (d)

11. (a) reflexive, symmetric, and transitive (b) reflexive only
    (c) reflexive and symmetric only
13. (a) {(1, 1), (1, 4), (2, 2), (2, 5), (3, 2), (3, 3), (4, 2), (4, 4), (5, 3), (5, 5)}
    (b) reflexive only
15. (a) {one}, {two, three, ten}, {four, five}, {six, seven}, {eight}, {nine}
    (b) {one, two, six, ten}, {four, five, nine}, {three, seven, eight}
    (c) {one, four, seven, ten}, {two, five, eight}, {three, six, nine}
17. (a) {(–1, 0), (0, –1), (0, 1), (1, 0)}
    (b) R is symmetric only.
19. (a) R is reflexive: If a is an integer, then $a - a = 0$, and 0 is a multiple of 4.
    R is symmetric: Suppose (a, b) is an element of R. Then $a - b$ is a multiple of 4, so
        $a - b = 4k$ for some integer k. Then $b - a = -4k = 4(-k)$, where $-k$ is an integer, so $b - a$ is a
        multiple of 4 and (b, a) is an element of R.
    R is transitive: Suppose (a, b) and (b, c) are elements of R. Then $a - b$ and $b - c$ are multiples of 4, so
        $(a - b) + (b - c) = a - c$ is also a multiple of 4 (see Theorem 7 from Section 3.1), and (a, c) is an
        element of R.
    (b) The equivalence classes are {..., –8, –4, 0, 4, 8, ...}, {..., –7, –3, 1, 5, 9, ...},
        {..., –6, –2, 2, 6, 10, ...}, and {..., –5, –1, 3, 7, 11, ...}.

Exercise Set 2.3

1. (a) {(3, 8), (5, 24), (2, 3), (1, 0)} (b) 15
   (c) The output is one less than the square of the input. (d) $f(x) = x^2 - 1$
3. (a) and (c). The others have two distinct outputs for the same input.
5. (a), (b), and (d)
7. (a) The domain is {1, 2, 3, 4, 5} for (a) and (b) and {1, 2, 4, 5} for (d).
   (b) Only (d) is one-to-one. (c) None of them are onto {1, 2, 3, 4, 5}.
9. (a) and (c)
11. (b) and (c)
13.

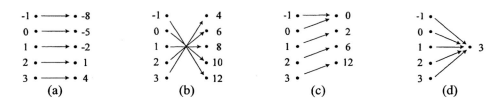

    (a)   (b)   (c)   (d)

15. (a) and (c)
17. (b)
19. (a) $50 + 15(x - 10) = 15x - 100$ cents if $x > 10$ (b) \$2; \$8   (c) 30 minutes

Exercise Set 2.4

1. (a) 5, 6, 7, 8, 9   (b) 2, 7, 12, 17, 22   (c) $1, \frac{1}{2}, \frac{1}{3}, \frac{1}{4}, \frac{1}{5}$

   (d) $\frac{1}{2}, \frac{1}{4}, \frac{1}{8}, \frac{1}{16}, \frac{1}{32}$ (e) 2, 5, 10, 17, 26   (f) 1, 3, 1, 3, 1

   (g) –2, 2, –2, 2, –2   (h) –2, –4, –8, –16, –32   (i) 10, 20, 40, 80, 160

   (j) 10, 9, 8, 7, 6

3. (a) Add 4; 23, 27, 31   (b) The counting numbers, alternating in sign; –6, 7, –8

   (c) Subtract 6; –22, –28, –34   (d) Multiply by 5; 6250, 31,250, 156,250

   (e) Multiply previous term by 1/3; 1/729, 1/2187, 1/6561

   (f) Each denominator is 3 more than the numerator; 6/9, 7/10, 8/11

   (g) Add 5, then subtract 1; 14, 13, 18   (h) multiply by –2; –352, 704, –1408

   (i) Add 2, add 3, add 4, etc.; 21, 28, 36

   (j) Multiply by 11; 161,051, 1,771,561, 19,487,171

5. (a) neither   (b) arithmetic, $d = -2$ (c) geometric, $r = -1$

   (d) geometric, $r = 5$ (e) neither

7. (a) arithmetic sequence: (d) $d = 7$ (b) geometric sequence: (e) $r = 2$

9. (a) $a_n = 10n - 3$ (b) $a_n = 9n - 13$ (c) $a_n = 6n - 4$ (d) $a_n = 23 - 3n$

11. The equation $y = ax + b$ on the calculator corresponds to the formula $a_n = a \cdot n + b$.

13. \$47,000

15. (a) $a_n = 3^{n-1}$ (b) $a_n = 32 \cdot \left(\frac{1}{2}\right)^{n-1}$ (c) $a_n = 2 \cdot (-20)^{n-1}$ (d) $a_n = 7 \cdot 2^{n-1}$

17. Yes; 3, 3, 3, 3, ..., for example, has a common difference of 0 and a common ratio of 1.

19. 2042

21. (a) the 28th square   (b) Approximately 92 quintillion dollars

Exercise Set 2.5

1. (a) 12, 10, 8, 6, 4   (b) 5, 10, 20, 40, 80   (c) 2, 5, 14, 41, 122

   (d) 3, 5, 8, 12, 17   (e) 4, –2, 1, –1/2, 1/4

3. (a) $a_n = a_{n-1} + 5$ (b) $a_n = \frac{1}{2} a_{n-1}$ (c) $a_n = a_{n-1} - 3$ (d) $a_n = -2 a_{n-1}$

5. (a) 1, 2, 6, 24, 120   (b) 1, 3, 10, 41, 206   (c) 1, 1, 2, 7, 34   (d) 1, 3, 8, 33, 164

   (e) 0, 1, 2, 9, 44

7. (a) $a_n = a_{n-1} + 5$ (b) $a_n = 3 a_{n-1}$ (c) $a_n = \left(a_{n-1}\right)^2$ (d) $a_n = a_{n-1} + (n-1)$

   (e) $a_n = 2 a_{n-1} - 1$

9. (a) 1, 2, 6, 24, 120   (b) 1, 2, 6, 24, 120   (c) They are the same.

11. (a) 1, 3, 5, 7, 9, 11, 13, 15, 17, 19   (b) 1, 1, 1, 2, 3, 4, 6, 9, 13, 19

13. (a) 1, 5, 7, 17, 31   (b) 1, 5, 7, 17, 31   (c) They are the same.

15. (a) 1   (b) 4   (c) 10   (d) 20   (e) 35

17. (a) It resembles the sequence in part (b).   (b) It resembles the sequence in part (c).

19. (a) 1, 2, 4, 7, 12, 20, 33, 54, 88, 143   (b) $f_1 + f_2 + \cdots + f_n = f_{n+2} - 1$

Exercise Set 3.1

1. (a) 9 is an integer with $27 = 3 \cdot 9$.   (d) -6 is an integer with $12 = -2 \cdot -6$.
   (b) 0 is an integer with $0 = 12 \cdot 0$.   (e) $6y$ is an integer with $6xy = x(6y)$.
   (c) 4 is an integer with $16 = 4 \cdot 4$.   (f) st is an integer with $rst = r(st)$.

3. (a) There is no integer k with $7 = 0 \cdot k$.   (b) -7 is an integer with $-42 = 6 \cdot -7$.
   (c) The prime numbers are numbers greater than 1.
   (d) The only positive divisors of 17 are 1 and 17.
   (e) The only divisors of 2 are 1 and itself. 2 is a divisor of all other even numbers.

5. (a) 28 is divisible by 7.   (b) 36 is not divisible by 5.

7. (a) $12 + 24 = 3 \cdot 4 + 3 \cdot 8 = 3 \cdot (4 + 8) = 3 \cdot 12$.
   (b) $25 + 15 = 5 \cdot 5 + 5 \cdot 3 = 5 \cdot (5 + 3) = 5 \cdot 8$.
   (c) Answers vary; for example, $7 \mid 14$ and $7 \mid 35$ and $7 \mid (14 + 35)$.
   (d) $b = ak$; $c = am$; $k + m$; integer   (e) Theorem 7

9. (a) $12 \cdot 5 = (4 \cdot 3) \cdot 5 = 4 \cdot (3 \cdot 5) = 4 \cdot 15$
   (b) $6 \cdot 11 = (3 \cdot 2) \cdot 11 = 3 \cdot (2 \cdot 11) = 3 \cdot 22$
   (c) Answers vary; for example, $4 \mid 8$ and $4 \mid 8 \cdot 5$.
   (d) $b = ak$; kn; $a \mid bn$ (e) Theorem 9

11.(a) Answers vary; for example, $3 \cdot 11 = 33$, which is odd.
   (b) If x and y are odd integers, then xy is an odd integer.
   (c) Since x is odd, there is an integer k with $x = 2k + 1$. Since y is odd, there is an
       integer j with $y = 2j + 1$. Then $xy = (2k + 1)(2j + 1)$. Using the properties of
       addition and multiplication of integers we can rewrite this as
       $xy = 4kj + 2k + 2j + 1 = 2(2kj + k + j) + 1$
       Since $2kj + k + j$ is an integer, xy is an odd integer.
   (d) Theorem 5

13.(a) $7(100) + 4(10) + 4$ (b) According to the Rule of 2, since $2 \mid 4$, $2 \mid 744$.
   (c) $744 = 2 \cdot 372$ (d) $10 = 2 \cdot 5$; Theorem 9; Theorem 8

15. z can be written in expanded notation as $z = a(100) + b(10) + c$.
    $4 \mid 100$ since $100 = 4(25)$. Therefore, $4 \mid [a(100)]$ by Theorem 9.
    By hypothesis, 4 divides the number formed by the rightmost two digits of z; that is,
    $4 \mid b(10) + c$. Since $4 \mid [a(100)]$ and $4 \mid [b(10) + c]$, then by Theorem 7, $4 \mid z$.

17. z can be written in expanded notation as $z = a(100) + b(10) + c$.
    $100 = 99 + 1$; $10 = 9 + 1$.
    Therefore $z = a(99 + 1) + b(9 + 1) + c = a(99) + b(9) + a \cdot 1 + b \cdot 1 + c = a(99) + b(9) + (a + b + c)$.
    $9 \mid 99$ since $99 = 9 \cdot 11$. Therefore, $9 \mid [a(99)]$ by Theorem 9.
    $9 \mid 9$ since $9 = 9 \cdot 1$. Therefore, $9 \mid [b(9)]$ by Theorem 9.
    $9 \mid (a + b + c)$ is the hypothesis of the "Rule for 9."
    It follows by Theorem 8 that 9 divides $a(99) + b(9) + (a + b + c)$ and hence $9 \mid z$.

19. (a) We would write the integers in expanded form as $a(1000) + b(100) + c(10) + d$ and then make the
        arguments looking at the divisibility properties of this decomposition.
    (b) If $a \mid b$, $a \mid c$, $a \mid d$, and $a \mid e$, then $a \mid (b + c + d + e)$.
    (c) Since $a \mid b$, $a \mid c$, $a \mid d$, and $a \mid e$, there are integers j, k, l, and m with $b = aj$, $c = ak$, $c = al$, and $e = am$. Thus $b + c + d + e = aj + ak + al + am = a(j + k + l + m)$. Since $j + k + l + m$ is an integer, $a \mid (b + c + d + e)$.
    (d) z can be written in expanded notation as $z = a(1000) + b(100) + c(10) + d$. $2 \mid 1000$ since $1000 = 2(500)$. Therefore, $2 \mid [a(1000)]$ by Theorem 9. $2 \mid 100$ since $100 = 2(50)$. Therefore, $2 \mid [b(100)]$ by Theorem 9. $2 \mid 10$ since $10 = 2(5)$. Therefore, $2 \mid [c(10)]$ by Theorem 9. "$2 \mid d$" is the hypothesis of the "Rule for 2." Since $2 \mid [a(1000)]$ and $2 \mid [b(100)]$ and $2 \mid [c(10)]$ and $2 \mid d$, then by (c) of this exercise, $2 \mid z$.

Exercise Set 3.2

1. (a) $\{v, w\}$ (b) Find in this section.   (c) v and w are the two elements in $A \cap B$, and each of these elements is also in A.

3. (a) $A \cup B = \{1, 2, 3, 5\}$. Since every element of A is an element of $A \cup B$, $A \subseteq A \cup B$.
   (b) See *Preparation* for Theorems 10 and 11.
   (c) The first sentence in our proof will be, "Let a be an arbitrary element of A." We will then try to show that a must be an element of $A \cup B$.
   (d) Let a be an arbitrary element of A. By the definition of $A \cup B$, a is also an element of $A \cup B$. Thus $A \subseteq A \cup B$.

5. Show that $A \cup (B \cap C) \subseteq (A \cup B) \cap (A \cup C)$.
   Let $a \in A \cup (B \cap C)$. Then, by definition of union, $a \in A$ or $a \in B \cap C$. If $a \in A$ then the definition of union ensures that a is an element of both $(A \cup B)$ and $(A \cup C)$. If $a \in B \cap C$, then by definition of intersection, $a \in B$ and $a \in C$. It follows that in either case, a is an element of both $(A \cup B)$ and $(A \cup C)$. Thus $a \in (A \cup B) \cap (A \cup C)$ and we have proved $A \cup (B \cap C) \subseteq (A \cup B) \cap (A \cup C)$. The proof can then be finished by showing that $(A \cup B) \cap (A \cup C) \subseteq A \cup (B \cap C)$ using a similar argument.

7. (a) Let $x = \dfrac{j}{k}$ and let $y = \dfrac{m}{n}$ where j, k, m, and n are integers and k and n are

   not equal to 0. Then $x + y = \dfrac{nj + km}{kn}$. Since $nj + km$ and kn are integers and

   kn is not equal to 0, $x + y$ is a rational number.

   (b) Let $x = \dfrac{j}{k}$ and let $y = \dfrac{m}{n}$ where j, k, m, and n are integers and k and n are

   not equal to 0. Then $xy = \dfrac{jm}{kn}$. Since jm and kn are integers and kn is not

   equal to 0, then xy is a rational number.

   (c) False. Let $x = \pi$ and let $y = -\pi$. Both π and $-\pi$ are irrational numbers, but $x + y = 0$ which is a rational number.

   (d) False. Let $x = \sqrt{2}$ and let $y = -\sqrt{2}$. Both $\sqrt{2}$ and $-\sqrt{2}$ are irrational numbers, but $xy = -2$, which is a rational number.

9. (a) $x - y = x + (-1)y$. Since $d \mid y$, then, by Theorem 9, $d \mid (-1)y$. It follows from Theorem 7 that if $d \mid x$ and $d \mid (-1)y$, then $d \mid (x + (-1)y)$ or $d \mid x - y$.
   (b) Since $a = bq + r$, then $r = a - bq$. Since $d \mid a$ and $d \mid b$, $d \mid bq$ by Theorem 9 and $d \mid r$ by the result in (a).

11. (a) 3, 6, 9, 12, 15, 18, 21, 24, 27, 30, 36; $3 \cdot 6 = 18, 3 \cdot 9 = 27, 3 \cdot 12 = 36, 9 \cdot 12 = 108$
    (b) If x and y are smooth, then xy is smooth.
    (c) Since x is smooth, there is an integer k such that $x = 3k$. Since y is also smooth, there is an integer j with $y = 3j$. Then $xy = (3k)(3j) = 9kj = 3(3kj)$. Since $3kj$ is an integer, xy is three times an integer, and hence xy is smooth.

13. (a) 5, 8, 11, 14, 17, 20, 23, 26, 29, 32, 35; $5 \cdot 8 = 40$, which is rough since $40 = 13(3) + 1$. $8 \cdot 11 = 88$ which is rough since $88 = 3(29) + 1$.
    (b) If x and y are abrasive, then xy is rough.
    (c) Since x is abrasive, there is an integer k such that $x = 3k + 2$. Since y is also abrasive, there is an integer j with $y = 3j + 2$. Then $xy = (3k + 2)(3j + 2) = 9kj + 6k + 6j + 4 = 9kj + 6k + 6j + 3 + 1 = 3(3kj + 2k + 2j + 1) + 1$. Since $3kj + 2k + 2j + 1$ is an integer, xy is rough.

15. Suppose $x = \sqrt{3}$ and x is rational. Then there are integers m and n ($n \neq 0$) with

$$x = \sqrt{3} = \frac{m}{n}$$

such that m and n have no common factors other than 1. Since $\sqrt{3} = \dfrac{m}{n}$, we can square both sides of the equation to learn that

$$3 = \frac{m^2}{n^2} \text{ or } 3n^2 = m^2.$$

Since m^2 is the product of 3 and the integer n^2, m^2 is smooth. By Exercise 14 this means that m is smooth, so there is an integer k with $m = 3k$. Thus $m^2 = (3k)^2 = 9k^2$. Hence $3n^2 = 9k^2$, and dividing by 3 reveals that $n^2 = 3k^2$. Using Exercise 14 once again indicates that since n^2 is smooth, n is smooth. Now we have shown that both m and n are smooth, that is, 3 is a divisor or factor of both m and n. This contradicts our assumption that m and n share no common positive divisors other than 1.

17. Since n^2 is even, by Theorem 12, n is even. Thus there is an integer k with $n = 2k$. It follows that $n^2 = (2k)(2k) = 4k^2$ and n^2 is divisible by 4.

Exercise Set 3.3

1. *Basis Step:* $3 = \dfrac{3(1)(1+1)}{2}$.

   Inductive Step: We wish to prove the statement "If $P(k)$ and $k \geq 1$, then $P(k + 1)$"

   Our hypothesis, then, is the statement $P(k)$: $3 + 6 + 9 + \ldots + 3k = \dfrac{3k(k+1)}{2}$

   Add $3(k + 1)$ to both sides of the equation given in the inductive hypothesis, and then perform algebraic manipulation on the right hand side of the resulting equation with these results.

   $$3 + 6 + 9 + \ldots + 3k + 3(k + 1) = \frac{3k(k+1)}{2} + 3(k + 1)$$

   $$3 + 6 + 9 + \ldots + 3k + 3(k + 1) = \frac{3k(k+1)}{2} + \frac{6(k+1)}{2}$$

   $$3 + 6 + 9 + \ldots + 3k + 3(k + 1) = \frac{(k+1)(3k+6)}{2} = \frac{3(k+1)(k+2)}{2}$$

   Since this last statement is $P(k + 1)$, we have shown that when $P(k)$ is true, $P(k + 1)$ is true.

3. *Basis Step:* $1^3 = \left[\dfrac{1(1+1)}{2}\right]^2$.

   Inductive Step: We wish to prove the statement "If $P(k)$ and $k \geq 1$, then $P(k + 1)$"

   Our hypothesis, then, is the statement $P(k)$: $1^3 + 2^3 + 3^3 + \cdots + k^3 = \left[\dfrac{k(k+1)}{2}\right]^2$

   Add $(k + 1)^3$ to both sides of the equation given in the inductive hypothesis and then perform algebraic manipulation on the right-hand side of the resulting equation with these results:

   $$1^3 + 2^3 + 3^3 + \cdots + k^3 + (k+1)^3 = \left[\frac{k(k+1)}{2}\right]^2 + (k+1)^3$$

$$1^3 + 2^3 + 3^3 + \cdots + k^3 + (k+1)^3 = \frac{k^2(k+1)^2}{4} + \frac{4(k+1)^3}{4}$$

$$1^3 + 2^3 + 3^3 + \cdots + k^3 + (k+1)^3 = \frac{(k+1)^2(k^2 + 4(k+1))}{4}$$

$$1^3 + 2^3 + 3^3 + \cdots + k^3 + (k+1)^3 = \frac{(k+1)^2(k^2 + 4k + 4)}{4} = \frac{(k+1)^2(k+2)^2}{4}$$

$$1^3 + 2^3 + 3^3 + \cdots + k^3 + (k+1)^3 = \left[\frac{(k+1)(k+2)}{2}\right]^2$$

Since this last statement is $P(k+1)$, we have shown that when $P(k)$ is true, $P(k+1)$ is true.

5. *Basis Step:* $1 < 2^1 = 2$.

   Inductive Step: We wish to prove the statement "If $P(k)$ and $k \geq 1$, then $P(k+1)$."
   $P(k)$ is the statement, $k < 2^k$. Multiplying both sides of this statement by 2 yields
   $2k < 2 \cdot 2^k = 2^{k+1}$
   Since $k \geq 1$, $k + 1 \leq k + k = 2k$ so $k + 1 \leq 2k < 2^{k+1}$. Since $k + 1 < 2^{k+1}$ is $P(k+1)$, we are finished.

7. *Basis Step:* Since $3^1 - 1 = 2$, $2 \mid (3^1 - 1)$.

   Inductive Step: Assume that k is a natural number and 2 divides $3^k - 1$. Show this
   assures that 2 divides $3^{k+1} - 1$. We observe that

$$\begin{aligned}
3^{k+1} - 1 &= 3 \cdot 3^k - 1 \\
&= (2+1) \cdot 3^k - 1 \\
&= 2 \cdot 3^k + 3^k - 1 \\
&= 2 \cdot 3^k + (3^k - 1)
\end{aligned}$$

Now we have the connection with $P(k)$ that we needed. By $P(k)$, the inductive hypothesis, 2 divides $3^k - 1$. Theorem 7 ensures that 2 divides $3^{k+1} - 1$.

9. (a) $\dfrac{1}{1 \cdot 2} = \dfrac{1}{2}$; $\dfrac{1}{1 \cdot 2} + \dfrac{1}{2 \cdot 3} = \dfrac{2}{3}$; $\dfrac{1}{1 \cdot 2} + \dfrac{1}{2 \cdot 3} + \dfrac{1}{3 \cdot 4} = \dfrac{3}{4}$

   Conjecture: $\dfrac{1}{1 \cdot 2} + \dfrac{1}{2 \cdot 3} + \dfrac{1}{3 \cdot 4} + \ldots + \dfrac{1}{n(n+1)} = \dfrac{n}{n+1}$

   (b) *Basis Step:* $\dfrac{1}{1 \cdot 2} = \dfrac{1}{2}$.

   Inductive Step: Assume that k is a natural number and that
   $\dfrac{1}{1 \cdot 2} + \dfrac{1}{2 \cdot 3} + \dfrac{1}{3 \cdot 4} + \ldots + \dfrac{1}{k(k+1)} = \dfrac{k}{k+1}$. Adding $\dfrac{1}{(k+1)(k+2)}$ to both sides of this
   equation yields

$$\begin{aligned}
\frac{1}{1 \cdot 2} + \frac{1}{2 \cdot 3} + \frac{1}{3 \cdot 4} + \ldots + \frac{1}{k(k+1)} + \frac{1}{(k+1)(k+2)} &= \frac{k}{k+1} + \frac{1}{(k+1)(k+2)} \\
&= \frac{k(k+2)}{(k+1)(k+2)} + \frac{1}{(k+1)(k+2)} \\
&= \frac{k^2 + 2k + 1}{(k+1)(k+2)} \\
&= \frac{(k+1)^2}{(k+1)(k+2)} = \frac{(k+1)}{(k+2)}
\end{aligned}$$

Since this last statement is $P(k+1)$, we have shown that when $P(k)$ is true, $P(k+1)$ is true.

11. *Basis Step:* If $a \mid b_1$, then a certainly divides b_1.

Inductive Step: Assume that k is a natural number and that if $a \mid b_1, a \mid b_2, a \mid b_3, \ldots,$ $a \mid b_k$, then $a \mid (b_1 + b_2 + \ldots + b_k)$.

If $a \mid b_1, a \mid b_2, a \mid b_3, \ldots, a \mid b_k, a \mid b_{k+1}$, then the inductive hypothesis asserts that $a \mid$ $(b_1 + b_2 + \ldots + b_k)$. Note that $b_1 + b_2 + \ldots + b_k + b_{k+1} = (b_1 + b_2 + \ldots + b_k) + b_{k+1}$. Using the fact that $a \mid b_{k+1}$ and Theorem 7, we see that $a \mid [(b_1 + b_2 + \ldots + b_k) + b_{k+1}]$, or $a \mid b_1 + b_2 + \ldots + b_k + b_{k+1}$.

13. *Basis Step:* When $n = 1$, the number consists of a single digit a. The number represented by the rightmost two digits of a single-digit number is that number. If $4 \mid a$, then certainly $4 \mid a$.

Inductive Step: Suppose k is a positive integer and the "Rule for 4" is true for numbers with k-digits. Consider any positive integer z with $k + 1$ digits, and suppose that 4 divides the number formed by the rightmost two digits of z. Observe that z can be written in expanded form as $z = a(10)^k + x$, where x is an integer with k digits. If $k = 1$, then z is a number with two digits, so $4 \mid z$. If $k > 1$, then the number formed by the right most two digits of z is the same as the number formed by the right-most two digits of x. By the inductive hypothesis, $4 \mid x$. Further, since $k > 1$, $4 \mid (10)^k$. Using Theorems 7 and 9, it follows that $4 \mid z$.

15. If in our arrangement of dominos we make sure that (a) the rth domino must fall and (b) for each $k \geq r$, when the kth domino falls, the $(k + 1)^{\text{st}}$ domino must fall, we can be sure that all the dominos from the rth domino onward will fall.

17. *Basis Step:* $5^2 = 25 < 32 = 2^5$.

Inductive Step: Suppose that k is an integer ≥ 5 and that $k^2 < 2^k$. We need to show that $(k + 1)^2 < 2^{k+1}$. Let us focus on the left-hand side of this desired inequality. $(k + 1)^2 = k^2 + 2k + 1$. We know something about k^2 so let us focus attention on $2k + 1$. Since $k \geq 5$, $2k + 1 \leq 2k + k \leq 3k < 5k$. Since $k \geq 5$, $k^2 = k \cdot k \geq 5k$. It follows that $2k + 1 < k^2$. Thus, putting all this together with the inductive hypothesis, we have
$$(k + 1)^2 = k^2 + 2k + 1 < k^2 + k^2 = 2k^2 < 2 \cdot 2^k = 2^{k+1}$$

19. *Basis Step:* We can produce 12 cents in postage using three 4-cent stamps.

Inductive Step: Let $k \geq 12$ be an amount of postage that we can form using 4-cent and 5-cent stamps.

Case 1: Suppose that in forming k cents of postage, we use at least one 4-cent stamp. Then if we replace this 4-cent stamp by a 5-cent stamp, we will form $k + 1$ cents in postage.

Case 2: Suppose that in forming k cents of postage, we use no 4-cent stamps. Since $k \geq 12$, we must have used at least three 5-cent stamps. Replace three 5-cent stamps by four 4-cent stamps to form postage for $k + 1$ cents.

Exercise Set 3.4

1. (a) (1) 1, 2, 3, 6; 1, 3, 5, 15 (2) 1, 3 (3) 3
   (b) (1) 1, 2, 3, 4, 6, 12; 1, 2, 3, 6, 9, 18, 27, 54 (2) 1, 2, 3, 6 (3) 6
   (c) (1) 1, 3, 7, 21; 1, 3, 7, 9, 21, 63 (2) 1, 3, 7, 21 (3) 21
   (d) (1) 1, 2, 4, 7, 14, 28; 1, 2, 3, 6, 9, 18, 27, 54 (2) 1, 2 (3) 2
   (e) (1) 1, 17; 1, 2, 3, 4, 6, 9, 12, 18, 36 (2) 1 (3) 1

3. (a) 2 (b) $\gcd(62, 14) = \gcd(14, 6) = \gcd(6, 2) = \gcd(2, 0) = 2$

5. (a) $15 = 2(6) + 3; 6 = 2(3) + 0, \gcd(6, 15) = 3$
   (b) $54 = 4(12) + 6; 12 = 2(6) + 0, \gcd(12, 54) = 6$
   (c) $63 = 3(21) + 0, \gcd(21, 63) = 21$
   (d) $54 = 1(28) + 26; 28 = 1(26) + 2; 26 = 13(2) + 0; \gcd(28, 54) = 2$
   (e) $36 = 2(17) + 2; 17 = 8(2) + 1; 2 = 2(1) + 0; \gcd(17, 36) = 1$

7. $2 = 9(14) + (-2)(62)$

9. (a) $3 = (1)(15) + (-2)(6)$ (b) $6 = (1)(54) + (-4)(12)$ (c) $21 = (0)(63) + (1)(21)$
   (d) $2 = (2)(28) + (-1)(54)$ (e) $1 = (17)(17) + (-8)(36)$

11.(a) Multiples of 8: 8, 16, 24, 32, 40, ...; Multiples of 12: 12, 24, 36, 48, ...; lcm(8, 12) = 24

   (b) 30   (c) 54   (d) 336

13.(a) 30   (b) 108   (c) 63

   (d) 756   (e) 612

15.Any integer k is a divisor of 0 since $0 = 0 \cdot k$. A is the greatest divisor of A and is also a divisor of 0. Thus $\gcd(A, 0) = A$.

17.(a) gcd(A, B) divides 9.

   (b) Let $d = \gcd(A, B)$. Then $d \mid A$ and $d \mid B$ so $d \mid MA$, $d \mid NB$, and $d \mid MA + NB$ by Theorems 7 and 9. Thus $d \mid 9$.

19.(a) $N = pk$.

   (b) $q \mid pk$, so $q \mid p$ or $q \mid k$.

   (c) Since p and q are distinct primes, q does not divide p. Therefore, $q \mid k$ and hence there is an integer m with $k = qm$. Thus $N = pk = pqm$. $pq \mid N$.

21.Let $d = \gcd(A, B)$. By Theorem 20, there are integers M and N with $d = MA + NB$.
   Since $D \mid A$ and $D \mid B$, then $D \mid MA + NB$ by Theorems 7 and 9. Thus $D \mid d$.

Exercise Set 3.5

1. (a)   (b)

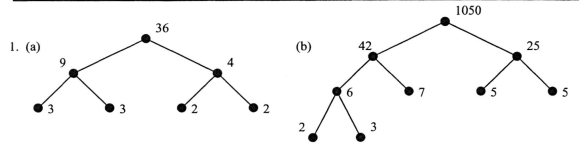

3. 2 3 5 7 11 13 17 19 23 29 31 37 41 43 47 53 59 61 67 71 73 79 83 89 97 101 103 107 109 113 127 131 137 139 149 151 157 163 167 173 179 181 191 193 197 199

5. (a) 34   (b) 74   (c) 114 (d) 153

7. (a) If a divisor is greater than 1, the prime factors of the divisor are either 2's and/or 3's, and there can be at most two 2's and three 3's as factors.

   (c) $1, 2^1, 2^2, 3^1, 3^2, 3^3, 2^13^1, 2^13^2, 2^13^3, 2^23^1, 2^23^2, 2^23^3$

9. (a) $1, 3^1, 3^2, 3^3, 3^4, 7^1, 7^2, 3^17^1, 3^17^2, 3^27^1, 3^27^2, 3^37^1, 3^37^2, 3^47^1, 3^47^2$

   (b) $1, 2^1, 5^1, 5^2, 13^1, 13^2, 2^15^1, 2^15^2, 2^113^1, 2^113^2, 5^113^1, 5^113^2, 5^213^1, 5^213^2, 2^15^113^1, 2^15^113^2, 2^15^213^1,$
        $2^15^213^2$

11. 5^37^2. Let d be the greatest common divisor. Since d divides 5^47^2, the prime factors of d must be 5's and 7's. Further, there are no more than four 5's and two 7's. Similarly, since d divides $2^3 5^3 7^3$, the factors must be 2's, 5's, and 7's, and there can be no more than three 2's, three 5's, and three 7's. Thus the greatest common divisor will have three factors of 5 and two factors of 7.

13. (a) Divisors of 96: 1, 2, 3, 4, 6, 8, 12, 16, 24, 32, 48, 96
       Divisors of 144: 1, 2, 3, 4, 6, 8, 9, 12, 16, 18, 24, 36, 48, 72, 144
       gcd(96, 144) = 48.

   (b) $144 = 1(96) + 48$; $96 = 2(48) + 0$; gcd(96, 144) = the last non-zero remainder = 48.

   (c) $96 = 2^53$, $144 = 2^43^2$, gcd(96, 144) = $2^43 = 48$

15. (a) 3 is the only prime that is one less than a square.

   (b) $n^2 - 1 = (n - 1)(n + 1)$. Thus $3 = 2^2 - 1 = (2 - 1)(2 + 1)$ is a prime. However, for larger values of n, $(n - 1)(n + 1)$ factors $n^2 - 1$ into factors other than 1 and itself.

17. Let d be a positive divisor of 3^{50} that is bigger than 1. Let $p_1 \cdot p_2 \cdot ... \cdot p_r$ be the prime factorization of d. Since d divides 3^{50}, there is a positive integer k with $3^{50} = d \cdot k$ or $3^{50} = p_1 \cdot p_2 \cdot ... \cdot p_r \cdot$ (prime factorization of k). Since the prime factorization of 3^{50} is unique by Theorem 23, each of the p's must be a 3, and thus d is a power of 3.

19. (a) 6, 10, 14, 15, 21, 22, 26

(b) n has exactly four positive factors if and only if n is a product of two distinct primes.

(c) *Show:* If n has exactly four positive factors, then n is a product of two distinct primes.

Since 1 and n are both factors of n, the prime factorization of n consists of at most two factors, $n = pq$. If $p = q$, then n has only three positive factors (See Exercise 18.) Thus p and q must be distinct.

Show: If n is a product of two distinct primes, then n has exactly four factors.

Suppose $n = pq$, where p and q are primes with p not equal to q. Clearly 1, p, q, and n are four factors of n. Suppose a is some other factor of n. Then there is a positive integer k with $n = ak$. Thus $pq = ak = $ (prime factorization of a)(prime factorization of k). Theorem 23 asserts that both a and k must be primes and that one must be p and one must be q. Thus there are only four positive divisors of n.

Exercise Set 4.1

1. Answer is not unique.

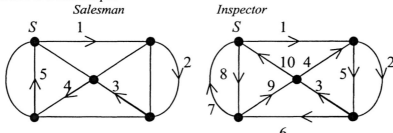

Salesman   *Inspector*

3.

Map	Number of Towns of Odd Degree	Number of Towns of Even Degree	Route for Inspector (Yes or No)
(a)	4	3	No
(b)	2	4	Yes
(c)	2	4	Yes
(d)	8	4	No
(e)	4	1	No
(f)	0	8	Yes

5. Since all destinations have odd degree, you cannot plan a walk that will take you over every bridge exactly once.

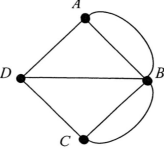

7. (a)

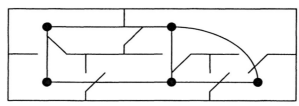

(b) In the road map in (a), each door is represented by a road. Since every door must be passed through, this is equivalent to passing over every road.

(c) Yes. By Theorem 5, since only two of the rooms have odd degree, a path through the house can be found.

(d)

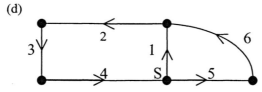

9. Yes, a salesman can visit each of the towns in this road system exactly once.

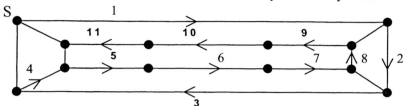

11. Answers will vary.

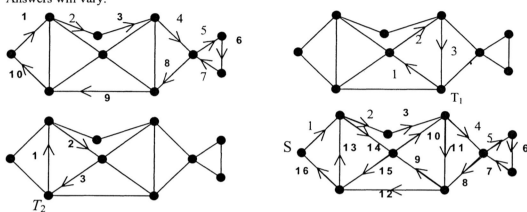

13. When the Salesman starts at a town, he can travel to either of two towns. But at each successive town there is a single road that will carry him to a town other than the one he just left. If this one choice takes him back to a town he has already visited, before he gets to all towns, then the road system is not connected. Any town not yet visited would have no roads to any of the towns on the trip he just completed.

15. By Theorem 1, we must begin and end our trip in a town of odd degree. With four towns, we can only begin in one, and end in one, leaving two towns with sections of road that have not been inspected.

Exercise Set 4.2

1. $V = \{ P, Q \}$ $E = \{\{P, Q\}\}$

3. $V = \{ w, x, y, z \}$ $E = \{\{w, x\}, \{w, y\}, \{w, z\}, \{x, y\}, \{x, z\}, \{y, z\}\}$

5. (a)   (b)

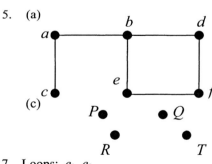

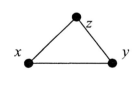

(c)

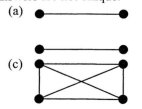

7. Loops: e_1, e_3 Parallel edges: e_5, e_6

9. Answers are not unique.

   (a)   (b)

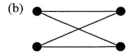

   (c)

11. Since there are exactly two vertices of odd degree, there exists an Euler path.

13. There does not exist a Euler path or circuit because there are four vertices of odd degree.

15. (a, c, e, d, b, a)

17. Suppose there is a Hamilton path or circuit. Any such path would have to both start and end at one of the two vertices of degree 1, else a vertex adjacent to a vertex of degree one would be visited twice. Assume that we have a Hamilton path that starts at one such vertex and ends at the other. As that path moves from vertices b and f to vertices c, d, and e, it necessarily visits vertices b and f a total of three times. Thus one of b and f is visited at least twice. Thus there is no Hamilton path.

19. Label the vertices in the following order:

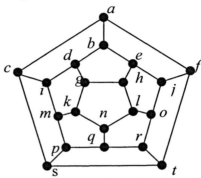

A Hamilton circuit is $(a, f, t, s, c, i, m, p, q, r, o, j, e, h, l, n, k, g, d, b, a)$.

Exercise Set 4.3

1. (a) 10   (b) 3   (c) 6

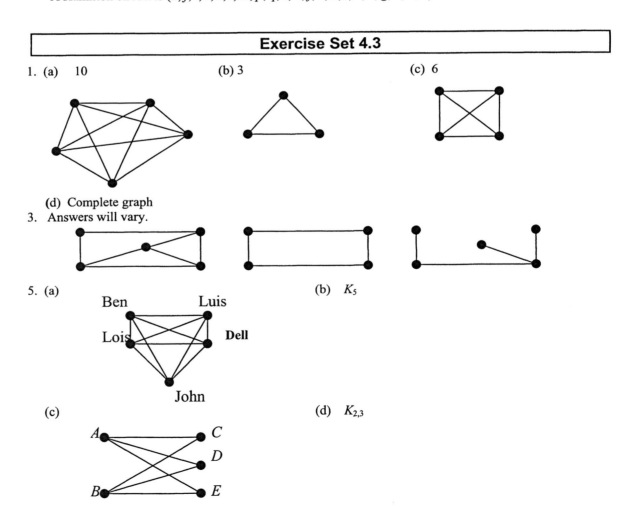

(d) Complete graph

3. Answers will vary.

5. (a)

Ben   Luis

Lois   Dell

John

(b) K_5

(c)

(d) $K_{2,3}$

7. (a)   No, they may not have the same number of edges.
    (b)   No, they may not have the same number of vertices.
9. In a complete graph on n vertices, all vertices have degree $n-1$. Therefore, all complete graphs with an odd number of vertices have an Euler circuit.
11. $a \to z$, $b \to x$, $c \to y$, $d \to w$
13. Vertex a has degree 1, and no vertex in the second graph has degree 1.
15. Vertex u has degree 2, and no vertex in the first graph has degree 2.
17.

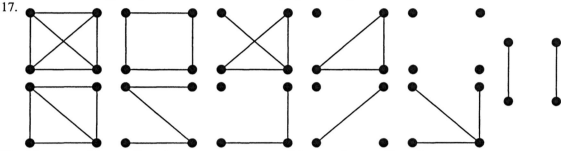

19. (a) a and b are adjacent.   (b) a and c are adjacent.
    (c) b and c are adjacent.
    (d)   G is not bipartite, because it is impossible to place a, b, and c in two disjoint sets.
21. (a)   No   (b)   Yes   (c)   No   (d)   No
    (e)   For all n except 2, there will be more than two vertices with odd degree. Thus, by Theorem 9, there will be no Euler path.
23. If m is odd, then for all n except 2, the number of vertices with odd degree will be some integer other than 2. By Theorem 9, there will be no Euler path.
25. (a)   Because $\{a_1, a_2, \ldots, a_k\}$ are all adjacent to u.
    (b)   Because the ismorphism function is one-to-one.
    (c)   Then $f(u)$ has the same number of neighbors as u, and hence the same degree.
    (d)   Because the isomorphism function is onto.
    (e)   By definition of isomorphism, $f(a)$ is adjacent to $f(u)$ if and only if a is adjacent to u.
    (f) $\{a_1, a_2, \ldots, a_k\}$ was defined to be the set of neighbors of u.

Exercise Set 4.4

1.

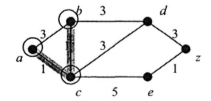

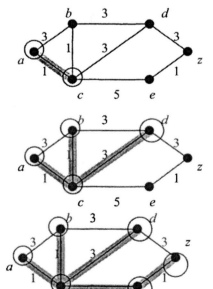

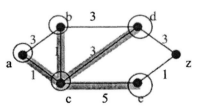

Thus a shortest path is $\{a, c, e, z\}$ with length 7.

3.

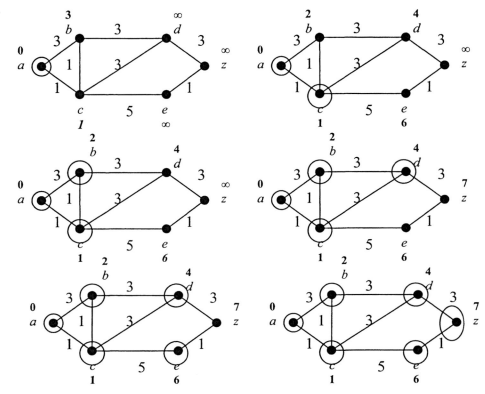

Thus a shortest path is (a, c, d, z) with length 7.

5. Length 530

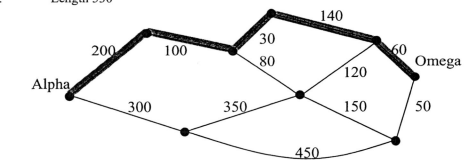

7. (a) and (b)   (c) The shortest path is $\{A, D, C, E, Z\}$ and has length 6.

Path	Length
$\{A, B, Z\}$	7
$\{A, B, C, E, Z\}$	8
$\{A, B, C, D, E, Z\}$	11
$\{A, D, E, Z\}$	7
$\{A, D, C, E, Z\}$	6
$\{A, D, C, B, Z\}$	9
$\{A, D, E, C, B, Z\}$	14

9.

Third Iteration		
Vertex	$L(u)$	$P(u)$
A	0	
B	4	C
C	2	A
D	6	C
E	8	C
F	10	D

Fourth Iteration		
Vertex	$L(u)$	$P(u)$
A	0	
B	4	C
C	2	A
D	6	C
E	8	C
F	10	D

Fifth Iteration		
Vertex	$L(u)$	$P(u)$
A	0	
B	4	C
C	2	A
D	6	C
E	8	C
F	10	D

The shortest path is (A, C, D, F) and has length 10.

11.

	α	β	ω	γ	κ	σ
α	-	120	250	200	210	50
β	120	-	320	220	90	170
ω	250	320	-	100	230	250
γ	200	220	100	-	130	150
κ	210	90	230	130	-	260
σ	50	170	250	150	260	-

13.   (a)

Vertex	Length of Shortest Path from a
a	0
b	2
c	6
d	5
e	7
z	10

(b)

Vertex	Length of Shortest Path from a
a	0
b	4
c	3
d	6
e	7
f	11
g	12
z	16

Exercise Set 4.5

1.

Graph	Vertices (v)	Edges (e)	Regions (r)	$v - e + r$
G_1	4	5	3	2
G_2	4	3	1	2
G_3	4	8	6	2
G_4	4	6	4	2

3.

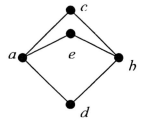

5.

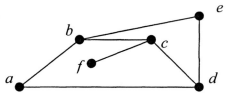

7. 6   **9.** 12

11. (a) (a, b, c, a) (b) 3   (c) (a, b, c, d, a) (d) 4

13.

Polyhedron	Vertices (v)	Edges (e)	Faces (f)	$v - e + f$
P_1	8	12	6	2
P_2	4	6	4	2
P_3	12	18	8	2
P_4	9	14	7	2

15. Edge $\{a, f\}$ was removed; vertex x and edges $\{a, x\}$ and $\{x, f\}$ were added. Edge $\{c, d\}$ was removed; vertex y and edges $\{c, y\}$ and $\{y, d\}$ were added.

17.

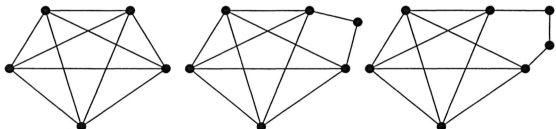

19. Each graph is formed by elementary subdivisions on $K_{3,3}$ or K_5, or it contains a subgraph isomorphic to such a graph.

Exercise Set 4.6

1.

5.

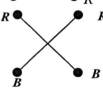

3.

7. (1) 3 (2) 2 (3) 3 (4) 4 (5) 2 (6) 3

9. (a)

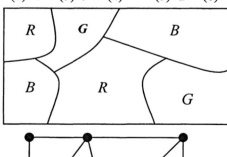

(b)

(c)

11. (a) 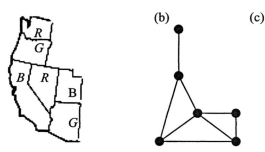   (b) 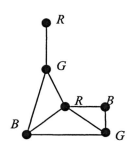   (c)

13. Three displays

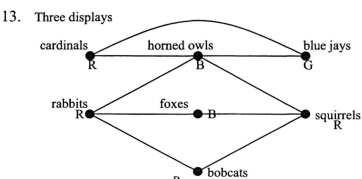

15. Three freight cars

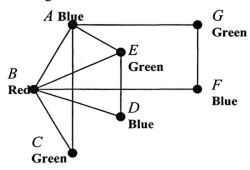

17. (a) 3   (b) 4   (c) 5   (d) 6   (e) K_5 and K_6 are not planar graphs.
19. (a) Because no pair of vertices in V_1 is adjacent, nor is any pair of vertices in V_2 adjacent.
    (b) Since the graph is connected and has more than one vertex, each vertex of the graph must be adjacent to at least one other vertex. Thus, we need at least two colors to color the graph.
    (c) In (a) we showed that two colors will color the graph. In (b) we showed that one color is not sufficient. Thus, the chromatic number is 2.

Exercise Set 5.1

1. $\{(p, q), (p, r), (p, s), (q, r), (r, q), (r, s)\}$; $\{p, q, r, s\}$
3. (a)   (b)

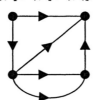 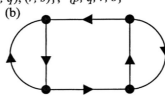

5.

Vertex	Indegree	Outdegree
p	2	2
q	2	2
r	3	1
s	2	1
t	1	4
u	1	1

7. (a) (O, H, I, J, G)
    (b) (G, K, L, O, A, B, D, C)
    (c) (G, K, L, O, H, I, J, G)
    (d) No, there are more than two vertices in which the indegree does not equal the outdegree.
9. No Euler path or circuit

11.

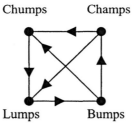

13. (a)   (b)   All teams may claim to be the winner.

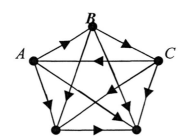

15. $(A, B, C, D), (A, D, B, C), (A, C, D, B)$

17. (a)   (b) (A, B, C, E, D)
                                                         (B, C, A, E, D)
                                                         (C, A, B, E, D)
                                              Note: If the edge connecting E and D is directed in the opposite direction, the paths would read (A, B, C, D, E) and so on.
                          (c) A, B, and C are winners.

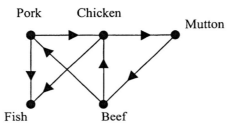

19.   Each team in S is a winner.

Exercise Set 5.2

1.   Tree
3.   Not a tree, this graph is not connected.
5.   Tree
7.   This confirms Theorem 6.
9.   (a)   Vertex Y
       (b)   Vertex W
       (c)   Vertex R
       (d)   Vertex W
       (e)   Vertices W, R, and Y
       (f)   Vertices Q and X

(g)
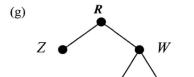

(b)
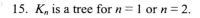

11.   (a)
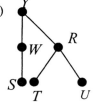

13.   (a)   14   (b)   10   15. K_n is a tree for $n = 1$ or $n = 2$.
17.
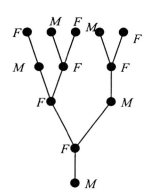

19.   (a)   The simple circuit C is a cycle in the graph.
       (b) $\hat{C}$ is a cycle in the graph.
       (c)   There are only a finite number of vertices in C.
       (d)   There are no repeated vertices in the simple circuit produced by the process other than the initial and terminal vertices.

Exercise Set 5.3

1.　Answers are not unique.

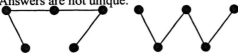 

3.　Answers are not unique.

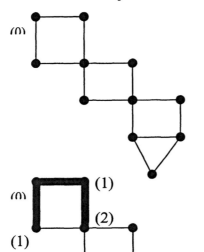

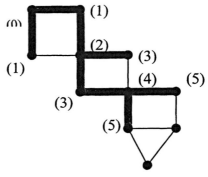

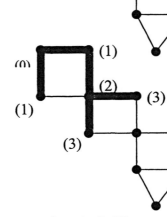

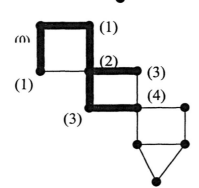

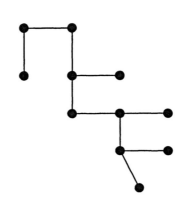

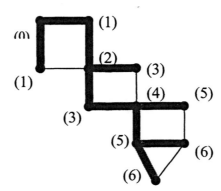

5. A spanning tree can be used to solve this problem. If the edges are labeled with weights representing the cost to upgrade the computer link, then a minimum spanning tree will minimize the cost to upgrade.

7. (a)   (b)   (c)

(d) The spanning tree resulting from a breadth-first search of K_n is $K_{1,n-1}$.

9. Answer is not unique.   11.

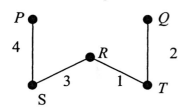

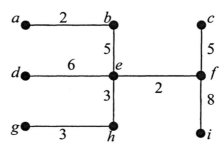

13. A minimum spanning tree would solve the problem. The minimum number of kilometers that need to be paved would be 197 kilometers.

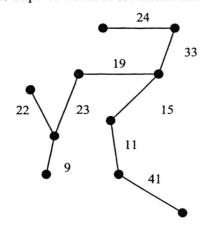

15. (a) u and v are vertices in G, and G is connected.
(b) C is a path from u to v in G'.
(c) Modify C by replacing (v_r, v_{r+1}) by $(v_r, v_{r-1}, \ldots, v_2, v_1, v_k, v_{k-1}, \ldots, v_{r+1})$ in C.
(d) The new path created in (c) is a path from u to v in G'. Thus G' is connected.

Exercise Set 5.4

1. (a) The left subtree of B consists of the vertices D and F and the connecting edge.
(b) The right child of E is the vertex G.
(c) F does not have a left child.
(d) The right subtree of D is the vertex F.

3.   (a)

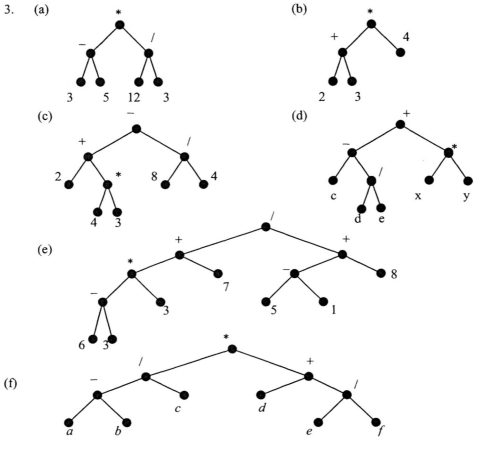

5. (a) *M, O, N, Q, P*   (b) *O, Q, P, N, M*
7. (a) *W, T, X, R, Y, U, Z, Q, V, S*   (b) *W, X, T, Y, Z, U, R, V, S, Q*
9. (a) 3 5 − 12 3 / *   (b) 2 3 + 4 *
   (c) 2 4 3 * + 8 4 / −   (d) *c d e* / − *x y* * +
   (e) 6 3 − 3 * 7 + 5 1 − 8 + /   (f) *a b − c* / *d e f* / + *

11. (a)

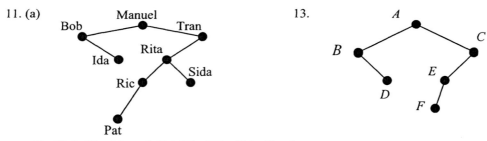

      (b) {Bob, Ida, Manuel, Pat, Ric, Rita, Sida, Tran}
15. (a) *M, N, P, O, Q, R*   (b) *A, B, D, H, E, I, C, F, J, G, K, L*
17. (a) 23   (b) 28

Exercise Set 6.1

1. (a) 6   (b)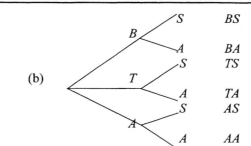

3. (a) 24   (b) 576
5. (a) 900   (b) 9000   (c) 90,000   (d) $9 \cdot 10^{n-1}$
7. 48
9. (a) 3! = 6   (b) *PQR, PRQ, QPR, QRP, RPQ, RQP*
11. (a) 2! = 2   (b) 3! = 6   (c) 4! = 24
13. 18
15. (a) *r*   (b) *k*(*k* − 1)   (c) *r*!   (d) *k*!/2   (e) *k*(*k* − 1)(*k* − 2)   (f) *k*!/6
17. (a) 10,000   (b) *P*(10,4) = 5040
19. (a) 5! = 120   (b) 3(5!) = 360

Exercise Set 6.2

1. (a) 336   (b) 56   (c) 1680   (d) 70
3. (a) 1   (b) *n*   (c) *n*   (d) 1
5. (a) 210   (b) 120   (c) 56   (d) 1013
7. 12,271,512
9. (a) 125,970   (b) 80,388
11. (a) 3060   (b) 1632   (c) 153   (d) 4845

Exercise Set 6.3

1. (a) 28   (b) 56   (c) 70   (d) 56   (e) 256
3. 210
5. (a) 1001   (b) 2002   (c) 3003
7. (a) 1001   (b) 2002   (c) 3003
9. (a) $x^6 + 6x^5y + 15x^4y^2 + 20x^3y^3 + 15x^2y^4 + 6xy^5 + y^6$

    (b) $x^6 - 6x^5y + 15x^4y^2 - 20x^3y^3 + 15x^2y^4 - 6xy^5 + y^6$

    (c) $x^7 + 7x^6y + 21x^5y^2 + 35x^4y^3 + 35x^3y^4 + 21x^2y^5 + 7xy^6 + y^7$

    (d) $x^7 - 7x^6y + 21x^5y^2 - 35x^4y^3 + 35x^3y^4 - 21x^2y^5 + 7xy^6 - y^7$

11. (a) $32a^5 + 80a^4b + 80a^3b^2 + 40a^2b^3 + 10ab^4 + b^5$

    (b) $a^5 - 15a^4b + 90a^3b^2 - 270a^2b^3 + 405ab^4 - 243b^5$

    (c) $x^5 - 10x^4 + 40x^3 - 80x^2 + 80x - 32$

    (d) $243s^5 - 810s^4t + 1080s^3t^2 - 720s^2t^3 + 240st^4 - 32t^5$

13. 15,736

15. $C(n,r)+C(n,r+1) = \dfrac{n!}{(n-r)!\,r!} + \dfrac{n!}{(n-r-1)!\,(r+1)!}$

$\quad = \dfrac{n!}{(n-r)!\,r!} \cdot \dfrac{r+1}{r+1} + \dfrac{n!}{(n-r-1)!\,(r+1)!} \cdot \dfrac{n-r}{n-r}$

$\quad = \dfrac{n!\,(r+1)}{(n-r)!\,(r+1)!} + \dfrac{n!\,(n-r)}{(n-r)!\,(r+1)!}$

$\quad = \dfrac{n!\,(r+1+n-r)}{(n-r)!\,(r+1)!}$

$\quad = \dfrac{n!\,(n+1)}{(n-r)!\,(r+1)!}$

$\quad = \dfrac{(n+1)!}{(n+1-(r+1))!\,(r+1)!}$

$\quad = C(n+1,r+1)$

Exercise Set 6.4

1. (a) 2,494,800   (b) 302,400   (c) 12,108,096,000   (d) 3,632,428,800
3. 1260
5. (a) 60   (b) 60
    (c) The coefficient in (a) is the number of arrangements of the letters in XXYYYZ.
7. (a) 256   (b) 56   (c) 20
9. $C(10,5) \cdot 9! \cdot 5! = 10,973,491,200$
11. (a) 2,821,109,907,456   (b) 98,425,600,000   (c) 35,152,000,000
13. (a) / / | | / / / | / | / / | /   (b) / | / / | | | / / / / / / | (c) | | / / | / / / / | | / / /
    (d) / | / / | / / / | / / | /
15. 286

Exercise Set 6.5

1. (a) the suits   (b) the cards   (c) 5
3. (a) colors   (b) the pencils   (c) 81
5. (a) 9   (b) 8   (c) $\frac{n+3}{2}$ (d) $\frac{n+1}{2}$
7. Number the 44 seats going across each row from left to right so that the front row has seats 1, 2, 3, and 4, the second row has seats 5, 6, 7, and 8, and the number of each seat (except those in the first row) is 4 more than the number of the seat directly in front of it. Let g_i represent the seat number of the ith girl, where $1 \le g_1 < g_2 < \cdots < g_{25} \le 44$ and consider the two lists: $g_1, g_2, \ldots, g_{25}$ and $g_1 + 4, g_2 + 4, \ldots, g_{25} + 4$. The numbers in each list must be positive integers no greater than $44 + 4 = 48$. With the numbers 1, 2, 3, ..., 48 as the pigeonholes and the 50 numbers in the two lists as the pigeons, at least two of the numbers in the two lists must be equal. Since no two of the numbers g_i are equal to each other, it must be the case that $g_i = g_j + 4$ for two distinct integers i and j. Thus the ith girl sits directly behind the jth girl.
9. Let h_i represent the total number of hits in the first i games of the streak. Since he had at least one hit in each of the 34 games, $1 \le h_1 < h_2 < \cdots < h_{34} = 52$. Consider the two lists $h_1, h_2, \ldots, h_{34}$ and $h_1 + 15, h_2 + 15, \ldots, h_{34} + 15$. The numbers in each list must be positive integers no greater than $52 + 15 = 67$. With the numbers 1, 2, 3, ..., 67 as the pigeonholes and the 68 numbers in the two lists as the

pigeons, at least two of the numbers in the two lists must be equal. Since no two of the numbers h_i are equal to each other, it must be the case that $h_i = h_j + 15$ for two distinct integers i and j. Thus the number of hits in the first i games is 15 more than the number of hits in the first j games, so the number of hits in games $j + 1, j + 2, \ldots, i$ must be 15.

11. 41

13. Divide the hexagon into 6 equilateral triangles, as shown at right, each with edges of length 5 cm. With the 7 points as pigeons and the 6 triangles as pigeonholes, at least two points must lie on or inside the same triangle. The maximum distance between these two points is 5 cm.

15. (a) Answers will vary. One possibility is $S = \{1, 3, 5, 7, 8\}$
    (b) Answers will vary. With $S = \{1, 3, 5, 7, 8\}$, one possibility is $\{5, 7\}$ and $\{1, 3, 8\}$.
    (c) Answers will vary. Other possibilities are $S = \{1, 3, 4, 6, 7\}$, $\{1, 3\}$ and $\{4\}$;
        $S = \{1, 2, 4, 6, 7\}$, $\{1, 2, 4\}$ and $\{7\}$; and $S = \{4, 5, 6, 7, 8\}$, $\{4, 7\}$ and $\{5, 6\}$.
    (d) There are at least two disjoint subsets of S for which the sums of their elements are equal.
    (e) Recall that S has a total of $2^5 = 32$ subsets, so it has 31 non-empty subsets. The sum of the elements of any non-empty subset of S must be a positive integer no greater than $4 + 5 + 6 + 7 + 8 = 30$. With the 31 non-empty subsets of S as the pigeons and the numbers $1, 2, 3, \ldots, 30$ as the pigeonholes, there must be at least two distinct subsets whose elements add up to the same number. If these two subsets are disjoint, then we are done. If not, remove any common elements from both subsets. Since the original subsets were distinct and one was not a subset of the other, the new subsets will be non-empty, they will be distinct, and the sums of their elements will still be equal.

Exercise Set 6.6

1. (a) 16   (b) 12   (c) 7   (d) 5   (e) 5
3. (a) $117 - 45 + 6 = 78$ (b) 25 (Note: 2, 3, 5, and 7 are prime, but 1 is not.)
5. (a) 125   (b)

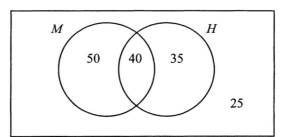

    (c) 25
7. (a) 30   (b) 5
9. 426
11. CCGGLL, CCGGLO, CCGGOO, CCGLLO, CCGLOO, CCLLOO, CGGLLO, CGGLOO, CGLLOO, GGLLOO

Exercise Set 7.1

1. (a) {0, 1, 2, 3, 4}   (b){*M, I, S, P*}
   (c) {(1, 1), (1, 2), (1, 3), (1, 4), (1, 5), (1, 6), (2, 1), (2, 2), (2, 3), (2, 4), (2, 5), (2, 6), (3, 1), (3, 2),
       (3, 3), (3, 4), (3, 5), (3, 6), (4, 1), (4, 2), (4, 3), (4, 4), (4, 5), (4, 6)}
   (d) {1, 2, 3, 4, ...}
3. (a) The event of getting heads an odd number of times = {1, 3}
   (b) The event of drawing a vowel = {*I*}
   (c) The event that the sum of the numbers landing face down is odd
       = {(1, 2), (1, 4), (1, 6), (2, 1), (2, 3), (2, 5), (3, 2), (3, 4), (3, 6), (4, 1), (4, 3), (4, 5)}
   (d) The event that it takes more than ten rolls to get doubles = {11, 12, 13, ...}
5. 1/24
7. (a) $P(1) = 105/600 = 7/40$, $P(2) = 87/600 = 29/200$, $P(3) = 82/600 = 41/300$, $P(4) = 65/600 = 13/120$,
       $P(5) = 125/600 = 5/24$, $P(6) = 136/600 = 17/75$
   (b) 288/600, or 48% of the time
   (c) 274/600, or approximately 46% of the time
9. (c), (d), (e), and (f)
11. (a) $P(A) = 1/4 = P(B)$; $P(C) = 1/12 = P(E)$; $P(D) = 1/3$ (b) 1/3   (c) 5/12
13. (a) $\frac{1}{n^2}$ (b) $\frac{n-1}{n^2}$ (c) $\frac{1}{n}$ (d) $\frac{1}{n^2}$

Exercise Set 7.2

1. (a) 0.17   (b) 0.83
3. (a) 165/2048   (b) 1981/2048
5. (a) 18/25   (b) 7/25   (c) 112/225
7. (a) 125/150 = 5/6   (b) 25/150 = 1/6
9. (a) 290/300 = 29/30 ≈ 0.967   (b) 45/300 = 3/20 = 0.15   (c) 10/300 = 1/30 ≈ 0.033

Exercise Set 7.3

1. 1/120 ≈ 0.0083
3. (a) 81/125 = 0.648   (b) 44/125 = 0.352   (c) 729/1000 = 0.729   (d) 271/1000 = 0.271
5. (a) 15/64   (b) 57/64   (c) 11/32
7. (a) 1/9,366,819   (b) 80/3,122,273
9. 688/1000 = 86/125 = 0.688
11. 9/64

Exercise Set 7.4

1. Because a probability cannot be greater than 1.
3. (a) $P(E \mid F)$ (b) $P(E \mid F)$ (c) $P(\overline{F} \mid E)$ (d) $P(\overline{E} \mid \overline{F})$ (e) $P(F \mid E)$
5. (a) 30/63 = 10/21   (b) 30/45 = 2/3
7. Yes; $P(E \cap F) = P(E) \cdot P(F)$.

9. (a)

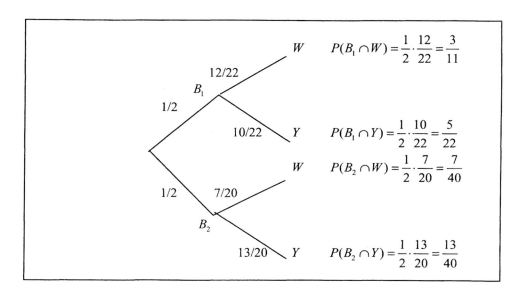

$$P(B_1 \cap W) = \frac{1}{2} \cdot \frac{12}{22} = \frac{3}{11}$$

$$P(B_1 \cap Y) = \frac{1}{2} \cdot \frac{10}{22} = \frac{5}{22}$$

$$P(B_2 \cap W) = \frac{1}{2} \cdot \frac{7}{20} = \frac{7}{40}$$

$$P(B_2 \cap Y) = \frac{1}{2} \cdot \frac{13}{20} = \frac{13}{40}$$

(b) 197/440   (c) 120/197

11.

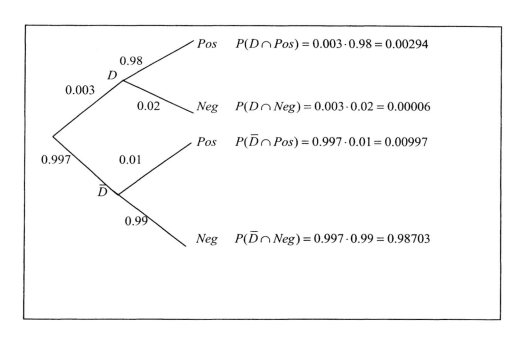

$$P(D \cap Pos) = 0.003 \cdot 0.98 = 0.00294$$

$$P(D \cap Neg) = 0.003 \cdot 0.02 = 0.00006$$

$$P(\overline{D} \cap Pos) = 0.997 \cdot 0.01 = 0.00997$$

$$P(\overline{D} \cap Neg) = 0.997 \cdot 0.99 = 0.98703$$

13. (a)

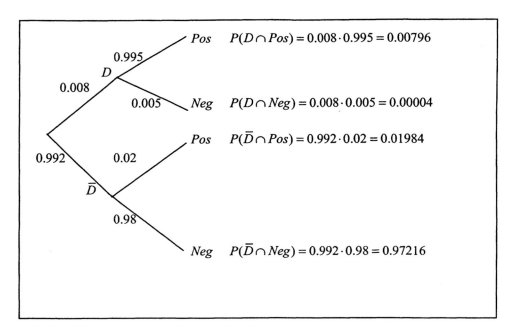

(b) $P(D \mid Pos) = \dfrac{P(D \cap Pos)}{P(Pos)} = \dfrac{0.00796}{0.0278} = 0.2863$

(c) $P(\overline{D} \mid Neg) = \dfrac{P(\overline{D} \cap Neg)}{P(Neg)} = \dfrac{0.97216}{0.9722} \approx 0.99996$

Exercise Set 8.1

1. (a)

	List 1	List 2	List 3	List 4	List 5
First	S	B	B	I	H
Second	I	S	I	H	I
Third	B	I	H	S	S
Fourth	H	H	S	B	B
Votes	2	3	1	2	1

(b) No   (c) B   (d) I

3. B   5. A   7. A

9. Answers will vary. Here is one example:

	List 1	List 2	List 3	List 4
First	A	B	C	D
Second	B	C	D	B
Third	C	A	B	C
Fourth	D	D	A	A
Votes	7	2	2	2

A has a majority of the votes; B has the highest Borda count.

11.   Answers will vary. Here is one example:

	List 1	List 2	List 3
First	A	C	D
Second	B	A	B
Third	C	D	C
Fourth	D	B	A
Votes	11	9	7

Exercise Set 8.2

1. (a) If a candidate has more than half of the votes, then the candidate has more votes than any other candidate and thus wins by the plurality method.
    (b) If a candidate has more than half of the votes, then the candidate will still have over half of the votes after votes are redistributed for a runoff.
    (c) If a candidate has more than half of the votes, then the candidate will defeat any other candidate in pairwise competition.
    (d) The Borda count does not satisfy the majority criterion, because A has a majority of the vote but B wins the Borda count.
3. (a) A   (b) B   (c) No   (d) C (e) No   (f) B   (g) Yes
5. If a candidate defeats all others in pairwise competitions, he or she certainly wins using the method of pairwise competitions.
7. (a) In the sequential runoff method, candidate Y is removed, so the 9 votes in List 2 are given to Z, giving Z 20 votes. This allows Z to win with the majority.
    (b) Y
    (c) No. The sequential runoff method does not satisfy the monotonicity requirement.

(d) If Candidate D is the winner by the plurality method and if every voter ranks D the same or higher in a second ballot (while not changing the order of the other candidates), then D would still have the most first-place votes.

(e) If Candidate D has the highest Borda score and if every voter ranks D the same or higher in a second ballot (while not changing the order of the other candidates), D's Borda score would improve while the Borda score of the others would fall or remain the same.

(f) Suppose Candidate D is the winner by the pairwise comparison method. If Candidate D defeats Candidate E in pairwise comparison and, if subsequently, every voter ranks D the same or higher in a second ballot, D still defeats Candidate E in pairwise comparison. Hence, in the second ballot, D wins as many or more pairwise comparisons as before. Further, if the relative order of other candidates is not changed, no other candidate can increase the number of pairwise comparisons he or she wins. Thus Candidate D is still the winner by the pair-wise comparison method.

9. (a) Z (b) Y (c) No (d) Y (e) X (f) No (g) Z (h) No

11.

A Comparison of Voting Methods and Fairness Criteria					
Criterion	Plurality	Single Runoff	Sequential Runoff	Borda Count	Pairwise Comparison
Majority	Y	Y	Y	N	Y
Condorcet	N	N	N	N	Y
Monotonicity	Y	N	N	Y	Y
Independence-of-irrelevant alternatives	N	N	N	N	N

Exercise Set 8.3

1. (a) 51   (b) Weight of $A = 52$; $B = 48$ (c) {A}   (d) A (e) A

3. (a) 3   (b) Weight of $A = 2$; $B = 1$; $C = 1$ (c) {A, B}   (d) None   (e) A

5.

(a) Winning Coalitions	(b) Critical Voters	(c) Dummies None
{A, B, C}	None	
{A, B}	A, B	
{A, C}	A, C	
{B, C}	B, C	

7.

(a) Winning Coalitions	(b) Critical Voters	(c) Dummies C, D
{A, B, C, D}	A, B	
{A, B, C}	A, B	
{A, B, D}	A, B	
{A, B}	A, B	

9. (a) 1/3, 1/3, 1/3   (b) 1, 0, 0   (c) 1/2, 1/2, 0, 0 (d) 6/12, 2/12, 2/12, 2/12 or 1/2, 1/6, 1/6, 1/6

11. (b) 26/72, 26/72, 4/72, 4/72, 4/72, 4/72, 4/72

13. (b) Each is 6/30, or 1/5.

15. Hempstead 1, Hempstead 2, and Oyster Bay each had a Banzhaf index of 1/3 while the other three voters had Banzhaf indices of 0.

Exercise Set 8.4

1. (a) 3.11
    (b)

Department	Faculty Members	Quota	Lower Quota	Upper Quota
Early Childhood Ed.	13	4.18	4	5
Secondary Ed.	8	2.57	2	3
Special Ed.	7	2.25	2	3

   (c) Early Childhood is represented by 4 committee members; Secondary Education, 3 committee members; and Special Education, 2 committee members.

3. Powerboats, 11 booths; personal water craft, 8 booths; sailboats, 4 booths; sea kayaks, 3 booths
5. (a) 5 (b) 6   (c) 5   (d) 1   (e) 2
7. Powerboats, 12 booths; personal water craft, 8 booths; sailboats, 4 booths; sea kayaks, 2 booths
9. Powerboats, 11 booths; personal water craft, 8 booths; sailboats, 4 booths; sea kayaks, 3 booths
11. (a)

	Number of Units	Least integer $\geq$ quotient $D = 11$	Least integer $\geq$ quotient $D = 12$	Least integer $\geq$ quotient $D = 11.7$
Building 1	44	4	4	4
Building 2	35	4	3	3
Building 3	24	3	2	3
Sum		11	9	10

   (b) 11.7
   (c) Building 1, 4 members; Building 2, 3 members; Building 3, 3 members

Exercise Set 8.5

1. (a) 633   (b) 802.9   (c) 1.27 (d) 27%
3. Route 24   5. Pipefitters, 3; electricians, 2; carpenters, 2
7. Arts and Sciences, 6; Education, 1; Health Professions, 2
9. Dismay, 11; Anxiety, 8; Distress, 6; Loneliness, 4; Distraction, 1
11. (a) The population of Anxiety increased from 7100 to 7400.
    (b) The number of representatives fell from 8 to 7.
    (c) The population of Distraction decreased from 1531 to 1525.
    (d) The number of representatives rose from 1 to 2.
    (e) Yes

INDEX